Smart Cities

This book discusses the integration of power electronics, renewable energy, and the Internet of Things (IoT) from the perspective of smart cities in a single volume. The text will be helpful for senior undergraduate, graduate students, and academic researchers in diverse engineering fields including electrical, electronics and communication, and computers. The book:

- Covers the integration of power electronics, energy harvesting, and the IoT for smart city applications.
- Discusses concepts of power electronics and the IoT in electric vehicles for smart cities.
- Examines the integration of power electronics in renewable energy for smart cities.
- Discusses important concepts of energy harvesting, including solar energy harvesting, maximum power point tracking (MPPT) controllers, and switch-mode power supplies (SMPS).
- Explores IoT connectivity technologies such as long-term evolution (LTE), narrow band NB-IoT, long-range (LoRa), Bluetooth, and ZigBee (IEEE Standard 802.15.4) for low data rate wireless personal communication applications.

The text provides knowledge about applications, technologies, and standards of power electronics, renewable energy, and IoT for smart cities. It will serve as an ideal reference text for senior undergraduate, graduate students, and academic researchers in the fields of electrical engineering, electronics and communication engineering, computer engineering, civil engineering, and environmental engineering.

Smart Cities

Power Electronics, Renewable Energy, and Internet of Things

Edited by

Ahteshamul Haque
Akhtar Kalam
Himanshu Sharma

CRC Press
Taylor & Francis Group
Boca Raton London New York

CRC Press is an imprint of the
Taylor & Francis Group, an **informa** business

Front cover image: metamorworks/Shutterstock

First edition published 2024
by CRC Press
2385 NW Executive Center Dr, Suite 320, Boca Raton, FL, 33431

and by CRC Press
4 Park Square, Milton Park, Abingdon, Oxon, OX14 4RN

CRC Press is an imprint of Taylor & Francis Group, LLC

ISBN: 978-1-032-31243-9 (hbk)
ISBN: 978-1-032-66978-6 (pbk)
ISBN: 978-1-032-66980-9 (ebk)

DOI: 10.1201/9781032669809

Typeset in Sabon
by SPi Technologies India Pvt Ltd (Straive)

I am thankful to Almighty Allah (SWT) for allowing us to complete this book.

This book is dedicated to last Prophet of Humanity Prophet Mohammad ﷺ, who has introduced Almighty Allah (SWT) to all of us.

Ahteshamul Haque

This book is dedicated to all my engineering students, all of whom are dedicated, conscientious, diligent and are giving their best to make this world sustainable, energy efficient and free of pollution.

Akhtar Kalam

I am thankful to almighty God and my teachers and parents who motivated me to contribute in this book.

Himanshu Sharma

Contents

About the authors

Dr. **Ahteshamul Haque** is Associate Professor at the Department of Electrical Engineering, Jamia Millia Islamia (A Central University) New Delhi. His area of research includes power electronics and its application in renewable energy, drives, electric control systems for artificial lighting, power quality improvements, smart grids, wireless power transfer, electric vehicles, electric traction, smart cities, and so on. He did his B.Tech in Electrical Engineering from AMU and his M.Tech from IIT-Delhi. He completed his Ph.D. from Jamia Millia Islamia in power electronics and renewable energy. Prior to Jamia Millia Islamia, he was working in power electronics R&D units of world-reputed multi-national companies. His inventions are patented and awarded in USA and Europe. He has published and presented his research papers at several international conferences and peer-reviewed journals. Since the inception of the Electrical Engineering Department, he received the maximum R&D grant in one project in an individual capacity from the Ministry of New and Renewable Energy (MNRE) of the Government of India. Dr. Haque has established the Advance Power Electronics Research Lab and installed a 1 kW solar PV-based energy conversion system, designed in this lab, and the load of Advance Power Electronics Research Lab gets its power from this installation. For the first time in the history of the Department of Electrical Engineering, B.Tech students working under his supervision have filed for a patent and had it awarded. He designed the course syllabus for both UG and PG levels. In his Advance Power Electronics Research Lab research work in the areas of solar-based energy conversion systems, embedded systems control of power electronics converters, electric control systems for artificial lighting, reliability of power electronics converter, AI-based control of power electronics converters, and so on are carried out. Dr. Haque is a senior member of IEEE PELS, IAS, PES, IES Smart Cities Society and Branch Counselor IEEE-JMI, and actively involved in IEEE activities at the Institution and in the Delhi

Section. Recently Dr. Haque has received an R&D grant under the reputed MHRD-SPARC scheme and has collaborated with Aalborg University Denmark. Dr. Haque has signed an MoU with National Institute of Solar Energy (NISE), MNRE, Govt of India. Dr. Haque was awarded with the Outstanding Engineer award 2019 by the IEEE Power & Energy Society for his research and development contribution in the area of Power Electronics and Renewable Energy. Dr. Haque has also won design contests from the Switzerland-based R&D company Typhoon and received an RTD emulator as the prize. He also won the most IEEE Smart Cities Best Newsletter Article Awards. He has also secured a place in the world's top 2% of scientists curated by the Meta Research Innovation center (METRICS), Standford University, USA.

Dr Haque works as Associate Guest Editor for the *IEEE Journal of Emerging and Selected Topics in Power Electronics* and the *IET Power Electronics Journal*, and is Associate Editor of Elsevier's e Prime journal, *Advances in Electrical Engineering, Electronics and Energy.*

Google Scholar Link: https://scholar.google.co.in/citations?user=fxFp6KAAAAAJ&hl=en

Webpage of Research Lab: https://apeel.eed.org.in/#/

Professor **Akhtar Kalam** has been at Victoria University (VU), Melbourne since 1984 and is a former Deputy Dean of the Faculty of Health, Engineering and Science and Head of Engineering and Head of External Engagement of the College of Engineering and Science. He is also the current Chair of the Academic Board, Texila College Australia, Melbourne, Australia and Deputy Chair of the Academic Board in the Engineering Institute of Technology, Perth, Australia. In addition, he is the Editor in Chief of the Australian Journal of Electrical & Electronics Engineering. Further, he has held the Distinguished Professorship position at various national and international universities. He has wide experience in educational institutions and industry across four continents. He received his B.Sc. and B.Sc. Engineering from Calcutta University and Aligarh Muslim University, India. He completed his MS and Ph.D. at the University of Oklahoma, USA and the University of Bath, UK. He has worked with Ingersoll Rand and other electrical manufacturers. He has held teaching appointments at the University of Technology, Baghdad, Iraq and Capricornia Institute of Advanced Education, Rockhampton, Queensland. He has had international and national recognition for his research. He is the first person to have received the John Madsen Medal from Engineers Australia in consecutive years 2016, 2019, and 2020. The award of the John Madsen Medal is for the best paper in Australia written by a current member of Engineers Australia and published in the

Australian Journal of Electrical and Electronic Engineering. Most recently, the prestigious ACPE-CIGRE Outstanding Academic Award (2021) has highlighted his outstanding impact. The award recognizes an exceptional Australasian academic for outstanding career-long contributions to industry, teaching, and research in electric power engineering. On invitation, he regularly delivers lectures, works on industrial projects, and examines external theses (national and international). His major areas of research interests are power system analysis, communication, control, protection, renewable energy, smart grid, cogeneration systems, and IEC61850 implementation. He has been actively engaged in the teaching of Energy Systems to undergraduates and postgraduates and in providing professional courses to the industry both in Australia and overseas. He regularly offers Continuing Professional Development and Master Class courses on Power System Protection, Renewable Energy, IEC61850, Cogeneration & Gas Turbine Operation, and PBL in engineering education to practicing engineers, the Energy Supply Association of Australia (ESAA), Instructor Development Course (IDC) Technologies, and the Australian Power Institute (API). He also runs a postgraduate distance education program on Power System Protection for the ESAA. He has conducted research, provided industrial consultancy, published over 600 publications on his area of expertise, and written over 29 books in the area. More than 49 higher degree research students have graduated under his supervision. He provides consultancy for major electrical utilities, manufacturers, and other industry bodies in his field of expertise. Professor Kalam is a registered Professional Engineer in the state of Victoria (PEV), Fellow of EA, IET, AIE, a senior member of IEEE, NER, APEC Engineer, IntPE (Aus), and a member of CIGRE AP B5 Study Committee.

Web address: www.vu.edu.au/institute-for-sustainable-industries-liveable-cities-isilc/research-programs/engineering-science-research/smart-energy-research

Google Scholar Link: https://scholar.google.com.au/citations?user=1ZnKlt0AAAAJ&hl=en

Dr. **Himanshu Sharma** is presently working as Head (R&D) and Associate Professor in Electronics & Communication Engineering (ECE) Department at the Noida Institute of Engineering & Technology (NIET), Greater Noida, Gautam buddh Nagar, Uttar Prasdesh (U.P.), India. Dr. Himanshu Sharma has worked as a Postdoctoral Researcher Fellow in the field of Industry 4.0 at Aalborg University, Denmark, Europe (QS world ranking #46) from the year 2022 to 2023. He contributed to two European Union (EU) funded projects titled the 5G-enabled Robotics Project and the Pipesense Project at the highest Technology Readiness Level (TRL-9). His areas of interest are Indoor Positioning & Navigation, Industrial IoT, 5G, Digital Twins, Wireless Sensor Networks (WSN), Artificial Intelligence, Machine Learning, and deep learning. Dr. Sharma also worked as a visiting researcher at Aarhus University, Denmark (QS world ranking #147). He worked as Assistant Professor (Research) in the Electronics & Communication Engineering (ECE) department of KIET Group of Institutions, Ghaziabad, Delhi-NCR, U.P., India from 2009 to 2021. He received his Ph.D. Degree in the field of Wireless Sensor Networks (WSN) & Internet of Things (IoT) from Jamia Milia Islamia (JMI) (a Central Government University), New Delhi, India in 2019. He completed his M.Tech. in Electronics & Communication (ECE) & B.Tech.(ECE) from the state government. Dr. APJ Abdul Kalam Technical University (AKTU), Lucknow, U.P., India in the year 2008 and 2014 respectively.

Dr. Himanshu Sharma has been awarded the IEEE Smart Cities Best Newsletter Article Award 2021 from IEEE Smart Cities Newsletter, USA in the 2022. He also received the prestigious Best Faculty Award-2020 from the state government. Dr. APJ Abdul Kalam Technical University (AKTU), Lucknow, U.P., India, in 2020. He has also received the Sir C. V. Raman Award, 2020 from the KIET Group of Institutions, Ghaziabad, U.P. in 2020 for the best research papers and patents. He won an award from Zero Investment Innovation for Education Initiative (ZIIEI) Awards (sponsored by Shri Aurobindo Society & HDFC Bank) for his innovative teaching-learning ideas in 2020.

Google Scholar Id: https://scholar.google.com/citations?hl=en&authuser=1&user=sszGRWUAAAAJ

Contributors

Mohammad Adnan
Department of Electrical Engineering
Faculty of Engineering and Technology
Jamia Millia Islamia (A Central University)
New Delhi, India

Mohammad Amir
Advance Power Electronics Research Lab
Department of Electrical Engineering
Faculty of Engineering and Technology
Jamia Millia Islamia (A Central University)
New Delhi, India

K. V. S. Bharath
Silicon Austria Lab
Austria

Dr. Manish Bhardwaj
Computer Science and Engineering
KIET Group of Institutions
Ghaziabad, India

Himanshi Chaudhary
Computer Science and Engineering
KIET Group of Institutions
Ghaziabad, India

Dr. Sourav Diwania
KIET Group of Institutions
Ghaziabad, India

Dhawal Gupta
Group Business Director, Public Policy
Chase India

Ahteshamul Haque
Advance Power Electronics Research Lab
Department of Electrical Engineering
Faculty of Engineering and Technology
Jamia Millia Islamia (A Central University)
New Delhi, India

Dr. Abhinav Juneja
KIET Group of Institutions
Ghaziabad, India

Akhtar Kalam
College of Sport, Health and Engineering
Victoria University
Melbourne, Victoria, Australia

Md Zafar Khan
Advance Power Electronics Research Lab

Department of Electrical Engineering
Faculty of Engineering and Technology
Jamia Millia Islamia (A Central University)
New Delhi, India

Dr. Maneesh Kumar
Yashwantrao Chavan College of Engineering (YCCE)
Nagpur, India

Azra Malik
Advance Power Electronics Research Lab
Department of Electrical Engineering
Faculty of Engineering and Technology
Jamia Millia Islamia (A Central University)
New Delhi, India

Junaid Ahmad Malik
Advance Power Electronics Research Lab
Department of Electrical Engineering
Faculty of Engineering and Technology
Jamia Millia Islamia (A Central University)
New Delhi, India

Suwaiba Mateen
Advance Power Electronics Research Lab
Department of Electrical Engineering
Faculty of Engineering and Technology
Jamia Millia Islamia (A Central University)
New Delhi, India

Shabana Mehfuz
Department of Electrical Engineering
Faculty of Engineering and Technology
Jamia Millia Islamia (A Central University)
New Delhi, India

Neirat Mohamad Fayez Mustafa
College of Sport, Health and Engineering
Victoria University
Melbourne, Victoria, Australia

Khushboo Pandey
KIET Group of Institutions
Ghaziabad, India

Prashant
Department of Electrical Engineering
Jamia Millia Islamia (A Central University)
New Delhi, India

Naila Shah
Advance Power Electronics Research Lab
Department of Electrical Engineering
Faculty of Engineering and Technology
Jamia Millia Islamia (A Central University)
New Delhi, India

Md Sarwar
Department of Electrical Engineering
Faculty of Engineering and Technology
Jamia Millia Islamia (A Central University)
New Delhi, India

Shrankhla Saxena
KIET Group of Institutions
Ghaziabad, India

Himanshu Sharma
Department of Electronics & Communication
Noida Institute of Engineering & Technology
Greater Noida, India

Anwar Shahzad Siddiqui
Department of Electrical Engineering
Faculty of Engineering and Technology
Jamia Millia Islamia (A Central University)
New Delhi, India

Shweta Singh
Computer Science and Engineering
KIET Group of Institutions
Ghaziabad, India

Mehwish Weqar
Department of Electrical Engineering
Faculty of Engineering and Technology
Jamia Millia Islamia (A Central University)
New Delhi, India

Zaheeruddin
Department of Electrical Engineering
Faculty of Engineering and Technology
Jamia Millia Islamia (A Central University)
New Delhi, India

Fatima Shabir Zehgeer
Advance Power Electronics Research Lab
Department of Electrical Engineering
Faculty of Engineering and Technology
Jamia Millia Islamia (A Central University)
New Delhi, India

Preface

Welcome to the world of smart cities – where technology, sustainability, and innovation converge to transform urban living into a seamless, intelligent experience. As we delve into the pages of this book, we embark on a journey to explore the fascinating realm of smart cities, focusing on the pivotal role of renewable energy, power electronics, and the Internet of Things (IoT).

The concept of smart cities has emerged as a response to the ever-growing challenges faced by urban areas around the globe. Rapid urbanization, increased energy demands, environmental concerns, and the desire for improved living standards have necessitated the integration of cutting-edge technologies into city infrastructures. This book aims to shed light on how renewable energy sources, harnessed through advanced power electronics systems, are shaping the sustainable future of urban landscapes, while the Internet of Things acts as the connective tissue that binds it all together.

OUR PURPOSE

The purpose of this book is to provide a comprehensive and insightful overview of the interplay between renewable energy, power electronics, and the Internet of Things in the context of smart cities. By delving into the fundamentals of each domain and showcasing real-world applications and case studies, we hope to equip readers with a deeper understanding of the profound impact these technologies have on urban development.

WHO SHOULD READ THIS BOOK

This book is intended for a wide audience, including students, researchers, urban planners, engineers, policymakers, and technology enthusiasts interested in exploring the dynamic landscape of smart cities and the key elements driving its transformation. Whether you are a seasoned professional in the field or a curious mind eager to learn about the future of urban living, this book offers valuable insights and knowledge.

WHAT TO EXPECT

Throughout the chapters of *Smart Cities – Renewable Energy, Power Electronics, and Internet of Things* (IoT), we will navigate through the essential concepts, advancements, and challenges that define smart cities. From the utilization of renewable energy sources like solar, wind, and hydroelectric power to the sophisticated control and conversion mechanisms provided by power electronics, we will explore how these technologies are paving the way for a sustainable, energy-efficient urban landscape.

Furthermore, we will delve into the transformative impact of the Internet of Things, which enables seamless communication, data exchange, and automation within smart cities. Through a series of case studies and examples, we will witness how IoT-driven solutions are enhancing various aspects of urban life, including energy management, transportation, healthcare, waste management, and more.

CONCLUSION

As we embark on this enlightening journey into the world of smart cities, we invite you to join us in envisioning a future where sustainable, intelligent, and connected urban environments create a better tomorrow for generations to come. The convergence of renewable energy, power electronics, and the Internet of Things opens the door to endless possibilities, and it is our hope that this book will inspire you to be an active participant in shaping the Smart Cities of the future.

Let's begin this transformative expedition together!

Chapter 1

Fundamentals of power electronics in smart cities

Ahteshamul Haque, Naila Shah, Junaid Ahmad Malik, and Azra Malik

Jamia Millia Islamia, New Delhi, India

1.1 INTRODUCTION

The idea of a "smart city" refers to the application of technology, data driven systems, and intelligent solutions to increase sustainability, improve the standard of life for its citizens, and optimize resource management. One of the major developments in today's technology is the smart city. The need for a "smart" and sustainable metropolis has arisen from the increase in the global population. The word "smart city" was first coined in 1990. At that time the main focus was given to infrastructure based on ICT (Information and Communication Technologies). Later on, this shifted to the concept of designing smart cities based on smart communities presented by California Institute [1]. There were many definitions given by authors regarding smart city. One definition given by Casio in [2] emphasizes the importance of ICTs by defining them as scalable systems to improve standards of life, minimize costs, and enhance efficiency. Another definition given by Chourabi stated that a smart city is a city that aims to achieve smartness by making itself more productive, democratic, enjoyable, and sustainable [3]. A city that is functioning well in terms of mobility, government, economics, population, and living standard of people can be summarized as smart city [4]. The main characteristics of smart cities include theme, quality, and structure, with theme being the prime support of smart cities. These attributes are dependent on four features that include quality of life, urbanization, sustainability, and smartness as shown in Figure 1.1. The quality of life aims to achieve welfare of citizens. Urbanization and smartness attributes include changing the infrastructure from rural to urban conditions. In sustainability the focus is given to ecosystems, pollution, energy, and climate change.

Smart cities employ optimization of resources, reduced operational costs, and improved overall efficiency through various techniques like data analytics, IOT (Internet of Things), and automation [5]. Integration of solutions is essential for a better and more regulated utilization of the resources due to their scarcity and high cost. Smart cities also focus on enhancing sustainability, governance, and control infrastructure and aim to reduce carbon footprint. Smart cities offer improved services to all of inhabitants using effective methods that minimize time and resource waste [6]. Smart cities

DOI: 10.1201/9781032669809-1

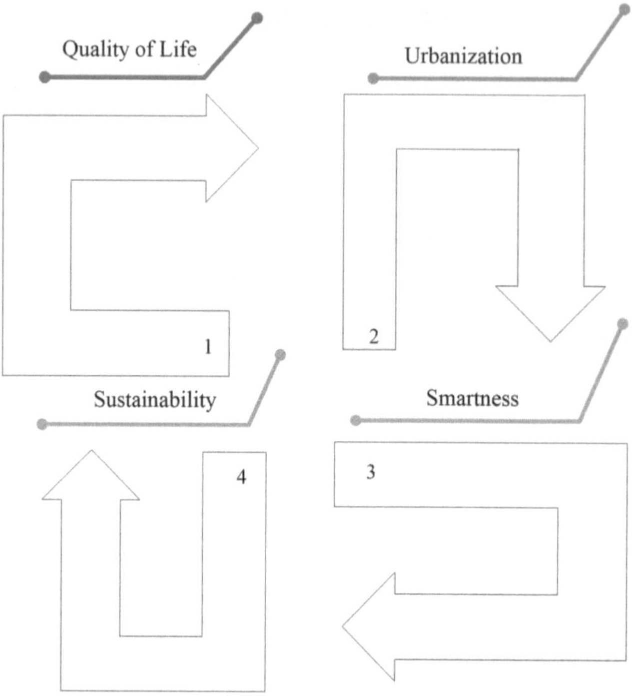

Figure 1.1 Characteristics of smart cities.

enhance the quality of life by providing comfort, convenience, security, and energy management for citizens. Remote healthcare monitoring systems help patients receive individual care, decreasing hospital visits. Smart cities deploy intelligent techniques and advanced technologies like surveillance systems, predictive analysis, and sensors to improve the safety and security of residents. Smart cities emphasize citizens participation in decision-making processes and involves in urban planning and policy making. Smart cities encourage innovation and attract investment [7]. They promote economic growth and job creation by fostering a climate that encourages technical development and entrepreneurship. By embracing technology and innovation, smart cities have the ability to overcome the urban issues and build healthier, more flexible, and successful communities.

Power electronics technology plays a vital role in the development and functioning of smart cities, revolutionizing the way we generate, distribute, and consume electrical energy [8]. As urban areas become increasingly complex and energy-intensive, power electronics offer a range of solutions that are essential for creating sustainable, efficient, and intelligent urban environments. Figure 1.2 shows the schematics of advanced power electronics technology in a smart city. Power electronics is essential for the operation and control of smart grids, which form the back bone of smart cities. The two

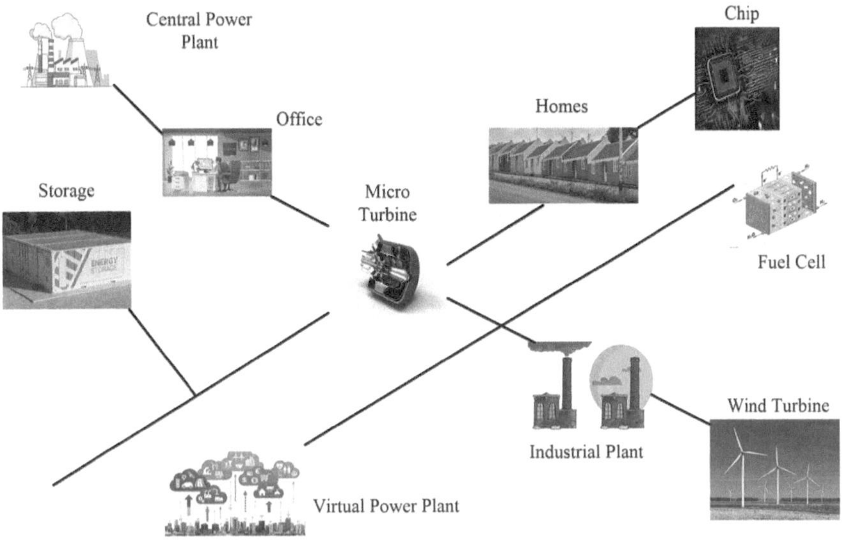

Figure 1.2 Block diagram of power electronics in smart city.

major functions of power electronics in a smart grid are (a) electrical power transmission and (b) electrical power distribution. Power can be transmitted through two ways: (a) HVDC transmission and (b) power transmission through FACTS devices [9]. Both techniques use power electronic devices like IGBT, GTO, and SCR. FACTS improve the reliability and stability of transmission systems by combining existing power electronic devices with modern ones. The quick control of FACTS technology may help China's energy supply and demand reach a more favorable equilibrium, boost electrical power transmission and distribution, and resolve the issue of voltage fluctuation [10]. The application of power electronic technology helps to maintain power quality and reliability in smart cities by setting up technical evaluation, economy assessment systems, rules, and regulations [11]. Solutions based on power electronics increase power grid stability, and dependability, and resilience also helps to reduce problems like power factor distortion, voltage fluctuations, and harmonics. Power electronics enable the seamless integration of renewable energy generation systems, such as solar photovoltaic panels and wind turbines, into the power grid. In a solar photovoltaic system, a DC/DC converter is used to link separate solar panels in series or parallel. Power electronics converters are used to interface two solar panels since they are not identical and direct interfacing is ineffective. Additionally, this increases the array's general effectiveness, since solar PV supplies DC and for grid integration this is converted into AC via a power electronic converter, that is, DC/AC. In the case of wind power plants, these power electronic converters perform the function of integrating wind power plants (WPP) to the grid as well as active and reactive power control.

With the help of power electronics, electricity can be converted, controlled, and managed properly, ensuring that there should be optimal energy usage and minimal waste in various sectors and industries. For sustainable usage of energy, integration of demand response systems, renewable energy sources, and energy storage systems are utilized to achieve this. In metropolitan regions, buildings are substantial energy users. LED lighting, for example, which greatly lowers electricity use, is made possible by power electronics. Additionally, it offers the effective regulation of heating, ventilation, and air conditioning (HVAC) systems, optimizing energy consumption depending on utilization, temperature, and environmental factors. In the event of power outage, energy storage systems are used to supply backup power to generators. Due to these energy storage technologies, the grid reliability is increased. Flywheel is the most significant device for energy storage as it can store the maximum amount of kinetic energy possible. Electric vehicle (EV) adoption and infrastructure development are supported by smart cities. In EV charging stations, power electronics is essential for effective power transfer, charging control, and grid integration. Fast-charging, two-way power flow and intelligent charging techniques that balance energy consumption and grid stability are made possible by power electronic converters. Additionally, power electronics aids in the effective management of vehicle performance, optimal energy usage, and electric drive trains in EVs.

In this chapter a comprehensive understanding and overview of the fundamental concepts, applications, and importance of power electronics in the context of smart cities is provided. This chapter also investigates the function of power electronics in the development and operation of various aspects of smart cities such as smart grids, smart buildings, and electric mobility.

1.2 BASICS OF POWER ELECTRONICS

Gh Power electronics refers to the technology that deals with power electronics devices like diodes, transistors, switches, converters, inverters, and rectifiers used for control and conversion of electrical power in order to meet a desired requirement. It can also be defined as the branch of electrical engineering that utilizes solid-state semiconductor apparatus to efficiently control and convert power from one form to another. The first power electronic apparatus that came into existence was a mercury arc rectifier in 1900 [12]. The first real power electronics revolution was started by Brattain, Bardeen, and Shockley in 1956 at the Bell Laboratory with the invention of the Silicon Controlled Rectifier (SCR) or PNP transistor [13]. The second revolution was started by the General Electric Company (GEC) in 1958 with the introduction of power devices or commercial thyristors like MOSFET, GTO, TRIAC, BJT, IGBTs, MCTs, and ICTs [14]. Due to this, the modern age of power electronics started and different power electronic devices with their control techniques are presented in Table 1.1. Power electronics plays a vital role in revolutionizing the power sector by controlling the generation, transmission, and distribution sectors. Examples include energy storage systems including supercapacitors and batteries, FACTs used to control power grids, inverters and converters used for integration of renewable energy resources with the grid, current limiting devices, transfer switches, and solid-state semiconductor devices [15].

1.2.1 Characteristics of power semiconductor devices

The characteristics of power semiconductor devices are: (i) fast switching speed, (ii) high-power handling capacity, (iii) low-power losses, (iv) gate drive requirements, (v) voltage blocking capability, and (vi) temperature stability. The three main classifications of power semiconductor devices are: (a) Diode, (b) SCR, and (c) Transistors.

Depending on the number of terminals in the device, power semiconductor devices are separated into two groups: (i) two-terminal device (Schottky diodes, Fast recovery diodes) and (ii) three-terminal device (SITs, MCT, MOSFET, IGBT, BJT, SCR) [6]. Based on the carrier used for conduction in semiconductor devices, these devices are classified as: (i) majority carrier semiconductor devices (MOSFET, SIT) and (ii) minority carrier semiconductor devices (MCT, GTO, IGBT, Thyristor, Power Diode). On the basis of the number of layers in semiconductor devices, semiconductor devices are grouped into categories: (i) two-layered device (Power Diodes), (ii) three-layered device (BJTs, MOSFETs), and (iii) four-layered device (SCR).

Table 1.1 Characteristics of power semiconductor devices

Semiconductor device	*Maximum frequency (KHz)*	*Voltage/ Current rating*	*Turn-on time*	*Turn-off time*
DIODE	1	5 KV/5 KA	50–100 μs	50–100 μs
SCR	1	10 KV/5 KA	2–5 μs	2–100 μs
MOSFET	100	1 KV/50 A	0.1 μs	1–2 μs
BJT	10	1400 V/400 A	2 μs	9–30 μs
IGBT	50	3.3 KV/2.500 A	0.2 μs	2–5 μs
GTO	2	5 KV/3 KA	3–5 μs	10–25 μs
TRIAC	0.5	1200 V/300 A	2–5 μs	200–400 μs
MCT	20	1200 V/100 A	0.2 μs	50–110 μs
SIT	100	1200 V/500 A	–	–

1.2.2 Power conversion techniques

A power electronic converter is used for the control or conversion of input power from one form to output power of another form by utilizing the switching characteristics of power devices. These power conversion tasks are effectively and efficiently executed by static converters. Solid-state devices with inductors and capacitors are used in power electronic switches. Comparing inductors and capacitors to resistors, they often show small power loss characteristics. The various power electronic converters are categorized as:

- *AC/DC Rectifiers (Diode Rectifier and Phase Controlled Rectifier)*
  The term AC to DC converter refers to a converter that converts an AC input signal into a DC signal. Such a device that performs this conversion is known as a rectifier. These rectifiers are further grouped into two categories as:
  - *Diode Rectifier*: These device circuits convert input AC voltage into fixed DC voltage. The input voltage applied could be either 1∅ or 3∅. These rectifiers are widely used in uninterruptible power supplies, welding, etc. They are employed in electromechanical processing like electroplating, electric transaction, battery charging, and many more.
  - *Phase Controlled Rectifier*: Phase Controlled Rectifiers transform a constant AC input voltage into a variable amount of DC output voltage in contrast to diode rectifiers. These converters are also known as naturally commutated converters or line commutated AC/DC converters because they operate on line voltages for the commutation process or turn-off process. In these rectifiers, the applied AC input voltage may be 1∅ or 3∅ AC source. These rectifiers are widely used in synchronous machines for excitation purposes, chemical, and metallurgical industries.
- *DC/DC Converters (Choppers)*
  These converters convert a fixed frequency DC input signal into a variable DC output signal. These converters have the ability to produce DC output voltage that may differ in amplitude from the source voltage. MOSFETs, power transistors, and thyristors are commonly employed semiconductor devices used for their construction. The output of DC choppers is controlled by utilizing low-power control signals and in this scenario, power transistors are used in place of thyristors. The choppers are categorized according to the nature of commutation and the direction in which power is delivered. Some significant applications for choppers include electric traction, battery-powered vehicles, SMPS, subway cars, and so on.
- *DC/AC Converters (Inverters)*
  An inverter transforms the input that is fixed DC voltage commonly sourced from batteries into an variable AC output voltage with adjustable voltage and frequency. Here, line commutation or

forced commutation is used to turn off the thyristor devices. Multiple semiconductor components including power transistors, MOSFETs, GTOs, IGBTs, and thyristors are used in the fabrication of inverters. These devices are designed to handle various power levels, with traditional thyristors utilized in high-power applications, whereas power transistors are utilized in low-power applications. Inverters are widely employed in synchronous motor drives, HVDC transmission, UPS, and flash photography cameras. Due to their versatile nature, they are adopted for effective power conversion over a wide range of power requirements.

- *AC/AC Converters*
  This converter transforms the fixed frequency AC input voltage into a variable AC output voltage. This converter is further divided into:
  - *Cycloconverters*: A cycloconverter converts fixed AC voltage at single frequency into a variable AC output voltage at another frequency employing a single-stage conversion process. The resulting output voltage signal has a lower frequency than the input voltage signal. These converters primarily utilize the line commutation technique. They are predominantly adopted for applications requiring low-speed speed, high-transaction AC drives such as multi-megawatt AC motor drives.
  - *AC voltage Controllers*: They are also known as AC voltage regulators as they convert a fixed voltage AC signal into a variable AC voltage signal with the same frequency as the input. These AC voltage controllers operate by using a pair of thyristors placed in an antiparallel arrangement and natural commutation is used to turn both devices off. The output voltage is effectively regulated by changing the delay angle. These controllers find wide application in tap changers, speed regulation of fans, lighting control, and so on.
- *Static Switch*
  The power devices can function as contractors or static switches and the source of power to these switches might be DC or AC, so these switches are referred to be DC switches or AC static switches respectively. Compared to mechanical and electromechanical circuit breakers, static switches have various benefits. The various advantages are:
  - i. The switching speed of a static switch is about 3 μs.
  - ii. The static switches don't bounce when turned on.
  - iii. There are reduced changes of wear and tear due to the absence of any moving part.
  - iv. Static switches possess longer operational life.

1.3 POWER ELECTRONICS IN SMART GRIDS

The smart grid is known as the modernization of the electrical grid through the use of advanced metering infrastructure (AMI), digital transmission substation, and current sensing technology by incorporating modern technology in relay protection devices. It is a new kind of power grid that integrates components of control technology, communication technology, information technology, and power electronics technology. The distinguishing features of smart grid are flexibility, efficiency, environmental safety, and reduced maintenance costs. In contrast to conventional grid, the smart grid offers continuous safety evaluation, fault detection, early warning system, reduced maintenance cost, and self-healing capabilities. Power electronic devices are essential for the development of the intelligent power grid and power systems in many ways, including energy storage systems (flywheel, batteries, super capacitors, water containers etc.), harmonics mitigation, integration of renewable energy resources, power quality improvement, bidirectional flow of power, load management, and so on. The importance of power electronic devices in power systems has substantially increased due to developments in modular converter units, intelligent control techniques, and high-voltage and high-power electronic devices. The growing significance of the power electronics technology in the field of smart grid development is highlighted by the rising global demand for power electronic devices in applications, including distributed independent power systems, power quality improvement, wind and solar power generation, large-scale industrial processes, and high-voltage inverters.

1.3.1 Role of power electronics in integration of renewable energy sources to grid

The main impediments to global sustainability are due to a shortage of raw materials and environmental contamination brought on by traditional energy sources such as coal and oil. The Paris Agreement (2015) stressed the need to achieve an energy transition through the development and use of renewable energy sources (RESs). As a result, several nations have taken significant steps to integrate RES, including wind, solar PV, ocean wave energy, and bioenergy into their energy systems in an effort to shift their energy paradigms. For example, Denmark intends to be 100% fossil fuel independent and 100% carbon neutral through the use of RESs in 2050 [16–18]. The increased penetration of RES is accompanied by two challenges: one is integrating RES into the electrical grid to guarantee stability during system outage and the other is how to utilize power electronics for energy conversion, transmission, distribution, and utilization efficiently. Therefore power electronic technologies have evolved quickly and grid integration standards are constantly being revised for RESs particularly in wind and PV systems [19, 20]. Recent years have seen fast advancements in power electronics, driven by two important aspects. The first is the development of very efficient high-power semiconductor switches and the second is the incorporation of real-time computer controllers with improved control algorithms. These two aspects made it possible to develop inexpensive and grid-friendly converters. Power converter topologies for low-power RESs focused on high-power density and high efficiency, with transformer-less topologies evolving as more efficient and less costly. In order to improve the grid integration of RESs, multi-port converters are being incorporated into energy storage systems [21]. Control algorithms for RES have first shifted from grid following to grid supportive and then to grid-forming capabilities in order to improve grid resiliency [22]. The consolidation of grid integration requires reliability-oriented control and power converters coopted with RES must be able to respond to individual end-user requests for uninterrupted power supply and system operators' global management directives.

1.3.2 Power quality and stability considerations in smart grids

The implementation of power quality technology in smart grids should strongly focus on development and enhancement of the power quality evaluation technique and hierarchical systems. It is also required to develop internal technical grade evaluation systems, user economic assessment systems and policies, and rules and regulations in order to accomplish "economically-efficient and high-quality operation" of smart grids. Table 1.2 shows the main technical parameter of power quality technology.

The use of power quality technology in the smart grid includes a number of related technologies such as unified power quality controller (UPQC), adaptive static reactive power compensation, high-quality power parks, and so on. UPQC plays a significant role in ensuring effective power quality for its users.

1.3.3 Demand-side technology in smart grids

Demand-side technology is largely concerned with satisfying the unique power quality and reliability demands of consumers or fulfilling the needs of customers with specialized power supply requirements. It involves the integration of power electronic technology and distribution automation technology to produce power supply solutions that closely match customer expectations. Power companies are faced with the dual challenge of balancing the need to satisfy rising power demand with the requirement of energy conservation due to the increasingly challenging landscape of power

Table 1.2 Main technical parameter of power quality technology

	Frequency test error	*Voltage deviation*	*Phase angle error*	*Power deviation*	*Power consumption*
Parameter	≤0.01 Hz	≤0.2%	≤0.2%	≤0.5%	<4 VA

load and power quality. Power companies must improve their technological capabilities in order to meet these problems. Demand-side technology is essential for ensuring a new energy grid and clean energy and offers creative solutions for problems like fault current limit prevention.

1.4 POWER ELECTRONICS IN SMART BUILDINGS

Power electronics is a rapidly advancing field that plays a crucial role in the development and implementation of smart buildings. With the rise of smart technologies and the increasing demand for energy-efficient solutions, power electronics has emerged as a key enabler in transforming traditional buildings into intelligent and sustainable infrastructures [1]. By effectively managing and controlling the flow of electrical power, power electronics systems in smart buildings optimize energy consumption, enhance power quality, and enable the integration of renewable energy sources, ultimately leading to improved comfort, cost savings, and environmental sustainability.

In smart buildings, power electronics technology is utilized across various applications, such as lighting, HVAC, energy storage systems, electric vehicle charging, and more. It facilitates efficient power conversion, voltage regulation, and power factor correction, enabling the seamless integration of diverse energy sources, energy storage systems, and energy management systems [2]. Moreover, power electronics devices, including inverters, converters, and controllers, enable bidirectional power flow, allowing for grid-to-building and building-to-grid energy exchange, promoting grid stability and resilience. The integration of power electronics into smart buildings not only enables advanced control and automation but also paves the way for intelligent energy management systems that adapt to the occupants' needs and preferences, optimizing energy consumption and reducing environmental impact.

1.4.1 Role of power electronics in energy-efficient systems

Power electronics plays a crucial role in enabling energy-efficient systems by effectively managing and controlling the flow of electrical power. It involves the conversion, conditioning, and control of electrical power to ensure optimal performance and reduced energy losses. Power electronic devices, such as converters, inverters, and controllers, provide the necessary tools to efficiently transform, regulate, and distribute electrical energy. One key aspect of power electronics is its ability to facilitate energy conversion between different sources and loads [3]. For example, in renewable energy systems, power electronics are used to convert the variable and often unpredictable output from sources like solar panels or wind turbines into a usable form of electricity. This conversion process ensures that the power generated

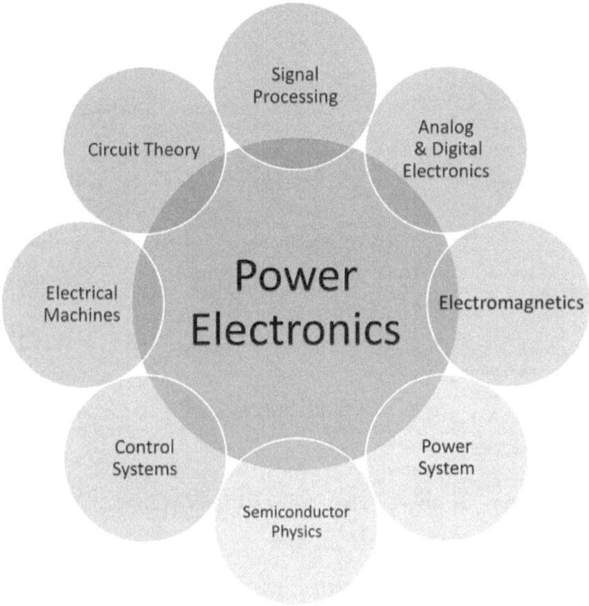

Figure 1.3 Power electronics role.

matches the requirements of the connected loads, enhancing overall system efficiency. Power electronics also enable bidirectional power flow, allowing energy to be efficiently stored in batteries and other energy storage systems, and then discharged when needed. As illustrated in Figure 1.3.

Moreover, power electronics devices contribute to energy efficiency by minimizing power losses during energy conversion. They employ advanced switching techniques, such as pulse-width modulation (PWM), to control the output voltage and current waveforms with high precision. This precise control helps in reducing energy losses through improved voltage regulation, reduced harmonic distortion, and efficient power factor correction. Additionally, power electronics systems incorporate features like soft switching and active clamping to further enhance energy efficiency by minimizing switching losses.

1.4.2 Power electronics applications in HVAC

Power electronics has significant applications in HVAC systems, contributing to their improved efficiency, control, and performance. One of the key areas where power electronics is employed in HVAC is in motor drives. Power electronic motor drives, such as variable frequency drives (VFDs), are utilized to regulate the speed of motors used in HVAC equipment like fans and pumps presented in Figure 1.4. By adjusting the motor speed based on the actual cooling or heating demands, VFDs enable the precise control of

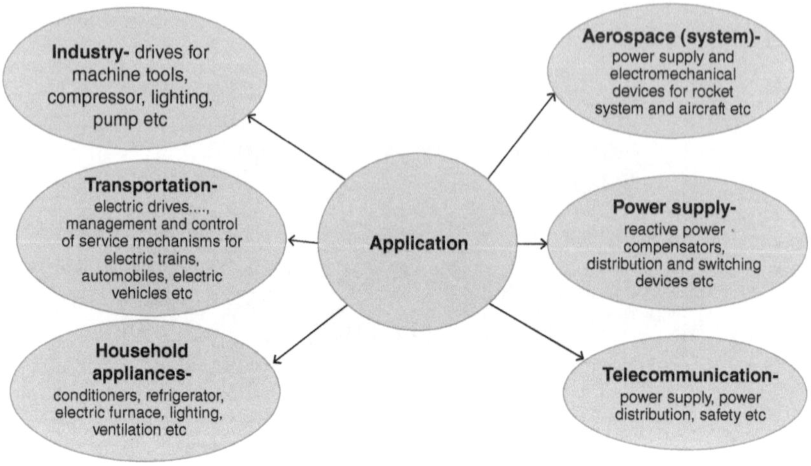

Figure 1.4 Applications of power electronics in HVAC.

airflow and water flow rates, resulting in reduced energy consumption and enhanced system efficiency [4]. Furthermore, VFDs offer soft-start capabilities, which minimize the inrush current during motor startup, extending the lifespan of the motor and reducing equipment wear and tear.

Another application of power electronics in HVAC is in the implementation of energy-efficient cooling systems. Air conditioning systems often employ compressors to provide cooling, and power electronics is employed to regulate and optimize their operation. Variable speed compressors, controlled by power electronic devices, allow for modulation of the cooling capacity based on the actual cooling load. This enables the system to operate at varying capacities, matching the cooling requirements precisely. As a result, energy wastage due to constant on/off cycling of the compressor is minimized, leading to significant energy savings and improved overall HVAC system performance.

1.4.3 Building automation and control systems enabled by power electronics

Power electronics plays a vital role in enabling building automation and control systems, enhancing energy efficiency, comfort, and overall operational performance. Building automation systems utilize power electronic devices to control and manage various building functions, including lighting, HVAC, security, and power distribution. By integrating power electronics, these systems can optimize energy usage, improve occupant comfort, and provide intelligent control and monitoring capabilities. Power electronic devices such as solid-state relays, electronic switches, and motor drives are employed in building automation systems to regulate and control electrical loads. These devices enable precise and efficient switching of electrical circuits, allowing

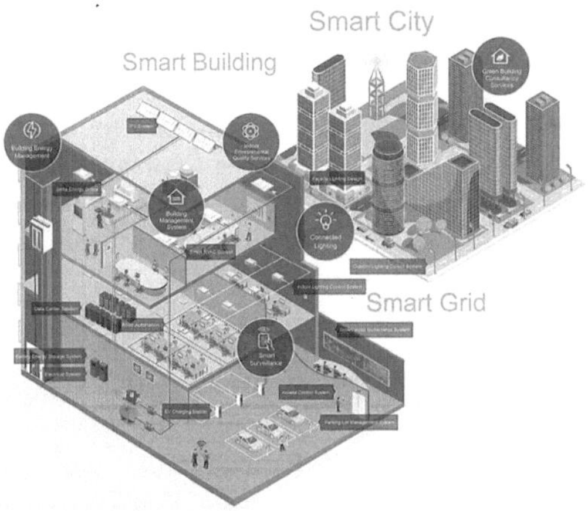

Figure 1.5 Building automation.

for intelligent control of lighting systems, motorized shades, and other electrical equipment. Additionally, power electronics facilitates power factor correction and voltage regulation, ensuring a stable and reliable power supply within the building, while reducing energy losses. Furthermore, power electronics enable advanced energy management strategies in building automation systems. Energy storage systems, such as batteries or supercapacitors, integrated with power electronics, can store excess energy during low-demand periods and release it during peak demand periods, optimizing energy usage and reducing utility costs [23] as represented in Figure 1.5. Power electronics also enable demand response capabilities, where the building's energy consumption can be dynamically adjusted based on real-time pricing or grid signals, contributing to load balancing and grid stability.

1.4.4 Power factor correction and load management in smart building

Power factor correction and load management are essential aspects of smart building systems, and power electronics plays a crucial role in implementing these functionalities. Power factor correction is the process of improving the power factor of a building's electrical system by reducing reactive power and bringing it closer to unity. Power electronics devices, such as capacitors and power factor correction controllers, are used to compensate for reactive power and adjust the power factor. By improving the power factor, smart buildings can minimize energy losses, optimize the use of electrical infrastructure, and reduce electricity costs. Power factor correction also enhances the overall efficiency of the electrical system and reduces stress on electrical equipment, leading to increased equipment lifespan and improved reliability. Load management in smart buildings involves the intelligent control and optimization of energy consumption based on demand, availability, and cost factors. Power electronics devices, such as programmable logic controllers (PLCs), sensors, [24] and actuators, enable real-time monitoring and control of loads in the building. These devices collect data on energy usage, occupancy patterns, and external factors to dynamically adjust the operation of electrical equipment and optimize energy consumption. Load management strategies can include load shedding during peak demand periods, load shifting to take advantage of off-peak electricity rates, and load balancing to ensure efficient distribution of energy within the building. By effectively managing loads, smart buildings can reduce peak demand, minimize energy waste, and enhance overall energy efficiency.

1.5 POWER ELECTRONICS IN ELECTRIC TRANSPORTATION

At the heart of electric transportation, power electronics devices such as inverters, converters, and motor controllers are utilized to convert DC power from the battery into AC power for the electric motor, providing smooth and precise control over the vehicle's propulsion. Power electronics also enable regenerative braking, where the kinetic energy during deceleration is converted back into electrical energy and stored in the battery for reuse [25]. Additionally, power electronics systems are integrated into EV charging infrastructure to regulate the charging process, providing safe and efficient charging for EV owners. This includes power conversion from AC to DC to charge the vehicle's battery, as well as communication and control functionalities to monitor charging status, manage power levels, and integrate with smart grid systems illustrated in Figure 1.6.

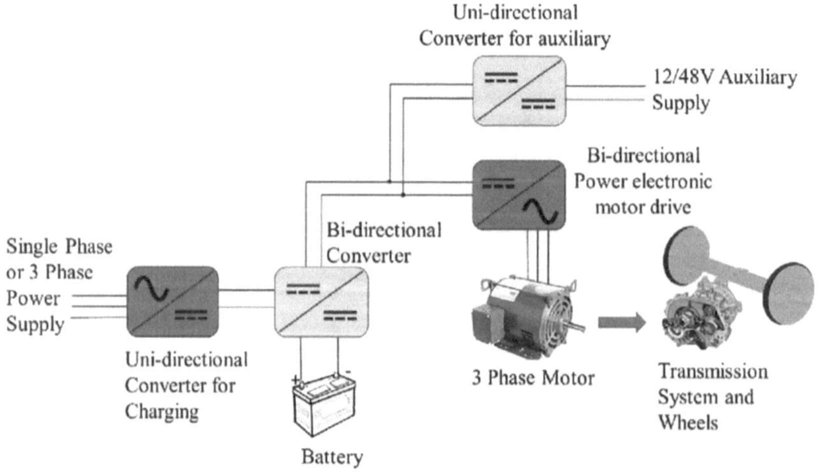

Figure 1.6 Electronics in transportation.

1.5.1 Overview of electric transportation in smart cities

Electric transportation plays a pivotal role in the development of electric cities, where sustainable and environmentally friendly modes of transportation are prioritized. Electric transportation encompasses various forms of EVs, including electric cars, electric buses, electric bikes, and electric scooters. These vehicles are powered by electricity instead of fossil fuels, resulting in reduced emissions and a cleaner urban environment. Electric transportation offers numerous benefits for electric cities. One of the primary advantages is the significant reduction in air pollution and greenhouse gas emissions. Electric vehicles produce zero tailpipe emissions, resulting in improved air quality and reduced carbon footprint. This contributes to creating healthier and more sustainable living environments for residents. Moreover, electric transportation contributes to noise reduction in cities [26]. Electric vehicles operate more quietly compared to conventional combustion engine vehicles, resulting in reduced noise pollution, especially in densely populated urban areas. This can lead to improved quality of life for residents, as well as quieter and more peaceful neighborhoods.

1.5.2 Overview of power electronics technology for electric transportation in smart cities

Power electronics plays a crucial role in the charging infrastructure of EVs, enabling efficient and reliable charging processes. Power electronic devices and systems are employed at various stages of the charging infrastructure to facilitate the conversion, control, and distribution of electrical power. At the charging station level, power electronics are used in AC/DC converters. These converters are responsible for converting the alternating current (AC) power from the grid to the direct current (DC) power required for charging the EV's battery. Power electronic converters ensure efficient power conversion, high-power quality, and compatibility with different charging standards [27]. Within the EV itself, power electronics are integral to the on-board charger. The on-board charger converts the AC power received from the charging station into the appropriate DC voltage and current levels for charging the EV's battery pack. Power electronic converters within the on-board charger ensure efficient power conversion, power factor correction, and control of charging parameters.

1.5.3 Overview of grid integration for electric transportation in electric cities

Integration of EV charging infrastructure with the power grid is a crucial aspect of the ongoing transition toward sustainable transportation. As the adoption of electric vehicles continues to grow, it becomes imperative to establish a seamless connection between the grid and EV charging systems to ensure efficient and reliable charging services. This integration allows for optimized charging strategies, grid management, and the utilization of renewable energy sources, facilitating the widespread adoption of electric mobility while minimizing the strain on the power grid. By integrating EV charging infrastructure with the power grid, several benefits can be realized. First, it enables the implementation of smart charging solutions that take advantage of off-peak hours and grid conditions, ensuring cost-effective charging for EV owners. Through communication protocols and advanced metering systems, charging stations can receive real-time data on electricity prices and grid demand, enabling them to adjust charging rates accordingly. This not only helps to balance the load on the grid but also encourages EV owners to charge their vehicles during periods of lower demand, reducing strain on the grid during peak hours. Additionally, this integration paves the way for Vehicle-to-Grid (V2G) technology, allowing electric vehicles to act as mobile energy storage units. During periods of high demand or grid instability [28], EV batteries can feed electricity back into the grid, providing support and flexibility to the overall power system. This bidirectional power flow not only enhances grid reliability but also enables the integration of renewable energy sources by absorbing excess generation and mitigating fluctuations in supply and demand. Overall, the integration of EV charging infrastructure with the power grid establishes a symbiotic relationship between electric vehicles and the energy system, contributing to a more sustainable and efficient transportation ecosystem.

1.6 POWER QUALITY AND RELIABILITY IN SMART CITIES

Power quality and reliability are critical factors in the development and operation of smart cities. As cities become increasingly connected and reliant on digital technologies, ensuring a consistent and high-quality power supply becomes paramount. Power quality refers to the characteristics of electrical power, including voltage stability, harmonic distortion, and frequency fluctuations, that can affect the performance of electronic devices and systems. In the context of smart cities, maintaining optimal power quality is essential for the smooth operation of various interconnected systems, such as smart grids, intelligent transportation systems, and smart buildings [29].

In smart cities, power quality issues can have far-reaching implications. For instance, voltage sags or fluctuations can disrupt the operation of sensitive equipment, leading to system failures and downtime. Harmonic distortions can cause interference and malfunctions in communication systems and electronic devices, affecting the overall efficiency and reliability of the city's infrastructure. Therefore, smart cities need advanced power quality monitoring and control systems that can detect and mitigate power quality

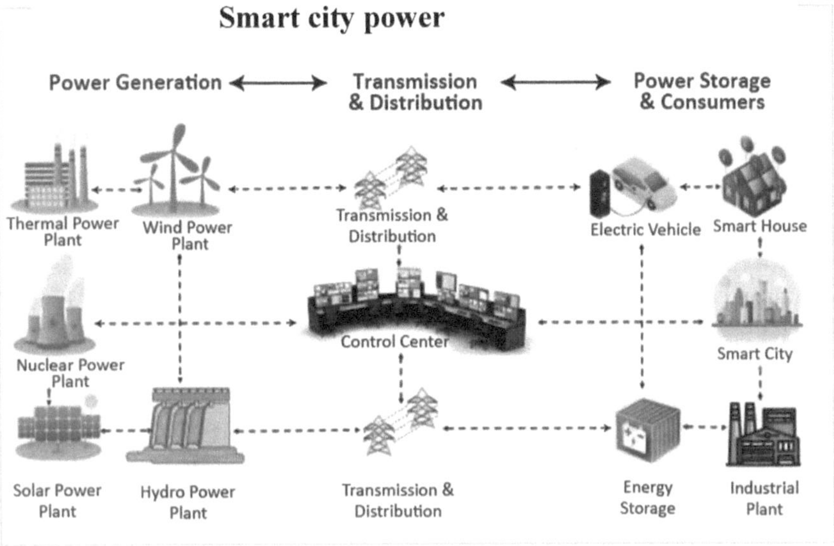

Figure 1.7 Smart grid power.

disturbances in real-time. Such systems employ smart sensors, data analytics, and automated control mechanisms to identify and address power quality issues promptly, minimizing their impact on critical systems and ensuring reliable operation as shown in Figure 1.7. Moreover, incorporating energy storage systems and microgrids into smart city infrastructure can provide backup power during outages, enhance grid stability, and improve power quality and reliability by regulating voltage fluctuations and minimizing disruptions. By prioritizing power quality and reliability in smart city planning and implementation, cities can ensure the uninterrupted and efficient operation of their interconnected systems, promoting sustainable growth and enhancing the overall quality of life for their residents.

1.6.1 Importance of power quality in smart cities

Power quality holds immense importance in the context of smart cities as it directly affects the performance, efficiency, and reliability of various interconnected systems and devices. In a smart city, where a multitude of critical infrastructure relies on digital technologies, maintaining optimal power quality is vital for uninterrupted operations and seamless connectivity. A high-quality power supply ensures that electronic devices, sensors, communication networks, and intelligent systems can function efficiently, accurately, and securely, leading to enhanced service delivery, improved safety, and overall productivity. One of the key reasons power quality is crucial in smart cities is its impact on the performance of sensitive electronic equipment and systems. Voltage sags, surges, and interruptions can cause disruptions, data loss, and system failures, leading to financial losses and service disruptions. In smart grids, for example, power quality issues can hamper the accurate monitoring and control of energy distribution, affecting the reliability and efficiency of the entire grid. Similarly, in smart buildings, poor power quality can result in malfunctions of automation systems, HVAC systems, and other critical infrastructure [30], impacting occupants' comfort and increasing energy consumption. By ensuring high-power quality, smart cities can mitigate these risks and optimize the performance of their interconnected systems, enabling seamless data exchange, efficient energy management, and effective decision-making.

Furthermore, power quality is essential for the successful integration of renewable energy sources and energy storage systems in smart cities. As cities strive to reduce carbon emissions and enhance sustainability, they increasingly rely on renewable energy generation. However, the intermittent nature of renewable sources like solar and wind can introduce power quality challenges, such as voltage fluctuations and harmonics. By addressing power quality concerns, smart cities can effectively integrate renewable energy into their grids, ensuring stability, grid resilience, and optimal utilization of clean energy. Additionally, power quality management facilitates the integration of energy storage systems, enabling cities to store excess energy and release it during peak demand periods, contributing to load balancing and grid stability.

1.6.2 Challenges and solutions for maintaining power quality in power electronic systems

Maintaining power quality in power electronic systems presents several challenges due to the complex nature of these systems and the impact of power disturbances on their performance. However, there are various solutions available to address these challenges and ensure optimal power quality. One major challenge in power electronic systems is the generation of harmonics and distortions that can result from the switching operations of power electronic devices. These harmonics can cause voltage and current distortions, leading to power quality issues. To mitigate this challenge, active power filters and passive filters can be employed to reduce harmonic distortions and maintain a clean power supply. Active power filters use advanced control algorithms to inject compensating currents that cancel out the harmonic components, while passive filters use inductors, capacitors, and resistors to filter out the harmonics. Implementing proper filtering techniques helps to minimize harmonic distortions and maintain power quality within acceptable limits [31].

Another challenge is the voltage fluctuations caused by varying load conditions and grid disturbances. Power electronic systems are susceptible to voltage sags, swells, and interruptions, which can disrupt their operation. Voltage regulators, such as voltage sag compensators and uninterruptible power supplies (UPS), can be utilized to address this challenge. Voltage sag compensators are devices that inject compensating voltages to mitigate the effects of voltage sags, ensuring a stable voltage supply to the power electronic systems. UPS systems, on the other hand, provide backup power during voltage interruptions, preventing disruptions and maintaining continuous operation [32]. By implementing voltage regulation solutions, power electronic systems can operate reliably and maintain power quality even in the presence of voltage disturbances [33].

1.6.3 Reliability considerations and fault-tolerant strategies

Reliability considerations and fault-tolerant strategies are of utmost importance in various systems and industries, particularly in critical applications where system failures can have severe consequences [34]. In the context of power electronics, ensuring reliability is crucial to maintain uninterrupted operation, mitigate downtime, and prevent potential hazards. Fault-tolerant strategies are implemented to detect, isolate, and accommodate faults or failures within power electronic systems, enabling continued operation or a graceful degradation to ensure system reliability. One key aspect of reliability considerations in power electronics is the design of robust and fault-tolerant systems. Redundancy is often employed to enhance reliability by duplicating critical components or subsystems. For example, redundant power modules or converters can be utilized in parallel to provide backup or alternate paths in the event of a failure. Fault detection and diagnostic techniques, such as current and voltage sensing, can be integrated into the system to monitor the health of components and detect abnormalities or faults [35]. Fault isolation techniques, such as circuit breakers or fault detection algorithms, are employed to isolate faulty components or subsystems, preventing the spread of faults and minimizing their impact on the overall system [36]. Additionally, fault-tolerant control strategies such as reconfigurable control algorithms or fault tolerant control hardware [37] can be implemented to ensure system stability and functionality in the presence of faults.

Furthermore, fault-tolerant strategies are designed to provide system resilience and recovery. This involves the implementation of backup power sources, such as UPS or energy storage systems, to ensure a continuous power supply during fault conditions or power outages. These backup systems are equipped with seamless switchover mechanisms that activate in the event of a fault, ensuring uninterrupted power to critical loads. Additionally, fault-tolerant strategies incorporate fault management protocols and automated diagnostic systems that can identify the root cause of faults, expedite troubleshooting, and facilitate timely repairs or replacements. By promptly addressing faults and failures, these strategies reduce downtime, increase system availability, and enhance overall reliability. In summary, reliability considerations and fault-tolerant strategies are essential in power electronics to ensure continuous operation, mitigate the impact of failures, and enhance system resilience. By employing redundancy, fault detection and isolation techniques, and fault-tolerant control strategies, power electronic systems can withstand faults, maintain functionality, and minimize disruptions. Incorporating backup power sources and fault management protocols further enhances system reliability and recovery. These reliability considerations and fault-tolerant strategies play a crucial role in critical applications, where system failures can have significant financial, operational, or safety consequences, promoting the overall dependability and performance of power electronic systems.

1.6.4 Mitigation of harmonics in voltage fluctuations and grid interactions

Mitigating harmonics in voltage fluctuations and grid interactions is essential to maintain power quality and ensure the reliable operation of power electronic systems. Harmonics, which are multiples of the fundamental frequency, can introduce distortion in voltage and current waveforms, leading to various issues such as equipment malfunctions, increased losses, and interference with communication systems. Several strategies can be employed to effectively mitigate harmonics and minimize their impact. One approach is the use of passive and active harmonic filters. Passive harmonic filters consist of passive components such as inductors, capacitors, and resistors that are designed to attenuate specific harmonic frequencies [38]. These filters are connected in parallel with the load to create a low-impedance path for the harmonics, diverting them away from sensitive equipment and preventing their propagation. Active harmonic filters, on the other hand, employ advanced control algorithms and power electronic devices to actively inject compensating currents that cancel out the harmonic components. These filters dynamically adapt to changes in harmonic content and offer precise control over the compensation process, resulting in effective harmonic mitigation [39].

Another approach to mitigate harmonics is through the implementation of grid-interaction standards and guidelines. Utility companies and regulatory bodies often define limits on harmonic distortion levels to ensure power quality and prevent adverse effects on the grid. Compliance with these standards helps to reduce the injection of harmonics into the grid by power electronic systems. It may involve the use of appropriate filtering techniques, compliance testing, and adherence to grid codes and regulations. By adhering to these standards, power electronic systems can minimize their impact on the grid and ensure compatibility with other connected loads and generation sources.

1.7 SUMMARY

The chapter on the fundamentals of power electronics in smart cities provides an overview of the crucial role that power electronics plays in the development and operation of smart cities. It covers the basic principles and key components of power electronics systems and explores their applications and benefits in the context of smart cities. The chapter emphasizes that power electronics is essential for efficient energy conversion, power management, and control in various smart city infrastructure, including smart grids, smart buildings, and electric transportation [40]. It highlights the significance of power quality and reliability considerations in smart cities, discussing the challenges faced in maintaining optimal power quality and the solutions available to address them. The chapter also delves into fault-tolerant strategies and the importance of resilience in power electronics systems, particularly in critical applications where system failures can have severe consequences. Furthermore, the chapter touches upon the integration of power electronics with renewable energy sources, energy storage systems, and grid management techniques, showcasing their role in enhancing sustainability and promoting clean energy utilization in smart cities. In summary, the chapter on the fundamentals of power electronics in smart cities provides a comprehensive overview of the essential aspects of power electronics in the context of smart city development. It highlights the significance of power quality, reliability, and fault-tolerant strategies, while emphasizing the integration of power electronics with renewable energy and energy storage systems. The chapter underscores the vital role that power electronics plays in the efficient and sustainable operation of smart city infrastructure, contributing to improved energy management, enhanced grid stability, and the promotion of clean and reliable power supply.

REFERENCES

[1] K. Dhibi et al., "Reduced Kernel Random Forest Technique for Fault Detection and Classification in Grid-Tied PV Systems," *IEEE J. Photovoltaics*, vol. 10, no. 6, pp. 1864–1871, 2020, doi: 10.1109/JPHOTOV.2020.3011068

[2] M. Manohar, E. Koley, Y. Kumar, and S. Ghosh, "Discrete Wavelet Transform and kNN-Based Fault Detector and Classifier for PV Integrated Microgrid," *Lect. Notes Networks Syst.*, vol. 38, pp. 19–28, 2018, doi: 10.1007/978-981-10-8360-0_2

[3] B. Cai, Y. Zhao, H. Liu, and M. Xie, "A Data-Driven Fault Diagnosis Methodology in Three-Phase Inverters for PMSM Drive Systems," *IEEE Trans. Power Electron.*, vol. 32, no. 7, pp. 5590–5600, 2017, doi: 10.1109/TPEL.2016.2608842

[4] D. E. Kim and D. C. Lee, "Fault Diagnosis of Three-Phase PWM Inverters Using Wavelet and SVM," *J. Power Electron.*, vol. 9, no. 3, pp. 377–385, 2009.

[5] A. S. Syed, D. Sierra-Sosa, A. Kumar, and A. Elmaghraby, "IoT in Smart Cities: A Survey of Technologies, Practices and Challenges," *Smart Cities*, vol. 4, no. 2, pp. 429–475, 2021, doi: 10.3390/smartcities4020024
[6] M. G. M. Almihat, M. T. E. Kahn, K. Aboalez, and A. M. Almaktoof, "Energy and Sustainable Development in Smart Cities: An Overview," *Smart Cities*, vol. 5, no. 4, pp. 1389–1408, 2022, doi: 10.3390/smartcities5040071
[7] T. Nam and T. A. Pardo, "Smart City as Urban Innovation," in *Proceedings of the 5th International Conference on Theory and Practice of Electronic Governance*, September 2011, pp. 185–194, doi: 10.1145/2072069.2072100
[8] O. Majeed Butt, M. Zulqarnain, and T. Majeed Butt, "Recent Advancement in Smart Grid Technology: Future Prospects in the Electrical Power Network," *Ain Shams Eng. J.*, vol. 12, no. 1, pp. 687–695, 2021, doi: 10.1016/j.asej.2020.05.004
[9] E. H. E. Bayoumi, "Power Electronics in Smart Grid Power Transmission Systems: A Review," *Int. J. Ind. Electron. Drives*, vol. 2, no. 2, p. 98, 2015, doi: 10.1504/IJIED.2015.069784
[10] E. B. Tchawou Tchuisseu, D. Gomila, and P. Colet, "Reduction of Power Grid Fluctuations by Communication between Smart Devices," *Int. J. Electr. Power Energy Syst.*, vol. 108, pp. 145–152, 2019, doi: 10.1016/j.ijepes.2019.01.004
[11] C. S. Lai et al., "A Review of Technical Standards for Smart Cities," *Clean Technol.*, vol. 2, no. 3, pp. 290–310, 2020, doi: 10.3390/cleantechnol2030019
[12] I. N. Jiya and R. Gouws, "Overview of Power Electronic Switches: A Summary of the Past, State-of-the-Art and Illumination of the Future," *Micromachines*, vol. 11, no. 12, p. 1116, 2020, doi: 10.3390/mi11121116
[13] N. Holonyak, "The Silicon p-n-p-n Switch and Controlled Rectifier (Thyristor)," *IEEE Trans. Power Electron.*, vol. 16, no. 1, pp. 8–16, 2001, doi: 10.1109/63.903984
[14] M. Guarnieri, "Solidifying Power Electronics [Historical]," *IEEE Ind. Electron. Mag.*, vol. 12, no. 1, pp. 36–40, 2018, doi: 10.1109/MIE.2018.2791062
[15] P. Vladimir and P. Iurii, "Supercapacitor Energy Storages in Hybrid Power Supplies for Frequency-Controlled Electric Drives: Review of Topologies and Automatic Control Systems," *Energies*, vol. 16, no. 7, p. 3287, 2023, doi: 10.3390/en16073287
[16] F. Danish, M. F. Shamsi, A. Kumar, Prashant, A. S. Siddiqui, and M. Sarwar, "Impact Assesment of Microgrid towards Achieving Carbon Neutrality: A Case Study," in *2023 International Conference on Power, Instrumentation, Energy and Control (PIECON)*, February 2023, pp. 1–5, doi: 10.1109/PIECON56912.2023.10085800
[17] "Power Electronics-the Enabling Technology for Renewable Energy Integration," *CSEE J. Power Energy Syst.*, vol. 8, no. 1, pp. 39–52, 2021, doi: 10.17775/CSEEJPES.2021.02850
[18] M. F. Shamsi, F. Danish, M. Sarwar, F. I. Bakhsh, and A. S. Siddiqui, "Transition towards Energy sufficient University Campus through Microgrid: Optimization and Configuration Analysis," in *2023 International Conference on Recent Advances in Electrical, Electronics & Digital Healthcare Technologies (REEDCON)*, May 2023, pp. 437–442, doi: 10.1109/REEDCON57544.2023.10150748
[19] IEEE, *Standard for Interconnection and Interoperability of Distributed Energy Resources with Associated Electric Power Systems Interfaces-2018*. 2018.

[20] M. Liserre, T. Sauter, and J. Hung, "Future Energy Systems: Integrating Renewable Energy Sources into the Smart Power Grid Through Industrial Electronics," *IEEE Ind. Electron. Mag.*, vol. 4, no. 1, pp. 18–37, 2010, doi: 10.1109/MIE.2010.935861

[21] L. Gevorkov, J. L. Domínguez-García, L. T. Romero, and À. F. Martínez, "Modern MultiPort Converter Technologies: A Systematic Review," *Appl. Sci.*, vol. 13, no. 4, p. 2579, 2023, doi: 10.3390/app13042579

[22] H. Emanuel, J. Brombach, R. Rosso, and K. Pierros, "Requirements for Control Strategies of Grid-connected Converters in the Future Power System," *IET Renew. Power Gener.*, vol. 14, no. 8, pp. 1288–1295, 2020, doi: 10.1049/iet-rpg.2020.0156

[23] W. Yuan, T. Wang, and D. Diallo, "A Secondary Classification Fault Diagnosis Strategy Based on PCA-SVM for Cascaded Photovoltaic Grid-connected Inverter," in *IECON Proc. (Industrial Electron. Conf.)*, vol. 2019-Octob, pp. 5986–5991, 2019, doi: 10.1109/IECON.2019.8927090

[24] Z. K. Hu, W. H. Gui, C. H. Yang, P. C. Deng, and S. X. Ding, "Fault Classification Method for Inverter Based on Hybrid Support Vector Machines and Wavelet Analysis," *Int. J. Control. Autom. Syst.*, vol. 9, no. 4, pp. 797–804, 2011, doi: 10.1007/s12555-011-0423-9

[25] W. Gong et al., "A Novel Deep Learning Method for Intelligent Fault Diagnosis of Rotating Machinery based on Improved CNN-SVM and Multichannel Data Fusion," *Sensors (Switzerland)*, vol. 19, no. 7, 2019, doi: 10.3390/s19071693

[26] M. Haque, M. N. Shaheed, and S. Choi, "Deep Learning Based Micro-Grid Fault Detection and Classification in Future Smart Vehicle," in *2018 IEEE Transportation Electrification Conference and Expo (ITEC) Conference*, pp. 201–206, 2018, doi: 10.1109/ITEC.2018.8450201

[27] Q. Sun, X. Yu, and H. Li, "Open-Circuit Fault Diagnosis Based on 1D-CNN for Three-Phase Full-Bridge Inverter," in *Proceedings of the 11th International Conference on Prognostics and System Health Management (PHM-2020 Jinan)*, pp. 322–327, 2020, doi: 10.1109/PHM-Jinan48558.2020.00064

[28] W. Gong, H. Chen, Z. Zhang, M. Zhang, and H. Gao, "A Data-Driven-Based Fault Diagnosis Approach for Electrical Power DC-DC Inverter by Using Modified Convolutional Neural Network with Global Average Pooling and 2-D Feature Image," *IEEE Access*, vol. 8, pp. 73677–73697, 2020, doi: 10.1109/ACCESS.2020.2988323

[29] S. Bolognani, M. Zordan, and M. Zigliotto, "Experimental Fault-Tolerant Control of a PMSM Drive," *IEEE Trans. Ind. Electron.*, vol. 47, no. 5, pp. 1134–1141, 2000, doi: 10.1109/41.873223

[30] D. Mohammadi and S. Ahmed-Zaid, "Active Common-Mode Voltage Reduction in a Fault-Tolerant Three-Phase Inverter," in *Annual IEEE Conference on Applied Power Electronics Conference and Exposition (APEC)*, vol. 2016-May, no. 1, pp. 2821–2825, 2016, doi: 10.1109/APEC.2016.7468264

[31] W. Wang, J. Zhang, and M. Cheng, "Common Model Predictive Control for Permanent-Magnet Synchronous Machine Drives Considering Single-Phase Open-Circuit Fault," *IEEE Trans. Power Electron.*, vol. 32, no. 7, pp. 5862–5872, 2017, doi: 10.1109/TPEL.2016.2621745

[32] A. Malik, A. Haque, V. S. B. Kurukuru, M. A. Khan, and F. Blaabjerg, "Overview of Fault Detection Approaches for Grid Connected Photovoltaic Inverters,"

e-Prime – Adv. Electr. Eng. Electron. Energy, vol. 2, p. 100035, 2022, doi: 10.1016/J.PRIME.2022.100035

[33] M. A. Khan, A. Haque, and K. V. S. Bharath, "Droop Based Low Voltage Ride through Implementation for Grid Integrated Photovoltaic System," in *2019 IEEE International Conference on Power Electronics, Control and Automation (ICPECA)*, Nov 2019. pp. 1–6, doi: 10.1109/ICPECA47973.2019.8975467

[34] A. Malik, A. Haque, and K. V. Satya Bharath, "Fault Tolerant Inverter for Grid Connected Photovoltaic System," in *2022 IEEE International Conference on Power Electronics, Smart Grid, and Renewable Energy (PESGRE)*, Jan. 2022, pp. 1–6, doi: 10.1109/PESGRE52268.2022.9715825

[35] V. S. Bharath Kurukuru, A. Haque, R. Kumar, M. A. Khan, and A. K. Tripathy, "Machine Learning Based Fault Classification Approach for Power Electronic Converters," in *9th IEEE International Conference on Power Electronics, Drives and Energy Systems (PEDES) 2020*, 2020, doi: 10.1109/PEDES49360.2020.9379365

[36] M. A. Khan, A. Haque, and V. S. Bharath Kurukuru, "Reliability Analysis of a Solar Inverter during Reactive Power Injection," in *9th IEEE International Conference on Power Electronics, Drives and Energy Systems (PEDES) 2020*, 2020, doi: 10.1109/PEDES49360.2020.9379776

[37] R. L. De Araujo Ribeiro, C. B. Jacobina, E. R. Cabral da Silva, and A. M. Nogueira Lima, "Fault-Tolerant Voltage-fed PWM Inverter AC Motor Drive Systems," *IEEE Trans. Ind. Electron.*, vol. 51, no. 2, pp. 439–446, 2004, doi: 10.1109/TIE.2004.825284

[38] V. S. B. Kurukuru, A. Haque, M. A. Khan, S. Sahoo, A. Malik, and F. Blaabjerg, "A Review on Artificial Intelligence Applications for Grid-Connected Solar Photovoltaic Systems," *Energies*, vol. 14, no. 15, p. 4690, 2021, doi: 10.3390/en14154690

[39] M. Moujahed, H. Ben Azza, K. Frifita, M. Jemli, and M. Boussak, "Fault Detection and Fault-Tolerant Control of Power Converter Fed PMSM," *Electr. Eng.*, vol. 98, no. 2, pp. 121–131, 2016, doi: 10.1007/s00202-015-0350-5

[40] Y. Song and B. Wang, "Analysis and Experimental Verification of a Fault-Tolerant HEV Powertrain," *IEEE Trans. Power Electron.*, vol. 28, no. 12, pp. 5854–5864, 2013, doi: 10.1109/TPEL.2013.2245513

Chapter 2

Fundamentals of renewable energy resources for smart cities

Maneesh Kumar
Yashwantrao Chavan College of Engineering (YCCE), Nagpur, India

Himanshu Sharma
Noida Institute of Engineering & Technology, Greater Noida, India

Sourav Diwania
KIET Group of Institutions, Ghaziabad, India

2.1 INTRODUCTION: SMART CITIES

Till 2020, urbanization has already increased at an unprecedented pace because of the migration of huge populations from rural areas. At the same time, the modernization of the cities has also expanded. The interconnection and connectivity between urban and rural areas are well established now as the online (internet) and offline (rail and road) routes are being developed in a positive way. Though the addition of human capital into the urban areas increases many economic activities such as the production of goods and services, on the other hand, the ever-increasing transfer of human capital toward unplanned cities continuously increases the burden in terms of food, education, healthcare services, transportation, sanitation, electricity, and so on [1].

To overcome these issues, with respect to the burden on cities and to be able to develop and manage them in a more appropriate way, the concept of smart cities emerged. The smart cities concept is also accommodating and incorporates the fundamentals of sustainable development. A number of objectives have been targeted under the smart cities concept, viz. smart buildings, smart sanitation and waste-management systems, smart healthcare systems, smart transport modes, and so on [2]. With a sustainable goal, smart cities also encourage the utilization of renewable energy to provide low-carbon, energy-based alternatives to the connected load. For this purpose, small-scale smart microgrids are being installed near or in combination with the smart buildings to provide a reliable power supply with reduced system losses. Microgrids inherently encourage the penetration of renewable energy sources (RESs) into the system. These are mainly of two types, the grid-connected and the isolated type [3]. The stability and control of grid-connected microgrids are comparatively easy compared to isolated microgrids, as these grids are supported by the main supply grid [4].

DOI: 10.1201/9781032669809-2

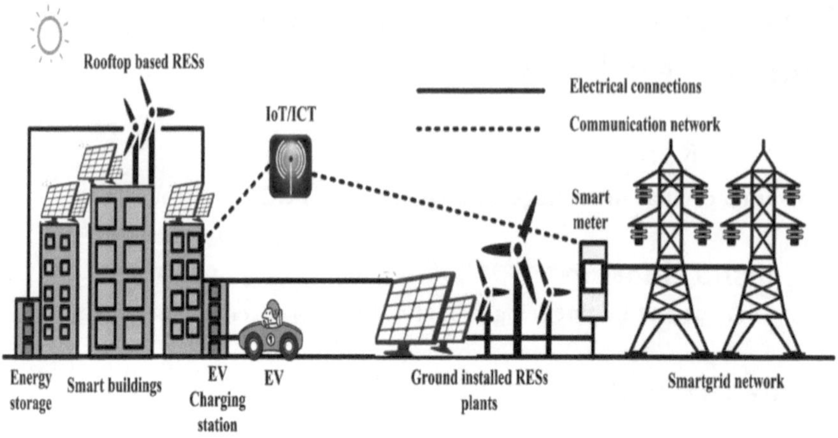

Figure 2.1 A typical RES-based smart city structure [6].

Nevertheless, the penetration of RESs reduces carbon footprint in an effective manner [5]. A typical smart city model with various features is shown in Figure 2.1.

The internet of things (IoT) or intelligent communication techniques (ICT)-enabled infrastructure are two of the key components of smart cities, which provide information and data flow between various nodes. These devices also encompass a number of sensors, software, and technologies to provide access to the data flow. These systems also enhance the monitoring and security of smart cities. Further, electric vehicles are an integral part of smart cities in providing ease of transport. These vehicles can also contribute to the peak load requirement of the city. There are multiple charging stations available at various locations in the smart cities to charge EVs. The energy storage systems (ESS) are also a part of smart city infrastructure. Nowadays, Li-ion-based batteries attract the attention of researchers because of its various advantages over other available batteries, such as high efficiency, low self-discharge, and acceptable performance with temperature variations.

2.2 ENERGY MANAGEMENT FOR SMART BUILDINGS

The buildings under the smart city's infrastructure are also smart systems and are capable of managing various connected loads such as lighting, Plug-in Electric Vehicles (PHEVs), charging stations in smart parking lots, Heating, Ventilation, and Air-Conditioning (HVAC), or comfort management system for residents, motors (elevator, escalators), pumps, and so on [6]. Furthermore, the integration of these buildings to the smart grid is done to increase the reliability of the power supply and for economic benefits by selling excess power when generation is surplus [7]. An appropriate energy management scheme of the smart building is required while integrating multiple technologies together. For serving this purpose, many optimization techniques are available. The objective of formulating an energy management problem shall be the minimization of the usage of grid power by supplying overall load through in-house generation and storage. This aspect of a smart building is known as the net-zero energy building (nZEB), which is capable of supplying all of its connected load. A multi-objective energy management approach for smart buildings considering conflicting objectives for the consumer's comfort is proposed [8]. These kinds of multi-objective problems also require a suitable demand response (DR) approach to match the generation-demand gaps; certain techniques, viz. automated DR approach, are proposed [9].

2.3 CONTROL TECHNIQUES FOR SMART BUILDINGS

The requirement for effective and efficient energy management of a smart building emerges because of the integration of multiple sources, load, and storage, and also uncertainty associated with renewable energy-based sources. In order to meet the generation demand, it is essential to model the whole building components individually and aggregate them to form the optimal energy management function. For the implementation, an objective or a cost function needs to be defined for the energy management of the smart building. The main objectives for an effective energy management scheme for a smart building can be to minimize the grid energy intake, maximize the RESs share, reduce the cost of energy required to feed the overall connected load to the smart building, implement an appropriate HVAC system, shifting the non-critical load demand to critical load demand during peak hours, control of power (active-reactive) and voltage for the smart buildings, temperature control of heating appliances available in the building, and so on. In this regard, a model predictive control (MPC) strategy is proposed [10] under a smart building infrastructure to control the voltage/VAR. A temperature control scheme for heating appliances connected to a smart building is proposed [11] using a discrete-event system-based approach. This system allows the modeling and design of the power control technique in an appropriate way for feasible scheduling. An MPC-based temperature control approach for the commercial building is proposed [12], considering comfort level as one of the parameters. A robust decentralized voltage control scheme for the isolated microgrid system proposed [13] can also be utilized for an isolated BIMG system with various uncertainties such as renewable energy intermittency, load contingencies, and so on. An intelligent autonomous load control approach with automatic features is proposed for a smart building [14]. A distributed real-time control of the HVAC system is proposed for a cost-effective commercial building under a smart grid environment [15]. For this purpose, the cost associated with energy

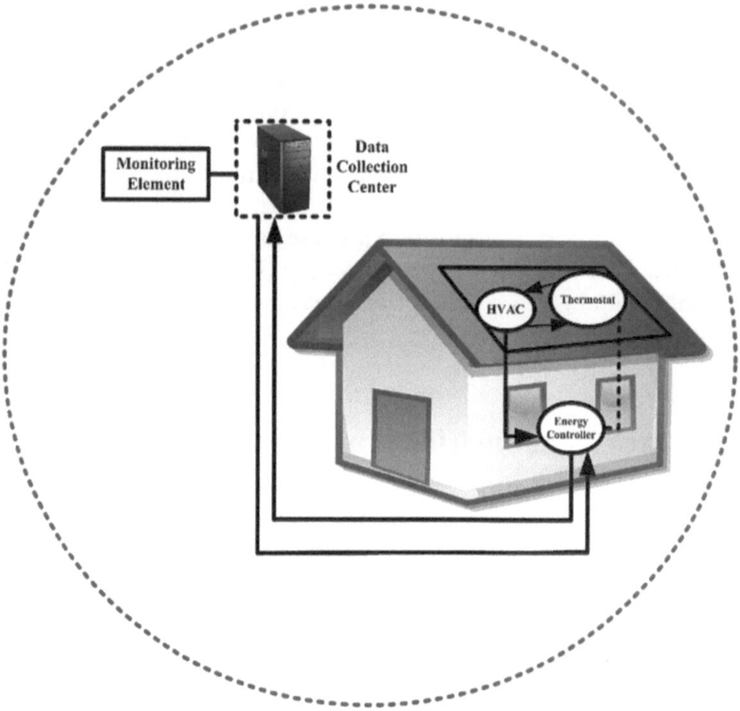

Figure 2.2 A distributed MPC-based control architecture for smart buildings [14].

use, outdoor temperature, comfortable temperature levels, and the external disturbance stochastic model is formulated. Figure 2.2 shows a distributed MPC for a smart building for thermal management. The data pertaining to the HVAC unit shares the data collection unit, which is monitored for corrective actions with respect to the predefined objectives such as desired comfort level and temperature control. The processed data is utilized by the energy controller for the required actions.

A smart building model also incorporates many stochastic factors such as RESs availability, load demand, and so on. The available state of charge of the battery storage and the energy pricing should also be incorporated such that utilization of the storage system can be improved with respect to the pricing of energy from the grid.

2.3.1 Fundamentals on MPC

Various complex electric networks and power electronics modules are required to be controlled while incorporating forecasted data and the available system constraints. The presence of RESs, with their intermittent nature, makes the overall system analysis and load balancing more challenging. This also makes the system a dynamic building energy management system (BEMS). The model of the overall dynamic system is analyzed through the MPC approach, which was developed in the 1980s. During that period, this approach is extensively used for multiple processes such as chemical industries, oil refineries, and so on. This approach provides optimization of a present action while considering future time slots.

The MPC theory relies on iterative and finite-horizon optimization of a system model. At time instant, 't' the current states of the plant are sampled and a cost minimization function control approach is calculated using a numerical algorithm for a relatively short future time-horizon $[t, t+T]$. In the majority of the cases, an online computation is utilized to find the state trajectories that originate from the present state and determine (using the solution of Euler–Lagrange equations) a cost-minimizing control approach until the predictive time-horizon $[t+T]$. With this method, we can forecast future events, which will be used for the current control actions. It is basically an optimal control approach that works in an iterative manner, that is, it optimizes the finite time-horizon over a predefined predictive horizon; this includes only the current timeslot over the control horizon and then repeats the optimization process.

2.4 RENEWABLE ENERGY TECHNOLOGIES (RETS) FOR SMART CITIES

Renewable energy is considered to be the illimitable energy that can be harnessed for overall sustainable development. Renewable energy technologies can be incorporated with many available RESs such as solar, wind, biomass, tidal, geothermal, and some of the mini/micro hydropower plants, and so on. Various available RETs are shown in Figure 2.3. Out of the available RESs, solar and wind are considered to be the maximum utilizing sources because of their broad availability which also gives them an edge over other sources for their utilization in smart cities. Now the big challenge comes when we look for the integration of these sources into our grid to extract electric power. One of the ways is to use multiple power electronic devices, but at the same time, excessive use of power electronics-based components reduces the system reliability by injecting some unwanted quantities into the system.

2.4.1 Types of renewable energy resource for smart cities

A renewable energy source means energy that is sustainable – something that can't run out, or is endless, like the sun. When you hear the term 'alternative energy' it usually refers to renewable energy sources too. It means

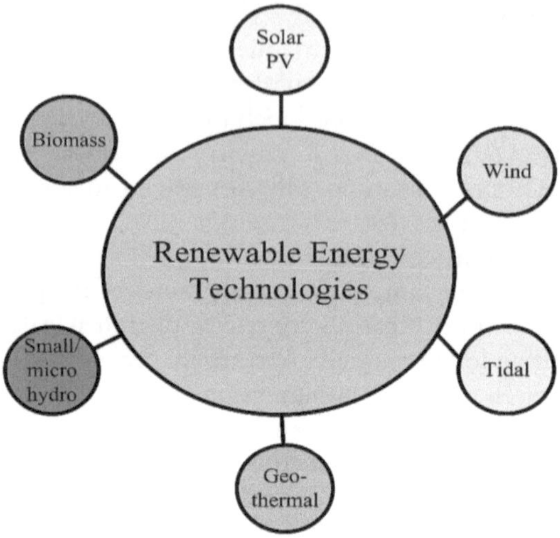

Figure 2.3 Various available RETs.

sources of energy that are alternative to the most commonly used non-sustainable sources – like coal. The most popular renewable energy sources currently are:

Solar energy
Wind energy
Hydro energy
Tidal energy
Geothermal energy
Biomass energy

- **Solar energy:** Sunlight is one of our planet's most abundant and freely available energy resources. The amount of solar energy that reaches the earth's surface in one hour is more than the planet's total energy requirements for a whole year. Figure 2.4 shows the PV panels to generate electrical energy.
- **Wind energy:** Wind is a plentiful source of clean energy. Wind farms are an increasingly familiar sight in the UK with wind power making an ever-increasing contribution to the National Grid. To harness electricity from wind energy, turbines are used to drive generators which then feed electricity into the National Grid. Figure 2.5 shows the wind turbine system for electricity generation.
- **Hydro energy:** As a renewable energy resource, hydropower is one of the most commercially developed. By building a dam or barrier, a large reservoir can be used to create a controlled flow of water that will drive a turbine, generating electricity. Figure 2.6 shows a typical hydro energy station.

Figure 2.4 Solar PV panels.

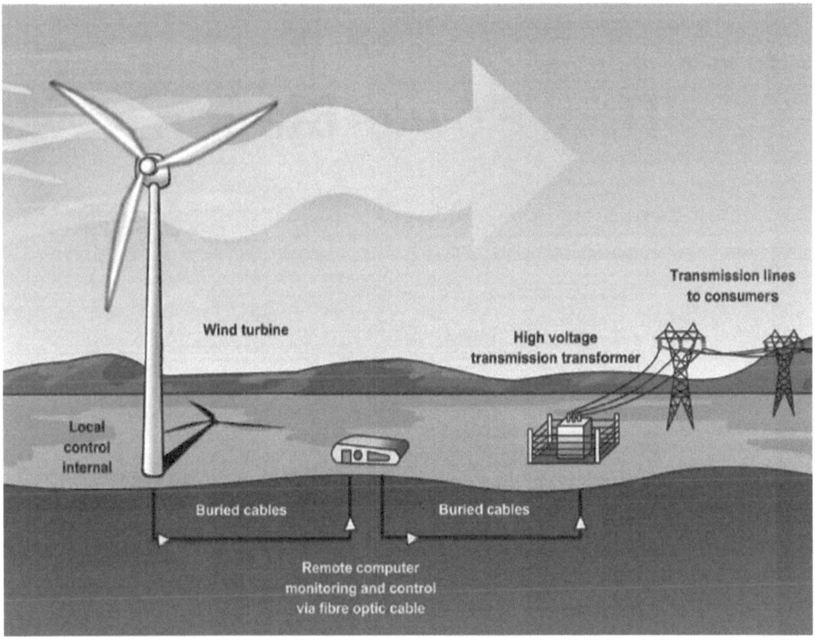

Figure 2.5 Wind turbine power generation and transmission.

- **Tidal energy:** This is another form of hydro energy that uses twice-daily tidal currents to drive turbine generators. Figure 2.7 gives a brief of tidal energy power station.
- **Geothermal energy:** By harnessing the natural heat below the earth's surface, geothermal energy can be used to heat homes directly or to generate electricity. Figure 2.8 shows a geothermal power station where the thermal energy is utilized to produce the electricity.

Figure 2.6 Hydro power station.

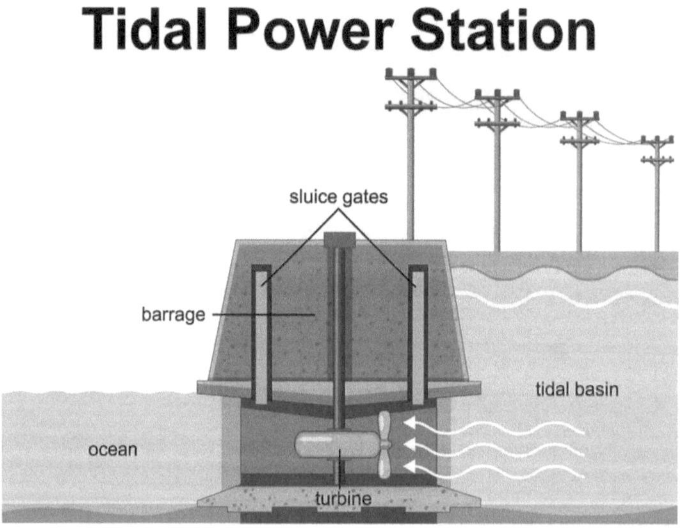

Figure 2.7 Tidal power station.

- **Biomass Energy:** This is the conversion of solid fuel made from plant materials into electricity. Figure 2.9 shows the biomass gasifier unit to produce the electricity from a biogas plant.

These integrations are more challenging in the isolated type of grid systems. Therefore, while adding renewable energy technology into the system, one has to look over the aforementioned issues. Solar photovoltaic (SPV) and wind turbines are installed in many ways to extract electrical energy from

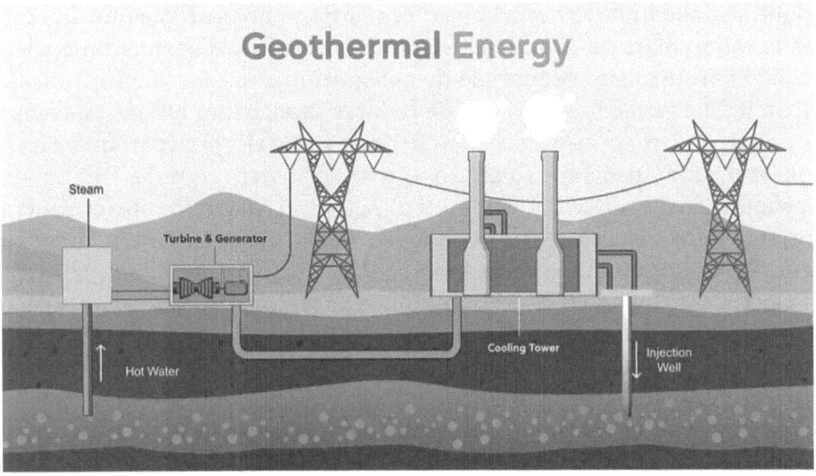

Figure 2.8 Geothermal power station.

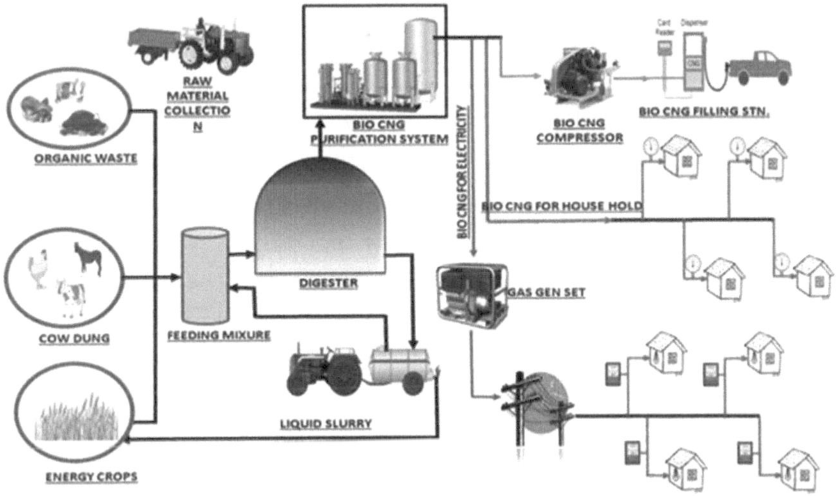

Figure 2.9 Biomass power generation.

them. The solar rooftop, the small-scale vertical axis (VAWT), and the horizontal axis (HAWT) wind turbine on the roof of the building as well as on the ground are examples of them. The establishment of smart cities also incorporates the goals of sustainable developments; therefore, the RESs can play a vital and undoubtedly crucial role in the smart cities' developments and plans.

The solar PV and the wind turbine units are vastly utilized for the production of electricity to reduce carbon emissions and to achieve sustainable energy goals. These sources can be used either in rooftop mode or as a ground-installed unit. The various types of RESs are available for integration into the smart city through a smart grid network. The intermittency of these RESs is the main issue while the integration into the system affects the dispatch. The grid-connected system reduces these issues but also increases the cost of electricity as we need to purchase more electricity from the grids to meet the load demand. To get an optimal dispatch from the RESs, more appropriate data modeling is required. A typical layout of power generation, transmission, and distribution along with storage systems is shown in Figure 2.10. The optimal dispatch of the RESs, which is required to provide the electrical supply to the smart buildings or the smart cities network, is of the most importance. For this, proper optimal energy management is required. This evolves a building-integrated microgrid system (BIMS) to be incorporated into the smart city, which will be capable of supplying required power and can be capable of selling the power in case of excess energy [16, 17].

The increasing energy demands in the modern buildings makes the system unavoidable to add supply from the smart grid network that makes the BIMS a grid-connected microgrid system. The various critical components of the BIMS are shown in Figure 2.11. These are rooftop renewable energy systems (such as small-scale solar and wind turbine unit), building-integrated energy storage systems, IoT-enabled devices, building-integrated EV charging stations, smart lighting systems, cyber security systems, smart camera systems, diesel engine generator (DEG) unit for backup purpose in case of isolated BIMS, smart heat sensors, and a proper energy management system (EMS) for integrating all the components as well as web connectivity. For the task of monitoring, protecting, and controlling BIMS, it is essential to deploy various sensors and relay devices at specific locations to provide the right feedback to the controller units [18].

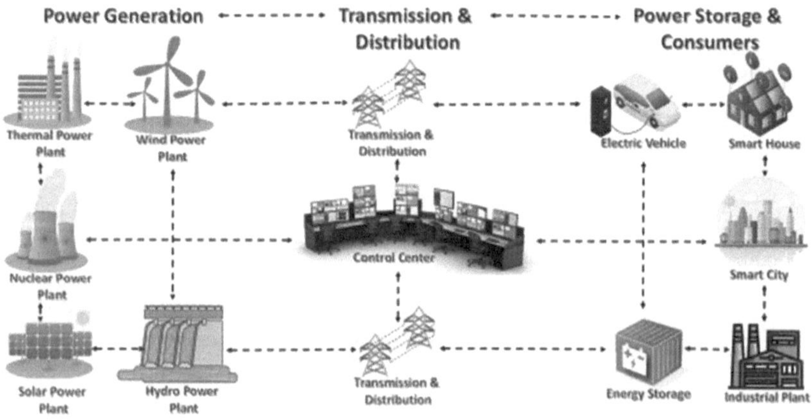

Figure 2.10 Power generation, transmission and distribution.

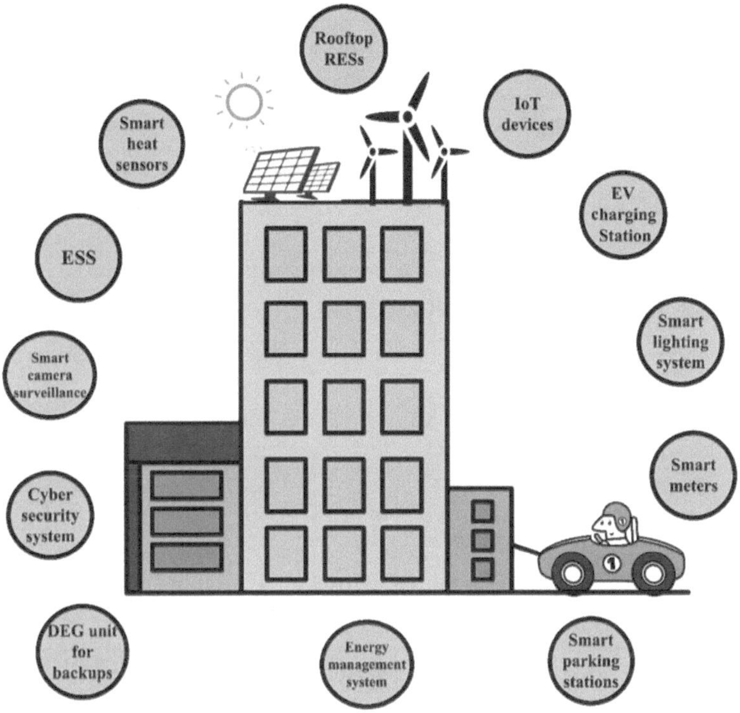

Figure 2.11 A typical BIMS system.

2.5 PSIM SIMULATION ON SPV-POWERED SYSTEM

2.5.1 DC subsystem requirements

The DC subsystem block diagram is shown in Figure 2.12. Here the photovoltaic cell provides input to the DC-DC boost converter. The boost converter shall accept a voltage from the photovoltaic cells.

- The input voltage shall be 48 Volts.
- The average output shall be 200 Volts +/– 25 Volts.
- The voltage ripple shall be less than 20 Volts.
- The open-loop boost converter shall operate above 65% efficiency.

The boost converter shall perform maximum power point tracking.

- The PWM of the boost converter shall be regulated based on current and voltage from the PV array.
- The efficiency of the MPPT system shall be above 80%.

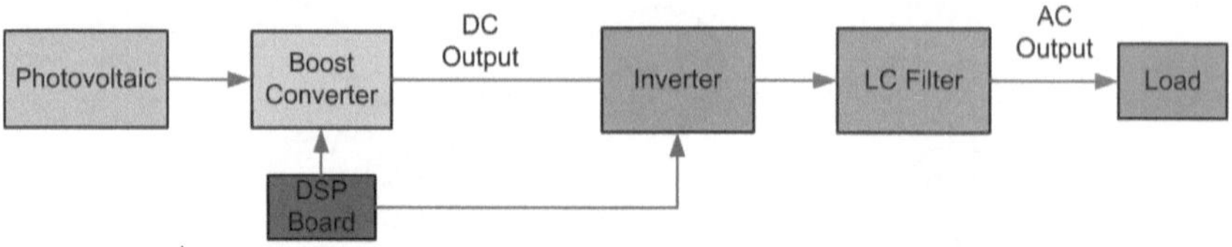

Figure 2.12 DC subsystem block diagram.

2.5.2 AC subsystem requirements

The AC side of the system shall invert the output of the boost converter.

- The output of the inverter shall be AC voltage.
- The output shall be 60 Hz +/– 0.1 Hz.

The inverter output shall be filtered by a LC filter.

- The filter shall remove high switching frequency harmonics.
- Total harmonic distortion of the output shall be less than 15%.

The AC subsystem block diagram is shown in Figure 2.13.

2.5.3 DC-DC converter

In solar energy harvesting the input voltage from the solar panel is an unregulated DC voltage. To regulate this dc voltage the DC-DC converters are used.

- **DC-DC converters control techniques:**
  Two control techniques for DC-DC converters have been analyzed as:
  - Time ratio control (TRC)
  - Hysteresis control or current limit control (CLC)

  The hysteresis control of DC-DC converter is better because:
  - The current never becomes discontinuous
  - Over current protection can be done.

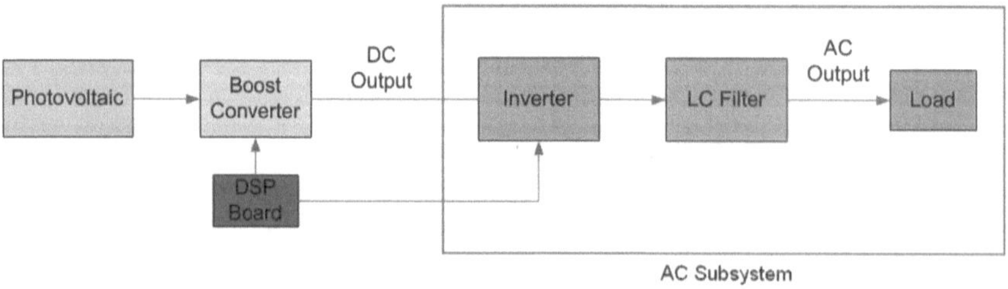

Figure 2.13 AC subsystem block diagram.

- **Power loss:**
  The total power loss of a DC-DC converter is modeled as:

$$P_{\text{Loss}} = P_{\text{MOSFET}} + P_{\text{Diode}} + P_{\text{inductor}} + P_{\text{capacitor}} + P_{\text{control}} \tag{2.1}$$

The power conversion efficiency of DC-DC boost converter is given as $(\eta) = P_o/P_{PV}$, where P_o is output power of DC-DC converter, and P_{PV} is called solar panel output power. The DC-DC boost converter efficiency (η) can further be increased using advanced control techniques.

Mathematical modeling of boost converter:

- ***I–V* Equation of ideal solar cell:**

$$I = I_L - I_0\left(\exp\left(\frac{qv}{kT}\right) - 1\right) \tag{2.2}$$

- ***IV* equation of equivalent circuit of solar cell:**

$$I = I_L - I_0\left(\exp\left(\frac{q(V + IR_s)}{nkT}\right)\right) - \left(\frac{V + IR_s}{R_{sh}}\right) \tag{2.3}$$

Where n = ideality factor of diode (1 = ideal diode, 2 = practical diode)
R_s = series resistance
R_{sh} = shunt resistance

- **Solar cell efficiency (η):**

$$\text{Solar cell efficiency}\,(\eta) = \frac{\text{FF}.V_{\text{oc}}.I_{\text{sc}}}{P_{\text{in}}} \tag{2.4}$$

Where V_{oc} = open circuit voltage, I_{sc} = short circuit current, FF = fill factor, P_{in} = incident optical power.

- **Fill factor (FF):**

$$\text{Fill factor}\,(\text{FF}) = \frac{I_m.V_m}{I_{\text{sc}}.V_{\text{oc}}} \tag{2.5}$$

- **Boost converter output voltage (V_o):**

$$V_{\text{out}} = \frac{D}{1 - D}.V_{\text{in}} \tag{2.6}$$

- **Boost converter efficiency (η):**

$$\text{Boost converter efficiency}(\eta) = P_o / P_{in} \tag{2.7}$$

Practically, boost converter efficiency = 99%

- **Rechargeable battery model:**

$$\text{NiCd battery model} = 1.5\text{v}, 600 - 1000\,\text{mAh}$$

Rechargeable battery is used as the load.

A 3.3v solar energy harvesting DC-DC boost converter is shown in Figure 2.14.

2.5.4 DC-DC converter simulation work using PSIM

Table 2.1 provides the solar panel and boost converter parameters.

Table 2.1 Solar panel & boost converter parameters

Name	*Value*
No. of cells in series (N_s)	2 (0.8v each)
Max. power ($P_{max.}$)	0.5 watts
Silicon material Energy band gap (E_g)	1.12 eV
Solar panel (2 solar cells in series connection)	V_{oc} = 1.6 V, I_{sc} = 2 A, V_m = 1.5 V, I_m = 1.8 A
V_{in}	1.5v dc (rippled)
V_o	3.3v dc
Light intensity	1000 watts/m^2 or 100 milli watts/cm^2
Temperature	25 degree Celsius
Capacitor, C1	100 uF, 5v
Diodes, D1, D2 Threshold voltage, Internal resistance	0.7 V, 25 ohm
Gating signal, G1	5 kHz
Inductor, L1	200 uH, 5v
MOSFET (MOS1)	2222 A
Load resistor, RL	100 ohm (WSN node)

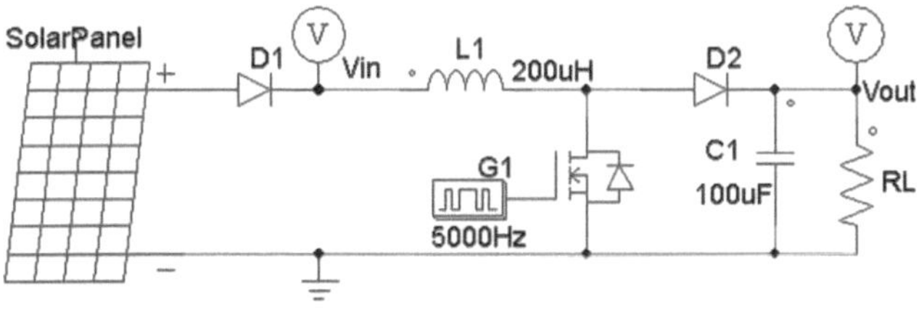

Figure 2.14 A 3.3v solar energy harvesting DC-DC boost converter.

2.5.5 Simulation results

Single-phase bridge inverter:

Figure 2.15 shows the input and output voltages of the boost converter in a single graph (Simulation Time = 10 sec). Figure 2.16 shows the input and output voltages of the boost converter in a single graph (Simulation Time = 0.1 sec).

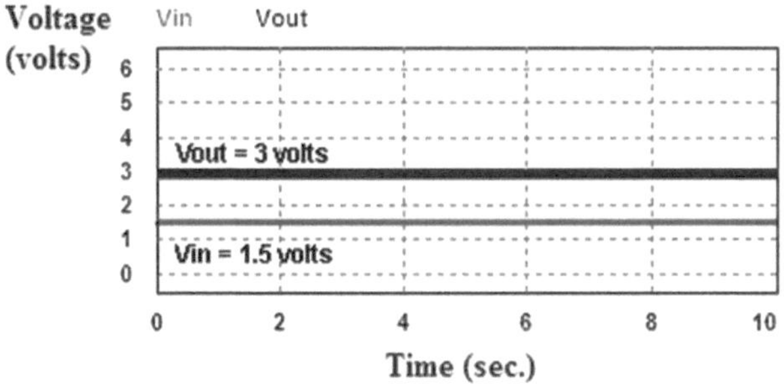

Figure 2.15 Input and output voltages of boost converter in a single graph (Simulation Time = 10 sec).

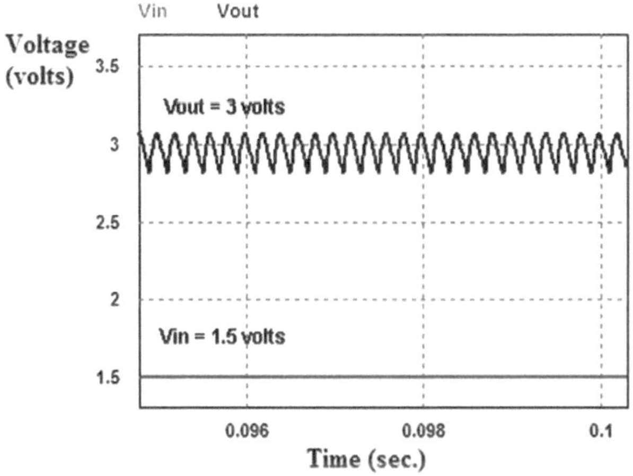

Figure 2.16 Input and output voltages of boost converter in a single graph (Simulation Time = 0.1 sec).

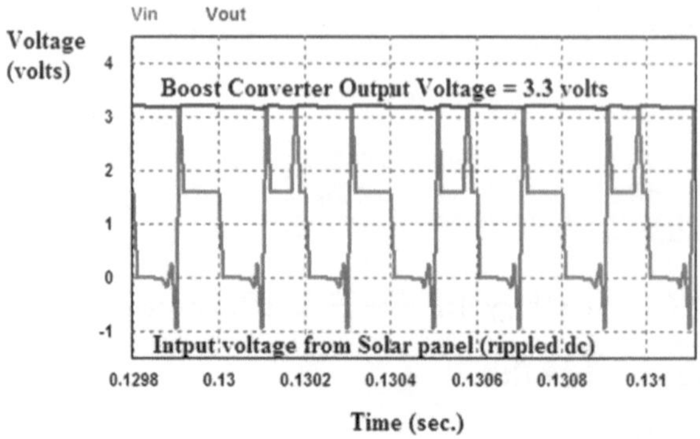

Figure 2.17 Final input and output voltages of boost converter after the capacitor C_1. (Simulation Time = 0.1 sec).

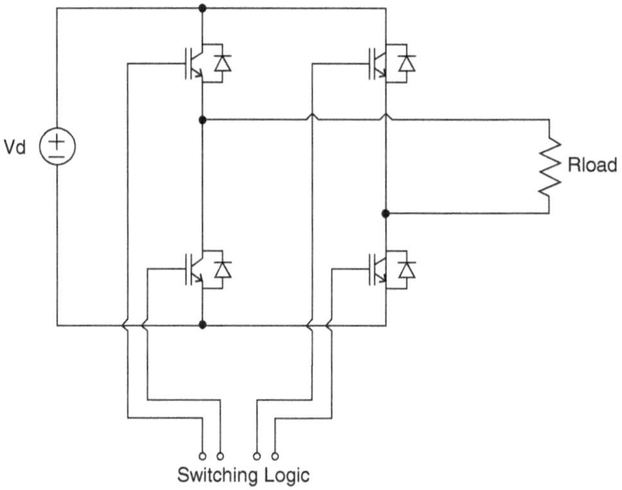

Figure 2.18 Single-phase bridge inverter.

Figure 2.17 shows the final input and output voltages of boost converter after the capacitor C1. (Simulation Time = 0.1 sec). Figure 2.18 shows the single-phase bridge inverter.

Design equations:

- Modulation index (m_i):

$$m_i = \frac{\hat{V}_{\text{control}}}{\hat{V}_{\text{triangle}}} \tag{2.8}$$

- **Frequency modulation ratio (m_f):**

$$m_f = \frac{f_{triangle}}{f_{carrier}} \tag{2.9}$$

- **Fundamental output magnitude:**

$$\left(\hat{V}_{out}\right)_1 = m_i{}^* V_d \tag{2.10}$$

- **Output frequency:**

$$f_{output} = f_{control} \tag{2.11}$$

The modulation index (m_i) can be used to control output magnitude (voltage).

$$\left(\hat{V}_{out}\right)_1 = m_i{}^* V_d \tag{2.12}$$

$$m_i = \frac{\hat{V}_{control}}{\hat{V}_{triangle}} \tag{2.13}$$

- Typically $0 < m_i \le 1$
- Overmodulation if $m_i > 1$ (nonlinear operation)

Figure 2.19 shows the Pulse Width Modulation (PWM) wave cycle.

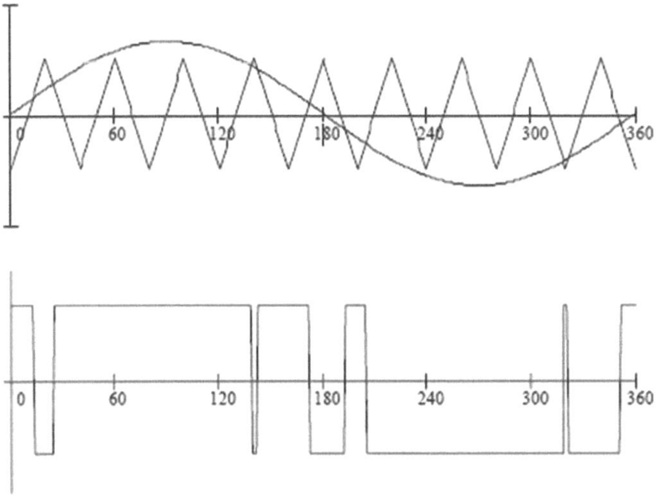

Figure 2.19 PWM wave cycle.

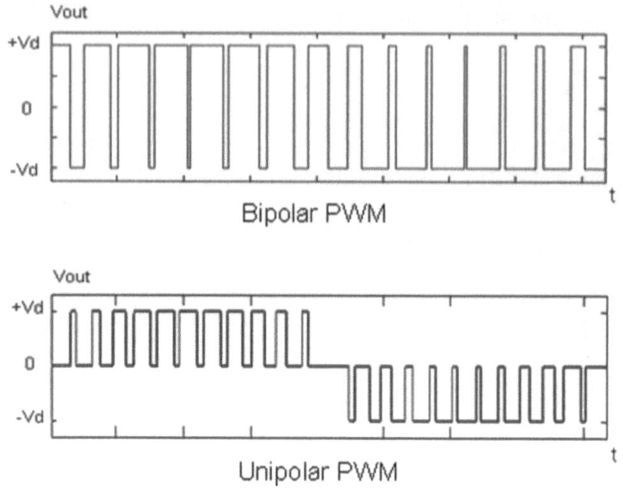

Figure 2.20 Bipolar and unipolar PWM.

Figure 2.20 shows the Bipolar and unipolar PWM.

The designer can select m_f to remove even harmonics from output spectrum

- For bipolar PWM, m_f = odd integer
- For unipolar PWM, m_f = even integer

$$m_f = \frac{f_{triangle}}{f_{carrier}} \tag{2.14}$$

The output isn't very sinusoidal; therefore, we use an LC filter.

LC filter:

Figure 2.21 shows the Inductor-Capacitor (LC) filter. Figure 2.22 shows the Magnitude response of LC filter.

- Goal: Smooth inverter output to smooth AC
- Second order LC filter transfer function:

$$G(s) = 1/\left(L^*C^*s^2 + 1\right)$$

- $f_{carrier}$ < cutoff frequency < $f_{carrier} \cdot m_f$

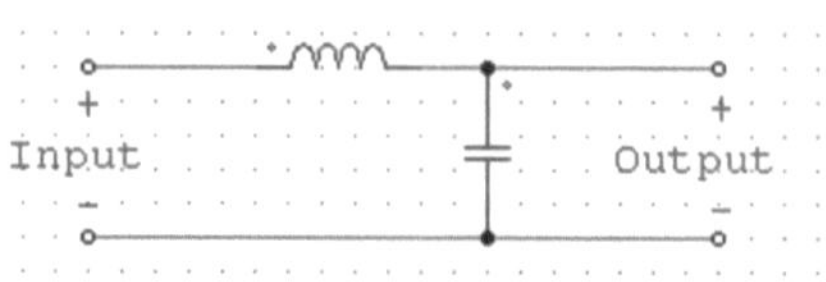

Figure 2.21 LC filter.

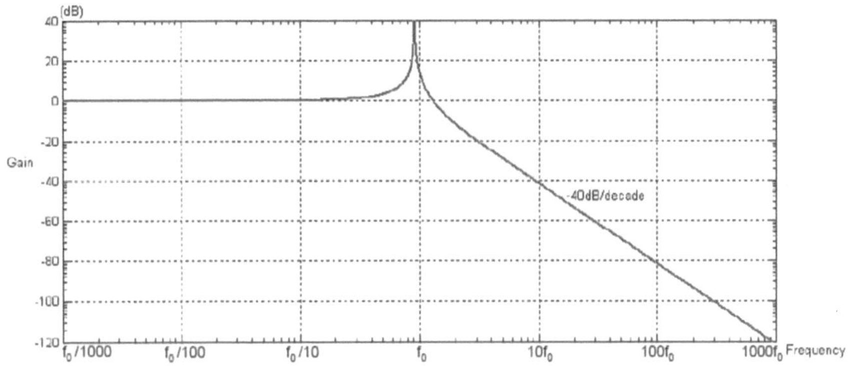

Figure 2.22 Magnitude response of LC filter.

2.6 POWER MANAGEMENT OF DIFFERENT RETS TO BE ELABORATED

2.6.1 Application of artificial intelligence (AI) for smart cities

In this section, the application of machine learning (ML), which is used to forecast some of the time-varying data such as load demand and renewable energy generation, is discussed in brief. The centralized control through a Field-Programmable Gate Array (FPGA) has also been described in the section.

As discussed earlier, the implementation of the RESs application for smart cities requires accurate and feasible data modeling and forecasting. The use of intelligent techniques, such as the use of machine learning, fuzzy logic system, evolutionary optimizations, multi-agent systems (MASs), and data mining, and so on, which are inherently the subsets of AI, are included at a very fast pace. Some of the AI application-based examples are web search engines such as Google, recommendation systems that are found on Amazon, Netflix, YouTube, Facebook, and so on, recognizing human tasks such as Siri and Alexa, for example. These techniques are not only capable of handling very large data but also provide more accurate and fast forecasting results. Today, many fast-speed computational algorithms are available and are capable of providing faster results. Therefore, the energy from the RESs can be utilized in smart cities in many ways.

One of the ways is to form a home energy management system (HEMS) where the daily demand of the residents can be met with the help of forecasting the historical demand data through ML and feeding through the BIMS connected sources in an optimal manner so that the generation demand can be matched. A simple HEMS structure is shown in Figure 2.23. Here, an FPGA-based central controller is used for controlling various critical and non-critical loads [19]. The demand-side management (DSM) can be

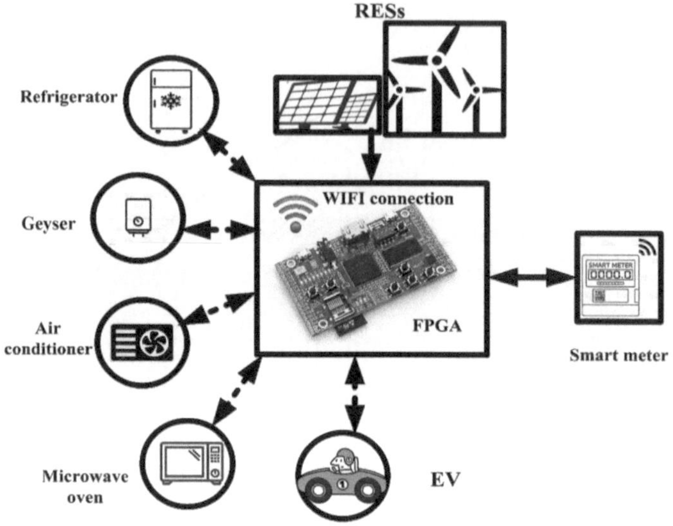

Figure 2.23 A simple HEMS for smart building.

easily implemented through such a centralized controller [20]. The multiple controllable smart devices such as refrigerator, geyser, air conditioner, microwave oven, and the data of EV vehicle can be communicated with a centralized FPGA-based controller, and the control actions with respect to load and the generation through RESs are being taken [21].

The FPGA-based centralized controller can also be used with multiple sensors for a smart HEMS. A basic layout of such an approach is shown in Figure 2.24. In this layout, various types of sensors are being used as a part of the smart home scheme. Multiple physical devices can be connected under this scheme. Real-time face recognition, biometric authentication, or any other such security feature can be interfaced through this mode. In this scheme, the wireless or wired channel of communication is adopted for the data transfer to the central controller to be processed. For the data analysis purpose, a cloud network is also proposed in this scheme, which can be used to develop an AI-enabled environment. Here, a cloud is proposed for the data analysis, and these data can also be used for the development of an artificial intelligence-enabled environment. It is a well-known fact that RESs technologies are the power resources for future global development in terms of achieving net-zero carbon goals and also integrating more and more intelligence techniques into the system. The AI-enabled technologies imply new ways to organize and implement these requirements of global development [22].

The AI-based approaches are being used in enabling the 5G network communications to implement biometric authentication technologies, smart banking, cyber-physical networking, and so on [23]. Renewable energy can be a part of these systems with a BIMS system.

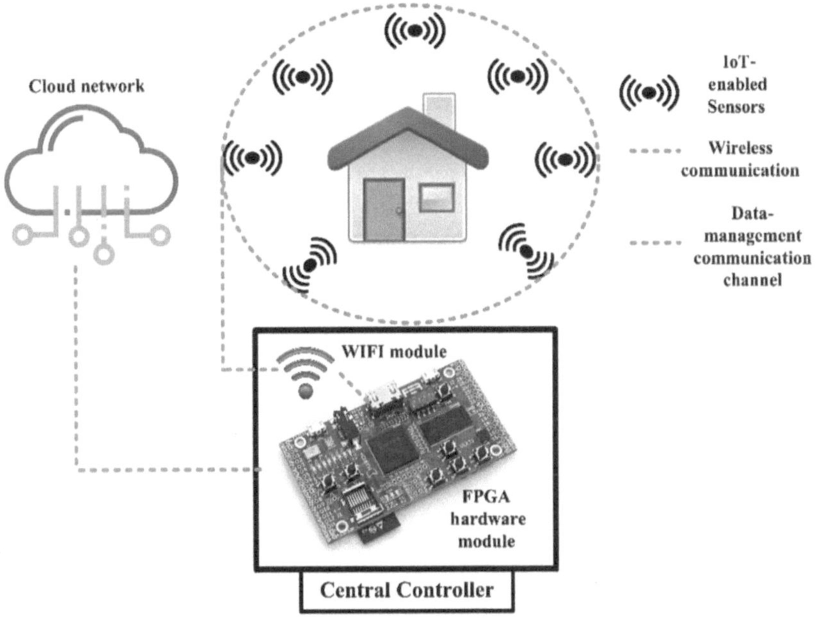

Figure 2.24 A basic layout of smart home scheme with multiple sensors.

Some basics of RESs used for the smart buildings under the smart city's infrastructure are provided with an example below.

Let us say a smart building has a total connected load of ***X*** kW, which needs to be fed through BIMS. For this, the data pertaining to solar energy and wind has been considered from past historical data. This data will be forecasted to get optimal dispatch of the RESs. The EMS for the BIMS can be solved by formulating an optimal dispatch problem, using various costs such as installation, operation, and maintenance associated with energy sources, losses in the battery storage system, DEG generator fuel cost, penalty on carbon emission on DEG unit, and so on.

Therefore, for obtaining an optimal energy dispatch for a smart BIMS with the RESs penetration, a cost-based objective function can be formulated as:

$$f(x) = \min\left(C^{\text{o\&m}} + C^{\text{tax}} + C^{\text{fuel}} + C^{\text{ESS}}\right) \tag{2.15}$$

subject to system's linearity and nonlinearity constraints.

Where

$f(x)$ is the objective function which need to be optimize (minimize).
$C^{\text{o\&m}}$: the operation and maintenance cost associated with the BIMS.

C^{tax}: tax or penalty imposed on the DEG unit for the CO_2 emission if any to reduce the carbon footprints (CFPs).
C^{fuel}: fuel cost associated with DEG unit.
C^{ESS}: the cost associated with the losses in the energy storage unit.

The data pertaining to RESs viz. solar irradiance and the wind speed for a specific geographical location can be forecasted. For this purpose, any of the learning models can be adopted which suits the type of available data set. The load demand data can also be forecasted, and an optimal dispatch problem can be solved. The BIMS can be developed with an optimal dispatch of the connected energy sources in the system for the fulfillment of *X* kW of the connected load demand. Since most of the real-world systems are nonlinear, any of the nonlinear optimization techniques such as sequential programming approach and other heuristic approaches can be implemented for the optimization.

Let's say the solar irradiance data (kW/m^2) as H_1, H_2, H_3, and so on for the per hour collected samples, and W_1, W_2, W_3, be the wind speed data (m/sec) for the same sample space. Now with the help of ML, this data can be forecasted to obtain a day-ahead data set as H_{f1}, H_{f2}, H_{f3}, and W_{f1}, W_{f2}, W_{f3} and so on, for solar irradiance and the wind speed, respectively, considering feasible inputs such as time of the day, temperature, humidity, and so on. The obtained data set can be used to get the optimal dispatch of the connected energy sources for the BIMS. For a grid-connected BIMS, the objective shall be the minimization of the cost of energy generation and maximizing the profile by selling the power to the grid. The other area where renewable energy can be explored in smart cities is solar thermal energy storage (STES) systems for heating as well as electrical energy production for the smart cities.

2.6.2 Future trends and challenges

The aforementioned discussion of smart buildings and the RESs penetration into the smart cities' infrastructure provides a way to explore many future trends and also certain key challenges. Some of them are discussed here.

- A robust control infrastructure is required to be implemented which incorporates sensing, monitoring, as well as accurate data processing with variable speeds.
- Use of fast-acting FPGA-based relays can be explored in the future to observe islanding and the mode of adaptability for the smart buildings.
- With emerging wireless communication techniques, viz. Wi-Fi, 5G network, etc., the emphasis shall be given on them rather than wired communication within the smart building architecture.
- IoT-based smart home appliances that can be utilized for health monitoring, fire safety, and user privacy can also be explored.

- More accurate load and renewable energy generation forecasting through AI-based approaches can also be explored extensively other than the present methodologies.
- The inclusion of solar thermal energy for power and heat can also be explored extensively for smart buildings in near future.

Some key technical challenges are also observed from the aforementioned discussions, and are discussed below.

- Since a large volume of data needs to be processed under the smart cities network, data privacy shall always be a point of concern.
- The coordination between various smart devices connected in a smart building system with high accuracy is always a challenging task.
- Since the optimal solutions are readily dependent on the type and the structure of the formulated problem, even with very fast computational algorithms, getting the corresponding global solution every time is difficult [24].
- For achieving net-zero carbon building, it is essential to utilize more and more renewable energy-based generation. Although presently many accurate and precious models of forecasting are available, this also raises the issues of system stability, more specifically for an isolated BIMS.

REFERENCES

[1] R. K. R. Kummitha, "Smart Cities and Entrepreneurship: An Agenda for Future Research," *Technological Forecasting & Social Change*, vol. 149, p. 119763, 2019.

[2] J. Shah, J. Kothari, and N. Doshi, "A Survey of Smart City Infrastructure via Case Study on New York," *Procedia of Computer Science*, vol. 160, pp. 702–705, 2019.

[3] Maneesh Kumar, and Barjeev Tyagi, "A State of Art Review of Microgrid Control and Integration Aspects," in *IEEE 7th India International Conference on Power Electronics (IICPE)*, Nov. 2016.

[4] D. S. Shafiullah, T. H. Vo, P. H. Nguyen, and A. J. M. Pemen, "Different Smart Grid Frameworks in Context of Smart Neighborhood: A Review," in *52nd International Universities Power Engineering Conference (UPEC)*, pp. 1–6, Aug. 2017.

[5] Maneesh Kumar, Sachidananda Sen, et al. "Emission-Averse Techno-Economical Study for An Isolated Microgrid System with Solar Energy and Battery Storage," *Electrical Engineering Journal*, 2023. https://doi.org/10.1007/s00202-023-01785-8

[6] M. Schmidta, and C. Åhlund, "Smart Buildings as Cyber-Physical Systems: Data-Driven Predictive Control Strategies for Energy Efficiency," *Renewable and Sustainable Energy Reviews*, vol. 90, pp. 742–756, 2018.

[7] J. Al Dakheel, C. Del Pero, N. Aste, and F. Leonforte, "Smart Buildings Features and Key Performance Indicators: A Review," *Sustainable Cities and Society*, vol. 61, p. 102328, 2020.
[8] Amjad Anvari-Moghaddam, Hassan Monsef, and Ashkan Rahimi-Kian, "Optimal Smart Home Energy Management Considering Energy Saving and a Comfortable Lifestyle," *IEEE Transactions on Smart Grid*, vol. 6, no. 1, pp. 324–332, 2015.
[9] Tariq Samad, Edward Koch, and Petr Stluka "Automated Demand Response for Smart Buildings and Microgrids: The State of the Practice and Research Challenges," *Proceedings of the IEEE*, vol. 104, no. 4, pp. 726–744, 2016.
[10] Surya Chandan Dhulipala, Raul Vitor Arantes Monteiro, Raoni Florentino da Silva Teixeira, Cody Ruben, Arturo Suman Bretas, and Geraldo Caixeta Guimarães, "Distributed Model-Predictive Control Strategy for Distribution Network Volt/VAR Control: A Smart-Building-Based Approach," *IEEE Transactions on Industry Applications*, vol. 55, no. 6, pp. 7041–7051, 2019.
[11] Waselul H. Sadid, Saad A. Abobakr, and Guchuan Zhu, "Discrete-Event Systems-Based Power Admission Control of Thermal Appliances in Smart Buildings," *IEEE Transactions on Smart Grid*, vol. 8, no. 6, pp. 2665–2674, 2017.
[12] Mahdi Ashabani, and Hoay Beng Gooi, "Multiobjective Automated and Autonomous Intelligent Load Control for Smart Buildings," *IEEE Transactions on Power Systems*, vol. 33, no. 3, pp. 2778–2791, 2018.
[13] Maneesh Kumar and Barjeev Tyagi, "A Robust Adaptive Decentralized Inverter Voltage Control approach for Solar PV and Storage based Islanded Microgrid," *IEEE Transactions of Industry Application*, vol. 57, no. 5, pp. 5356–5371, 2021.
[14] Giancarlo Mantovani and Luca Ferrarini, "Temperature Control of a Commercial Building with Model Predictive Control Techniques", *IEEE Transactions on Industrial Electronics*, vol. 62, no. 4, pp. 2651–2660, 2015.
[15] Yu Liang, Di Xie, Tao Jiang, Yulong Zou, and Kun Wang, "Distributed Real-Time HVAC Control for Cost-Efficient Commercial Buildings Under Smart Grid Environment," *IEEE Internet of Things Journal*, vol. 5, no. 1, pp. 44–55, 2018.
[16] H. Fontenot and B. Dong, "Modeling and Control of Building-Integrated Microgrids for Optimal Energy Management – A Review," *Applied Energy*, vol. 254, p. 113689, 2019.
[17] Maneesh Kumar, Sachidananda Sen, et al. "Optimal Planning for Building Integrated Microgrid System (BIMGS) for Economic Feasibility with Renewable Energy Support," in *IEEE International Conference PIICON*, Nov. 2022.
[18] Jinsoo Han, Chang-sic Choi, Wan-Ki Park Ilwoo Lee, and Sang-Ha Kim, "Smart Home Energy Management System Including Renewable Energy Based on ZigBee and PLC," *IEEE Transactions on Consumer Electronics*, vol. 60, no. 2, pp. 198–202, 2014.
[19] S. Sharma and R. Deokar, "FPGA Based Cost Effective Smart Home Systems," in *International Conference on Advances in Communication and Computing Technology*, pp. 397–402, 2018.

[20] Maneesh Kumar and Barjeev Tyagi, "Optimal Energy Management and Sizing of a Community Smart Microgrid Using GA with Demand Side Management and Load Uncertainty," *ECTI Transactions on Computer and Information Technology*, vol. 15, no. 2, pp. 186–197, Apr. 2021.
[21] M. Casini, "Active Dynamic Windows for Buildings: A Review," *Renewable Energy*, vol. 119, pp. 923–934, 2017.
[22] Andreea Claudia and Miltiadis D. Lytras, "Artificial Intelligence for Smart Renewable Energy Sector in Europe-Smart Energy Infrastructures for Next Generation Smart Cities," *IEEE Access*, vol. 8, pp. 77364–77377, 2020.
[23] Ahmed Sedik, Ahmed A. Abd Al-Latif, et al, "Deep Learning Modalities for Biometric Alteration Detection in 5G Networks-Based Secure Smart Cities," *IEEE Access*, vol. 9, pp. 94780–94788, 2021.
[24] Maneesh Kumar and Barjeev Tyagi, "Multi-Variable Constrained Nonlinear Optimal Planning and Operation Problem for Isolated Microgrids with Stochasticity in Wind, Solar and Load Demand data", *IET Generation, Transmission & Distribution*, vol. 14, no. 11, pp. 2181–2190, 2020.

Chapter 3

Fundamentals of internet of things (IoT) for smart cities

Mohammad Adnan, Ahteshamul Haque, Md Sarwar, and Anwar Shahzad Siddiqui
Jamia Millia Islamia, New Delhi, India

Prashant
Department of Electrical Engineering, Jamia Millia Islamia (A Central University), New Delhi, India

3.1 INTRODUCTION

Due to the fast-growing population of cities and urban areas, it becomes difficult for citizens to have access to decent services. Several initiatives are being taken by the government to improve the living conditions of its citizens. One such step toward the quality of services is taken by the government in the form of a smart city. "A smart city is a framework, predominantly composed of Information and Communication Technologies (ICT), to develop, deploy, and promote sustainable development practices to address growing urbanization challenges." Its main objectives are to provide improved quality of life, digital connectivity, increased efficiency of urban operations and services, and enable better decision-making, all while meeting the economic, social, and environmental needs of current and future generations [1]. In the last few years, the concept of a smart city has drawn the attention of the world toward transforming existing cities in order to address the urbanization challenges such as health, education, traffic, waste, energy, and so on. Several initiatives and policies are being taken and framed by the government for the smooth transition of an urban city into a smart city. However, these objectives cannot be met without the use of information and communication technologies as revealed from the definition of a smart city. The application of the internet of things (IoT) can prove to be a panacea for making the smart city a reality [2]. IoT refers to the network of physical objects that are implanted with sensors and software along with other technologies to connect and exchange data with devices over the internet. It provides such a powerful medium to interact with different components of smart cities in order to meet its objectives [3]. It has limitless applications in smart cities [4]. Some of the examples of its applications are intelligent transportation services which provide better traffic management, safety, and space availability for parking. Smart energy services are the other examples that provide efficient energy management [5–7]. However, there are some

 DOI: 10.1201/9781032669809-3

issues that need to be addressed for the implementation of IoT in smart cities. There are some existing works that address the network and communication issues that arise. To address this issue, Zanella et al. [8] presented a work which analyzed the available current technologies for implementing IoT in a smart city. Taking into account all these factors in this chapter, the significance and application of IoT in the context of a smart city has been discussed. Also, the challenges of implementing IoT in a smart city are presented. The explanation starts from the fundamentals of IoT and smart city. Later on, a detailed explanation of application of IoT to address the challenges of smart cities is presented.

The chapter is organized as follows: Section 3.2 presents the basics of smart cities, Section 3.3 presents the basics of IoT which is followed by detailed application of IoT in smart cities in Section 3.5. Finally, the challenges of implementing IoT in smart cities are discussed in Section 3.6.

3.2 SMART CITY

A smart city is an innovative city that uses information and communication technologies and other means to improve quality of life, provide digital connectivity, increase the efficiency of urban operations and services, and enable better decision-making, all while meeting the economic, social, and environmental needs of current and future generations [9]. This technology uses various types of sensors, voice activation methods, and CCTV surveillance systems for collecting the data. The information is then utilized to manage the city's resources and services; in return, the data is used to improve the various operations throughout the city. A smart city may include a smart lighting system, smart traffic management, and smart parking, as well as improved data privacy and security.

Smart cities focus on the most pressing demands and research the potential to improve people's lives. To make a difference, they use a variety of strategies and best practices in digital and information technologies, planning, policy, and so on. They prioritize the interests of the people.

The goal of the Smart Cities Mission [10] launched by the government of India in 2015 is to develop cities to provide better infrastructure and improved life quality to their residents, along with a clean and sustainable environment. Its focus is on sustainable and equitable development, with the goal of creating a repeatable model that will serve as a beacon for other aspiring cities [11]. Three models of area-based development are the key initiatives of the Smart City Mission which are redevelopment, green field, retrofitting, and developing areas step-by-step.

According to the Smart Cities Mission, a smart city's core infrastructure includes adequate water supply, guaranteed electricity supply, proper sanitation, and solid waste management, robust digitalization with IT connectivity, affordable housing, efficient urban mobility and public transportation, sustainable environment, good governance (particularly e-governance) and citizen participation, citizen safety and security, health and education, and so on [12]. Governments, corporations, software providers, device makers, energy providers, and network service providers all have a role to play in integrating solutions that meet basic security goals.

3.3 INTERNET OF THINGS (IOT)

It is evident from the name that IoT covers things that are connected to the internet. The IoT is a platform that connects various sensors and devices without requiring human involvement to perform a specific function as shown in Figure 3.1 [1]. This has made cities smarter and smarter day by day all over the world. Everything is connected through internet protocol (IP) and interacts on the pre-defined logic. Anyone can access things from different parts of the world or it can be obtained automatically. IoT is an emerging and accelerating trend. With the help of this, things can be done more speedily and accurately, and most importantly without the intervention of humans. An IoT system enables better integration, automation, and analysis of different systems. It increases the reach of the application areas with improved accuracy. It exploits the recent advances in information and communication technologies as well as the innovations in smart devices. Its cutting-edge components and mechanisms have proven to be vital in bringing major changes in various services which would impact the social, environmental, and economic conditions [13]. How IoT can help cities become smarter is discussed in further sections of this chapter.

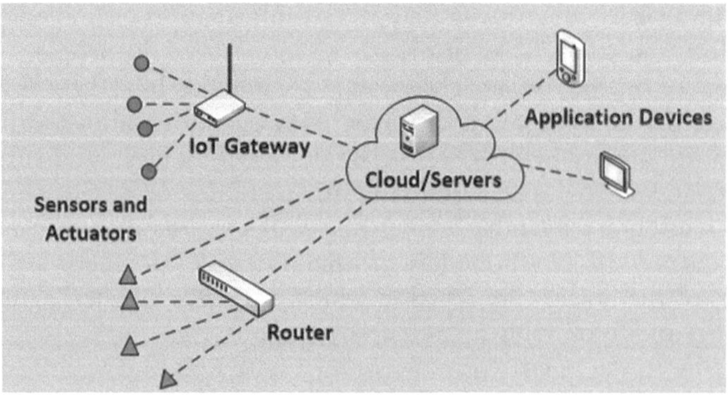

Figure 3.1 Components of IoT.

3.4 IOT ARCHITECTURE AND NETWORK PROTOCOLS IN SMART CITY

There is no universally accepted common architecture for an IoT environment. Different architectures are proposed by different researchers from time to time in their own perspective. Some of the commonly identified architectures make use of different layers such as data collection, data processing, data integration and reasoning, device control layers, and so on. There are three-layer, four-layer, and five-layer architectures [14] and these layers are: perception network, processing, and application. The raw data acquired from sensors is kept at the perception layer which is also called the physical layer for later processing. The data collected at the data perception layer before being transmitted, analyzed, and merged at higher levels, utilizing advanced algorithms at the processing layer and network layer. The primary goal of these levels is to convert the obtained heterogeneous data into a standardized format. The services to the user are delivered by the application layer.

Implementation of IoT architecture is based on the different network protocols being used for transforming urban cities into smart cities [15]. The short-range communication application such as smart apartments, smart water management, and so on, may use IEEE 802.15.4 (Zigbee) and 801.15.1 (Bluetooth) which fall into the category of personal area network (PAN) class protocols. However, the application of IoT in smart cities requires communication for longer ranges which can utilize IEEE 802.11 (Wi-Fi) which comes under the category of local area network class. Its applications include intelligent traffic management or logistics. Besides these, wide area network (WAN) class protocols are used for wide-range communications such as smart grids. The WAN class protocols include IEEE 802.16 (WiMAX), cellular, and satellite. All of these protocols have provisions to support asynchronous and synchronous data connections.

3.5 IOT APPLICATIONS IN SMART CITY

Smart cities are intelligent, responsive, connected, and sustainable. It is made up of several components as shown in Figure 3.2 [11]. Typically, the application of IoT to integrate the different components of smart cities in order to achieve its objectives has four major key drivers which make it work toward achieving sustainable development [16]. The first one is analytics, that is, putting data to work; second is smart transportation solutions; third is building healthier communities, and fourth is storage. Data gathering is application-dependent, and it has been a major driver of sensor development in a variety of fields. The smart city concept is facilitated by fast-developing technology and sensors. There is a massive influx of data being

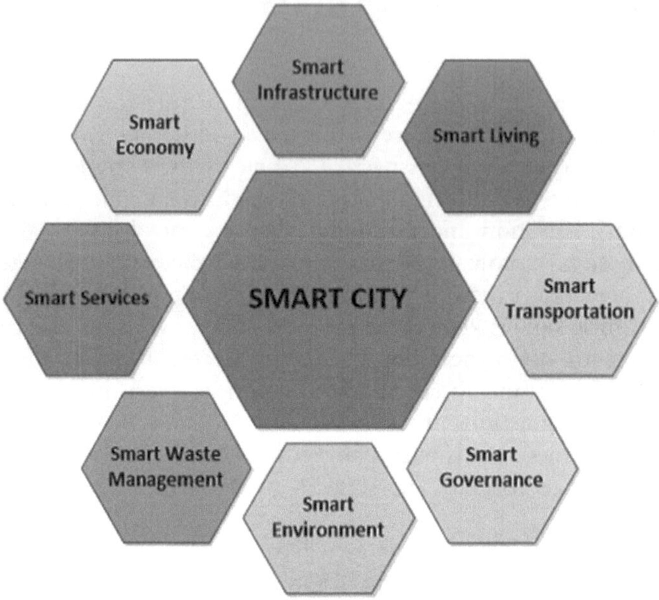

Figure 3.2 Typical components of a smart city.

generated and made available from gadgets starting from the vehicle, home appliances, and so on, to the sensors [17]. Furthermore, people, public transportation, vehicles, infrastructure and so on all work together in a smart city to increase mobility and safety through a connected transportation system [18]. Data storage can be easily accomplished with storage in the cloud [16]. It facilitates data storage with ease of access from anywhere. The application of IoT in smart cities is shown in Figure 3.3. It can be categorized into three major parts: public utilities, transportations, and services for residents which are discussed in Figures 3.2 and 3.3.

Due to problems such as the complexity of components in the IoT ecosystem, deploying IoT is tough. According to surveys, an IoT project's average time to market is 18–24 months, and approximately 70% of IoT ventures fail. As a result, it is advised that enterprises follow standard IoT implementation phases and best practices. The number of IoT active connections statistics in smart cities in the European Union is shown in Figure 3.4 for the years 2016, 2019, 2022, and 2025 [19]. The number of IoT smart cities' active connections is expected to increase through the years.

3.5.1 Public utilities

Public utilities are the key component of a smart city. It makes the life of its residents pleasant and comfortable. It includes a number of services such as smart lighting, smart waste management, smart meters, and so on.

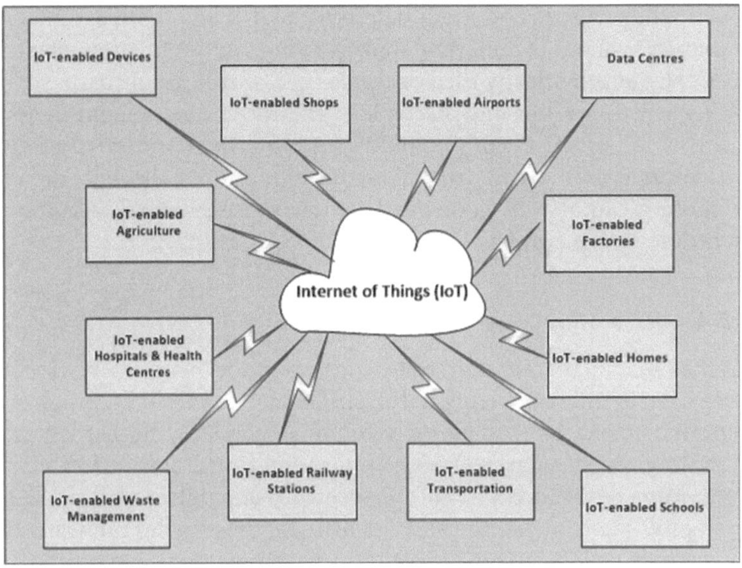

Figure 3.3 A typical diagram showing the application of IoT in smart city.

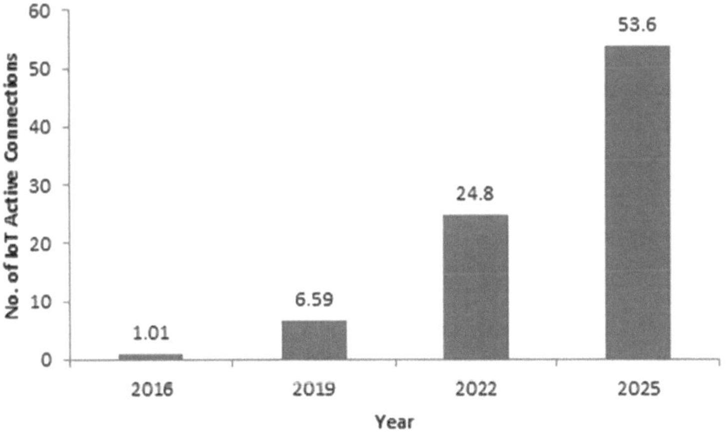

Figure 3.4 Number of IoT active connections in smart cities in the EU.

The application of IoT to these components in order to achieve the objectives of smartness is discussed below.

3.5.1.1 Smart lighting

Automatic lighting control and dimming in the streets and residential areas is adapted to weather conditions and occupancy based on data collected from luminance sensors and motion detectors [20]. Smart street lights work automatically when it is dim, they start glowing, and in the day when there is bright, they automatically turn off. This type of smart lighting system not only saves electricity but also needs less attention. Smart lighting systems make use of IoT-enabled sensors, bulbs, and internet connectivity to allow users to control their lights from their mobile phones through the cloud. Smart lighting can also be controlled remotely using Google Assistant, set on a schedule system, and actuated by sound or motion.

3.5.1.2 Smart meters

One of the most important milestones in the development of citywide smart grid networks is smart metering [21, 22]. Through meter data insights, IoT-enabled smart metering may improve customer service. Another key advantage is the low cost of smart metering implementation. Using an IoT platform gives you immediate access to all the necessary capabilities, allowing focus on developing smart metering use cases instead, thus saving time and avoiding the dangers of in-house IoT development. Smart metering acquires data from smart meters using various connectivity options, representing data on a dashboard, saving data for reporting and past analysis, sending processed smart metering data into third-party accounting and billing applications, and analyzing data to derive actionable insights.

3.5.1.3 Waste collection

Waste containers are monitored for their fill levels to optimize waste collection schedule and routes. Vehicles arrive with a full container load, eliminating the clearance of unfilled containers and overfilling of containers. A microcontroller with a Wi-Fi system, ultrasonic sensors, and a web server is used in the proposed system. In this method, the bins are fitted with ultrasonic sensors that measure the waste level and transfer the data to a server via the internet using a microcontroller and Wi-Fi protocol. The server keeps track of the bins that are spread across the city in various areas. Based on the waste level in the bin, the system tells the garbage truck driver when the garbage has to be removed. The server delivers an SMS to the given cellphone number, which includes a route for the driver based on all the data collected from bins. In metropolitan cities, this smart waste management system makes the management of waste more efficient [23].

3.5.1.4 Municipal vehicles monitoring

Municipal and sweeping vehicle fleets equipped with sensors are tracked in real-time, allowing for real-time control of their optimal route, operation mode, and fuel consumption, as well as the detection of violations and technical concerns. Monitoring gas levels and consumption, tracking driver behaviors for unsafe driving or drowsiness, tracking vehicle location, and monitoring overall vehicle performance are all part of fleet management solutions [24]. Municipal fleet management ensures the safety of drivers while increasing operating efficiency, optimizing maintenance schedules, and lowering insurance rates.

3.5.1.5 Utility infrastructure monitoring

To identify divergences, overheating, response times to accidents, and perform predictive maintenance, water supply pipes and heating network indicators like temperature, pressure, vibrations, and so on, are monitored. Some of the applications of IoT utility infrastructure monitoring are as follows:

(i) *Asset Tracking and Logistics*: Keep track of crucial goods in real-time to avoid supply chain delays.
(ii) *Pole Tilt Monitoring*: By deploying sensors to monitor utility pole infrastructure, pole maintenance, storm hardening, and recovery can be streamlined.
(iii) *Gas Infrastructure Monitoring*: Improve reaction times and eliminate human testing requirements by remotely detecting natural gas leaks and voltage at corrosion potential gas piping test points.
(iv) *Water Distribution Monitoring*: Connect sensors to routinely monitor chlorination, leak detection, turbidity, and other parameters in existing water distribution infrastructure.
(v) *Environmental Monitoring*: At facility discharge locations, keep an eye on radiation, particles, and other pollutants.

3.5.2 Transportation

Another major component of a smart city is smart transportation. Densely populated areas and traffic congestion create havoc in the daily life of residents. Their daily rituals are affected by inefficient transportation systems. Therefore, smart transportation is key to the success of the smart city. IoT plays a vital role in dealing with these challenges and helps in providing a cutting-edge solution to the transportation systems which are discussed below.

3.5.2.1 Intelligent transportation system

Adaptive vehicle traffic management is achieved based on dynamic traffic modeling, and real-time data from detectors and cameras in order to prevent congestion. The IoT-based intelligent transportation system [25] aids in the automation of trains, motorways, airways, and marine transportation, improving customer satisfaction with the way goods are transported, tracked, and delivered. The Intelligent Traffic Management System based on IoT and big data is used for smart traffic solutions in smarter cities. The goal of an intelligent and smart transportation system is to establish structures for managing transportation in an orderly, clean, and hassle-free way. The ever-increasing traffic congestion, high levels of pollution, and increase in the number of accidents are the major factors throughout the world that have forced the adoption of intelligent transportation systems and, as a result, have reduced these crucial issues.

3.5.2.2 Smart parking

To make a city smart, we must accomplish a variety of things that can be produced using the internet of things, one of which is smart parking [19]. In practically every metropolitan city, parking is an issue. Smart parking can be created via the internet of things. A person may reserve a parking spot using a mobile device and avoid having to look for it. Anyone can pay the parking fees in addition to making a reservation. If there is a violation at a parking area or an unnecessary crowd formation, sensors will detect it and alert security.

3.5.2.3 Connected public transport

This allows for smart fare analysis and payment, as well as knowing which bus is scheduled to arrive at a specific location at a specific time [19]. Furthermore, with the help of an intelligent transportation system, the availability of seats, the condition and kind of bus, and its final destination can all be easily determined, allowing passengers to choose their route more intelligently. It can be determined via mobile apps and electronic information boards on public transport.

3.5.2.4 Video surveillance of traffic violations

Nowadays, road traffic is a major problem. Using live video monitoring, various concerns such as vehicle collision or accident detection, and related issues, may be easily resolved. An IoT-based accident detection and traffic monitoring system comprised of Raspberry Pi and a Pi camera as hardware will take live video as input and process it to obtain information about traffic in real-time. Also, magnetic loops are currently utilized to count and categorize automobiles on the road. The magnetic coil collects data on passing cars as well as other parameters such as vehicle speed, traffic congestion, and accidents. The wireless sensor-based network performs admirably in the traffic control system, which is a real-time application [26]. The sensor-based system also provides energy efficiency, dependable data transmission, and precise operations.

In addition to the above, the IoT-enabled smart airports and smart railway stations are also used in contemporary day. IoT has begun to reshape the railway and airline industry, offering higher efficiency, improved passenger experience, and increased safety.

IoT-driven advances in preventive maintenance methods have helped to restore the dependability of even the oldest assets. By incorporating IoT sensors into crucial components such as brakes, wheelsets, and engines, trains become more sensitive to their operations. Train track maintenance can also benefit from IoT. Operators can keep track of track stress and conditions, temperatures, and other data that have predictive values for maintenance teams by placing sensors across track systems. In the internet of things era, predictive and preventative maintenance is more practical and effective. Rail systems may be remotely monitored and fixed using smart sensors and analytics on the train engine, carriages, and tracks before a minor issue becomes a major problem. By tracking trains across networks and analyzing the data with analytics, the operators can better manage the trains. Some companies also use IoT to monitor passenger flow such as waiting at stations, movement in each railway coach, and the peak times for passenger traffic. Weather has an impact on railway schedules, and with IoT, one can now predict future weather conditions and change train schedules accordingly. Using IoT technology to understand the customer experience and make enhancements for more comfortable and convenient travel, rail operators can increase passenger loyalty. For example, using IoT, an operator may determine a passenger's frequency of travel, and services can be charged differently for various travelers according to their frequency of travel.

Similarly, many airports have implemented IoT-enabled smart baggage trolleys to track and maintain baggage trolley availability for passengers in real-time across the airport. This technology aids the operations staff in tracking the number of trolleys available at various locations throughout the airport and planning with real-time dashboards to ensure that passengers may use the trolleys at the appropriate time and location.

Many adjustments may be made to improve the travel experience by adopting IoT technologies. The person entering the airport can be identified and admitted using advanced biometric systems. Digital IoT beacons strategically placed across the airport can help with navigation and reduce waiting times. The beacon employs Bluetooth localization to deliver real-time information on parking, airport shuttle time tables, restaurant table availability, and store locations.

3.5.3 Health services

Apart from public utilities and smart transportation, other components are of greater significance in achieving the objectives of smart cities. One of the key services is providing a smart health service [27]. This provides safety as well as security to the residents of the smart city. The term "smart health" refers to the use of information and communication technology to improve the availability and quality of health care. With an increasing population and rising healthcare expenses, researchers and healthcare providers have been concentrating their efforts in this field. The smart health system includes virtual hospitals, wearable biosensors, smart thermometer, smartly connected inhalers, and smart watching monitors like smart watches and automated insulin delivery systems. IoT enables the provision of better governance of health services. Some of the IoT-enabled smart services are discussed below.

3.5.3.1 *Telemedicine*

Several studies have focused on using IoT technologies to make telemedicine smarter [28]. IoT has the potential to improve a variety of medical applications, including remote health monitoring, exercise programs, rehabilitation, chronic disease management, and elderly care. As smart devices, a variety of medical equipment, sensors, and imaging devices are critical components of the telemedicine architecture. Many devices, such as the smart inhaler, smart wheelchair, and smart dialysis machine, have made our lives easier. As a result, the global healthcare industry of the future should be prepared for widespread remote healthcare monitoring via IoT and telemedicine.

3.5.3.2 *Support for emergency services*

Emergency teams are automatically assigned to calls based on task and location. Photoelectric sensors are embedded inside and outside of hospitals, offices, and buildings, so if they catch fire the sensors detect and give the information to the microcontroller. The microcontroller will then activate the alarm and send a text or email message to the fire department.

3.5.4 Miscellaneous services

In addition, there are other components that without IoT, certain objectives could not be met. IoT enables these components to make them smarter in the appropriate sense so that they contribute to the sustainable development of smart cities. Some of the components are discussed below.

3.5.4.1 *Smart schools*

Smart schools [29] connect students and teachers with connected devices, allowing for more meaningful interactions and the improvement or expansion of lesson plans. IoT and its connected gadgets, on the other hand, can assist administrators and staff in maintaining student security. Teachers can take student attendance using apps and sensor wristbands. A notice can be sent when students sign into a learning application or scan their wristband at the door to notify the teacher or attendance recorder which students are there and which are not. Also, if a student skips class and hangs out somewhere else, his parents will receive an immediate notification along with his GPS location.

In addition to wristbands, schools can use IoT to connect wireless locks with video capabilities that register and recognize those who are permitted to enter a specific class, lab, or building.

For schools, IoT is opening up new possibilities. Connected technologies improve and enhance the classroom experience in a variety of ways, from assisting teachers to securing institutions.

3.5.4.2 Smart homes

These are at the core of a smart city, and in order to make the city smart, we must first accept smart technologies in our own houses [30]. And, in order to create a smart house, a home automation system is required. Home automation enables us to control all electronic equipment over the cloud while sitting in another part of the world. Boards like ESP8266 and Raspberry Pi allow us to connect the device to the cloud using a Wi-Fi module. Also nowadays, there are smart televisions, smart refrigerators, and smart inverters, too, making our homes smarter and easier.

3.5.4.3 Monitoring of public safety

Any escaped prisoner can be readily detected and monitored by facial recognition cameras thanks to IoT technologies, ensuring the city's safety [31]. Any unnecessary crowd formation or violence on the street will be reported to the nearest police station in real-time by the microcontroller, allowing prompt action to be taken.

Officials can use data generated by IoT devices (sensors) to drive artificial intelligence (AI) and machine learning (ML) algorithms to gain insights into city infrastructure and identify damage by measuring shifts and vibrations in constructions that can indicate cracks, extensions, as well as strain. This enables officials to discover potentially life-threatening problems ahead of time and carry out preventative maintenance to ensure that minor concerns do not become major difficulties.

3.5.4.4 Public security

Security systems are being used at an exponential rate in our daily lives. Security at a commercial place, organization, or bank locker is now crucial to everyone. As discussed earlier, CCTV cameras are used to protect and secure places, but they need a human being all the time to function. The smart IoT security system has the potential to interact in real-time with the device [31]. The system consists of a camera, Wi-Fi module, motion sensors, and voice sensors, which can easily detect and gather all the information and, depending on the situation, allow the actuators to react. An IoT-based smart locker has various facilities and provides better security. One cannot easily break the system, however, if someone tries to break it then with the help of a camera all its activity can be recorded and officers are alerted in real-time. The capacity to monitor one's property remotely 24/7 is the key benefit of this technology. With this IoT-based smart locker, users can monitor, receive notifications, and inform in the event of an emergency from anywhere in the world using a mobile application connected to the cloud 24/7. To be more specific, we want to create a wireless smart security system that is lightweight, low cost, extendable, and versatile, and that uses IoT to integrate multiple cutting-edge technologies.

3.6 CHALLENGES IN IMPLEMENTATION OF IOT IN SMART CITY

The success of the smart city depends on the involvement of every individual. Citizens, government, businesses, and so on, all have a role in its development. Once the equilibrium is established, it provides enormous opportunities for improved quality of life and livelihood for its residents. However, there are so many components to consider and various challenges are also being faced in developing smart cities which are discussed below [32, 33].

3.6.1 Infrastructure

Sensor technology is used in smart cities to collect and analyze the data which improves quality of life. The installation and upkeep of these sensors necessitate a complicated and expensive infrastructure. There is a limited amount of money available for new infrastructure projects, and its approval processes might take years. As a result of the installation of new sensors and other advancements, residents in these cities face temporary but still frustrating problems.

3.6.2 Security

As the use of IoT and sensor technology grows, the threat to security has also increased. Creating a city that can be controlled from a central system creates severe security concerns. With thousands of devices dispersed around a metropolis, all it takes is a single sensor to give an access point for an attacker. The attacker may then try to break into the central control system, which could allow them to alter traffic systems, street lights, train barriers, and even emergency services. Creating a city that can be controlled from a central system creates severe security concerns. With thousands of devices dispersed around a metropolis, all it takes is a single sensor to give an access point for an attacker. The attacker may then try to break into the central control system, which could allow them to alter traffic systems, street lights, train barriers, and even emergency services.

3.6.3 Privacy

There is a proper balance in big cities between quality of life and violation of privacy. Everyone wants to live their life independently and wants freedom; no one wants to feel like anyone is watching them all the time. Although cameras help in the detection of crime, nobody wants transparency in their lifestyle.

3.6.4 Costs and implementation

Infrastructure costs and implementation will be a problem for smart cities. It may take several years for a city to upgrade its hardware, and the cost to taxpayers may be significant. Also once it is installed, after some time it will become outdated, hence require upgradation and regular maintenance. And these will raise costs.

3.6.5 Employment

Although IoT has made our lives easier, it has increased unemployment since it eliminates the need for middlemen. Because everything is managed remotely and many processes are automated, the need for labor is reduced, resulting in unemployment. In developing countries literacy rate is very low, meaning less employees know how to work with sensors and digital devices.

3.7 CONCLUSIONS

The goal of an IoT-powered smart city is to improve the quality of life of its residents in a variety of ways, including measures that promote an environmentally friendly, sustainable environment and the delivery of linked healthcare services to inhabitants at home and when outside the home. IoT has made our lives easier and more efficient in a variety of areas, including education, security, healthcare, and transportation. The first credit for making the world smarter goes to IoT because it is the only thing that connects the physical world with the virtual world. Sensors collect data from the physical world, send it to the microcontroller, and the microcontroller gives instructions to actuators to carry out the tasks. Actuators are the devices that control the operation of machines and other equipment. Smart cities powered by IoT rely on an increasing number of sub-technologies and subsystems that must be seamlessly networked and interfaced in real-time. This can only be accomplished by implementing appropriate measurements, communication, integration, and control standards and protocols.

REFERENCES

[1] Silva, B.N., Khan, M., Han, K. 2018. Towards sustainable smart cities: A review of trends, architectures, components, and open challenges in smart cities. *Sustainable Cities and Society*, 38:697–713.

[2] Hammi, B., Khatoun, R., Zeadalli, S., Fayad, A., Khoukhi, L. 2018. IoT technologies for smart cities. *IET Networks*, 7(1):1–13.

[3] Syed, A.S., Sierra-Sosa, D., Kumar, A., Elmaghraby, A. 2021. IoT in smart cities: A survey of technologies, practices and challenges. *Smart Cities*, 4:429–475.

[4] Ahlgren, B., Hidell, M., Ngai, E.C. 2016. Internet of things for smart cities: Interoperability and open data. *IEEE Internet Computing*, 20:52–56.

[5] Mohamed, N., Lazarova-Molnar, S., Al-Jaroodi, J. 2017. Cloud of things: Optimizing smart city services. In *Proceedings of the IEEE International Conference on Modeling, Simulation and Applied Optimization*, Sharjah UAE, pp. 1–5.

[6] Erol-Kantarci, M., Mouftah, H.T. 2012. Suresense: Sustainable wireless rechargeable sensor networks for the smart grid. *IEEE Wireless Communications*, 19(3):30–36.

[7] Gutiérrez, J., Villa-Medina, J.F., Nieto-Garibay, A., Porta-Gándara, M.Á. 2014. Automated irrigation system using a wireless sensor network and GPRS module. *IEEE Transactions on Instrumentation and Measurement*, 63(1):166–176.

[8] Zanella, A., Bui, N., Castellani, A., Vangelista, L., Zorzi, M. 2014. Internet of things for smart cities. *IEEE Internet of Things Journal*, 1(1):22–32.

[9] Sánchez-Corcuera, R., Nuñez-Marcos, A., Sesma-Solance, J., Bilbao-Jayo, A., Mulero, R., Zulaika, U., Azkune, G., Almeida, A. 2019. Smart cities survey: Technologies, application domains and challenges for the cities of the future. *International Journal of Distributed Sensor Networks*, 15(6):1–36.

[10] Smart Cities Mission: https://smartcities.gov.in/about-the-mission

[11] Rong, W., Xiong, Z., Cooper, D., Li, C., Sheng, H. 2014. Smart city architecture: A technology guide for implementation and design challenges. *China Communications*, 11(3):56–69.

[12] Sharma, N., Solanki, V.K., Davim, J.P. 2019. Basics of the internet of things (IoT) and its future. In *Handbook of IoT and Big Data*, eds. V. K. Solanki, V. G. Díaz, J. P. Davim, 1–22. CRC Press, Boca Raton.

[13] Lee, I., Lee, K. 2015. The internet of things (IoT): Applications, investments, and challenges for enterprises. *Business Horizons*, 58(4):431–440.

[14] Palaniappan, R., John, T.J., Nagaraj, V. 2020. IoT solutions for smart cities. *IOP Conference Series: Materials Science and Engineering*, 955(1):012004.

[15] Jawhar, I., Mohamed, N., Al-Jaroodi, J. 2018. Networking architectures and protocols for smart city systems. *Journal of Internet Services and Applications*, 9(26):1–16.

[16] Ejaz, W., Anpalagan, A. 2019. Internet of things for smart cities: Overview and key challenges. In *Internet of Things for Smart Cities: Technologies, Big Data and Security*, eds. W. Ejaz, and A. Anpalagan, 1–15. Springer, Cham.

[17] Khan, Z., Anjum, A., Soomro, K., Tahir, M.A. 2015. Towards cloud based big data analytics for smart future cities. *Journal of Cloud Computing*, 4(2):1–11.

[18] Wang, Y., Ram, S., Currim, F., Dantas, E., Sabóia, L.A. 2016. A big data approach for smart transportation management on bus network. In *Proceedings of the IEEE 2nd International Smart Cities Conference: Improving the Citizens Quality of Life, ISC2 2016*, Italy, 1–6.

[19] IoT active connections in smart cities EU 2016-2025|Statista: https://www.statista.com/statistics/691843/smart-city-iot-active-connections-in-the-eu/
[20] Park, D.M., Kim, S.K., Seo, Y.S. 2019. S-mote: SMART home framework for common household appliances in IoT Network. *Journal of Information Processing Systems*, 15(2):449–456.
[21] Shirazi, E., Jadid, S. 2018. Autonomous self-healing in smart distribution grids using multi agent systems. *IEEE Transactions on Industrial Informatics*, 15(12):6291–6301.
[22] Garcia, F.C.C., Creayla, C.M.C., Macabebe, E.Q.B. 2017. Development of an intelligent system for smart home energy disaggregation using stacked denoising autoencoders. *Procedia Computer Science*, 105:248–255.
[23] Pardini, K., Rodrigues, J.J., Kozlov, S.A., Kumar, N., Furtado, V. 2019. IoT-based solid waste management solutions: A survey. *Sensor and Actuator Networks*, 8(1):5.
[24] Desdemoustier, J., Crutzen, N., Giffinger, R. 2019. Municipalities' understanding of the smart city concept: An exploratory analysis in Belgium. *Technological Forecasting and Social Change*, 142:129–141.
[25] Farag, S.G. 2019. Application of smart structural system for smart sustainable cities. In *Proceedings of the 2019 4th MEC International Conference on Big Data and Smart City (ICBDSC)*, Muscat, Oman, 1–5.
[26] Al-Turjman, F., and Malekloo, A. 2019. Smart parking in IoT-enabled cities: A survey. *Sustainable Cities and Society*, 49:101608.
[27] Sevillano, X., Màrmol, E., Fernandez-Arguedas, V. 2014. Towards smart traffic management systems: Vacant on-street parking spot detection based on video analytics. In *Proceedings of the FUSION 2014—17th International Conference on Information Fusion*, Salamanca, Spain, 1–8.
[28] Machorro-Cano, I., Alor-Hernández, G., Paredes-Valverde, M.A., Ramos-Deonati, U., Sánchez-Cervantes, J.L., Rodríguez-Mazahua, L. 2019. PISIoT: A machine learning and IoT-based smart health platform for overweight and obesity control. *Applied Sciences*, 9(15):3037.
[29] Andreão, R.V., Athayde, M., Boudy, J., Aguilar, P., de Araujo, I., Andrade, R. 2018. Raspcare: A telemedicine platform for the treatment and monitoring of patients with chronic diseases. In Alejandro Rafael Garcia Ramirez and Marcelo Gitirana Gomes Ferreira, *Assistive Technologies in Smart Cities*. IntechOpen, London, UK, DOI: 10.5772/intechopen.76002
[30] Zhu, Z.T., Yu, M.H., Riezebos, P. 2016. A research framework of smart education. *Smart Learning Environments*, 3:4.
[31] Risteska Stojkoska, B.L., Trivodaliev, K.V. 2017. A review of internet of things for smart home: Challenges and solutions. *Journal of Cleaner Production*, 140(3):1454–1464.
[32] Janssen, M., Luthra, S., Mangla, S., Rana, N.P., Dwivedi, Y.K. 2019. Challenges for adopting and implementing IoT in smart cities: An integrated MICMAC-ISM approach. *Internet Research*, 29(6):1589–1616.
[33] Elmaghraby, A.S., Losavio, M.M. 2019. Cyber security challenges in smart cities: Safety, security and privacy. *Journal of Advanced Research*, 5(4):491–497.

Chapter 4

Role and applications of power electronics, renewable energy and IoT in smart cities

Ahteshamul Haque
Jamia Millia Islamia, New Delhi, India

K. V. S. Bharath
Silicon Austria Labs, Graz, Austria

Mohammad Amir and Md Zafar Khan
Jamia Millia Islamia, New Delhi, India

4.1 OVERVIEW

The development of power electronics (PE) technology has received growing attention from the power system as a result of the growth of the nation's economy and the continued development and renewal of computer technology in the new era. Informational, digital, and interactive goals are included in the development of the intelligent power grid. The whole power generation, transmission, distribution, and other power generation and generation activities are where the application of sophisticated PE technology in smart grids (SGs) is most evident. PE technology supports the development of SGs. Experts are thus investigating improved PE technology. Power electronics technology aids in the creation of SGs. The use in SGs is quite important. This chapter combines the author's professional experience with his or her personal experiences, principally to analyze the use of PE technology throughout the power system with an eye toward its potential development in the future [1–3]. Electricity exchange is the main application for advanced PE technology, a PE with electrical innovation in the area of power systems. More specifically, PE is the technique that uses power electrical devices to efficiently control and convert electric energy. PE technology developed in China is currently widely used in a number of industries. It enables power utilization to reach an acceptable level, optimize electricity utilization [4], and contribute to the development of mechanical building as well as the transformation of conventional industries. The crucial function. Parallel to this, the intelligent development of PE technology has led to ground-breaking methods of power and information processing integration. Therefore, SGs have to take advantage of current PE technologies [5–7]. The acceptance of renewable energy as a competitive substitute for conventional

 DOI: 10.1201/9781032669809-4

energy sources powered by fossil fuels is gradually growing. The deployment of sustainable energy sources including the sun's rays, the wind's, water-based, and geothermal sources of energy has significantly increased during the past ten years. Several factors have aided the expansion of renewable energy. To begin with, worries about climate change and environmental damage have increased awareness of the need to switch to cleaner, more sustainable energy sources. Furthermore, technological developments and lower costs have made renewable energy economically competitive with fossil fuels in many locations [8].

The growth of emerging communication technologies (ECT) has been exponential. These technologies include, among others, detectors, sensor networks, the Internet of Things (IoT), and intelligent systems [9]. These technologies have had a significant impact on every aspect of our lives, and it is expected that this trend will continue. Advances in technology and creative applications are changing society more and more. These advancements have had the greatest influence on smart healthcare, smart buildings, smart infrastructure, smart energy, and many other comparable systems [10].

The growing use of sensors across multiple sectors is generating a very productive platform for data collection across a wide range of applications. The platform for gathering data is greatly enhanced by the IoT, a broad type of sensor network with good connectivity. Data is effectively used by artificial intelligence (AI), leading to intelligent systems and applications. SCs result from a wider application of this technique to deal with various social needs and city functions. The same process might be used to develop intelligent regions, states, countries, the intelligent planet (an IBM project), and intelligent universe [11].

The concept of the Internet of Things makes use of several ubiquitous services to enable the creation of smart cities all around the world. The IoT creates new opportunities, including the ability to remotely manage devices, analyze, and take action on data from various real-time traffic data streams. By delivering more effective municipal services, improving transit services by reducing traffic congestion, and strengthening public safety, IoT technologies are revolutionizing cities. Instead of offering a separate smart city component, smart city architects and suppliers recognize that cities must provide flexible and safe IoT solutions that integrate efficient IoT systems in order to achieve the full potential of IoT [12]. The following are the key objectives of this chapter.

- The conversion and management of electrical power are the main goals of power electronics.
- Enhancing energy efficiency, maximizing power utilization, and increasing the dependability of power systems are the goals of smart cities.
- In order to diminish dependency on fossil fuels, lower greenhouse gas emissions, and promote the production and use of sustainable energy, smart cities employ renewable energy.

- To generate energy in smart cities sustainably, emphasize the need to employ renewable sources like sunlight and wind energy.
- The goal of the IoT in SCs is to create a network of interlinked objects, devices such as sensors, and systems that gather and analyze data in order to raise urban efficiency, security, and quality of life.
- Identify the challenges and barriers associated with integrating power electronics, renewable energy, and IoT in smart cities and discuss potential solutions.

4.2 POWER ELECTRONICS, IOT, RENEWABLE ENERGY IN SMART CITIES

4.2.1 Role of power electronics

Forward, Push-pull, Flyback, Resonant, and Bridge DC-DC converters are examples of isolated converters, and Boost, Buck-Boost, Cuk, SEPIC, and so on are examples of non-isolated type converters, as illustrated in Figure 4.1. The benefits, duty cycle characteristics, and input-output relationships of non-isolated converters are displayed in Table 4.1.

4.2.1.1 Safety and security in smart grid

India, China, the USA, Australia, and Russia possess extensive power networks, which may increase the risk of natural disasters. These disasters can have varying degrees of impact on the regular operation of SGs. To enhance the safety and reliability of SGs, it is crucial to incorporate advanced power electronics technology. Therefore, the adoption of modern PE technology within SGs is essential [1, 5, 13].

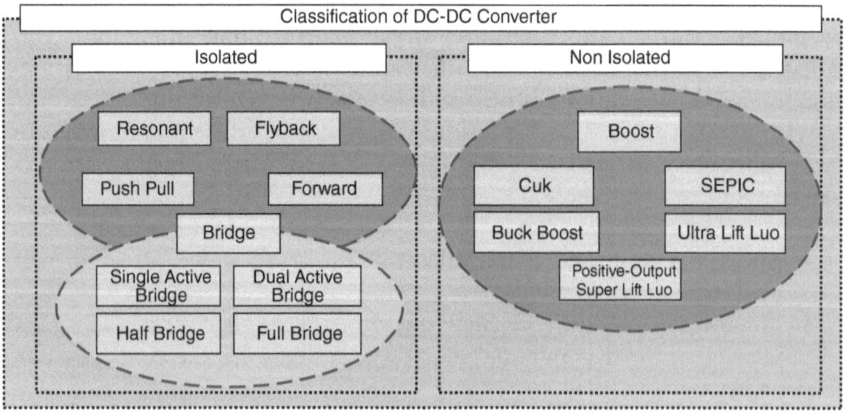

Figure 4.1 Classification of DC-DC converters.

Table 4.1 Power electronics circuit with specification

Circuit topology	*Duty cycle and characteristic equation*	*Input and output relationship*	*Advantages in smart city*
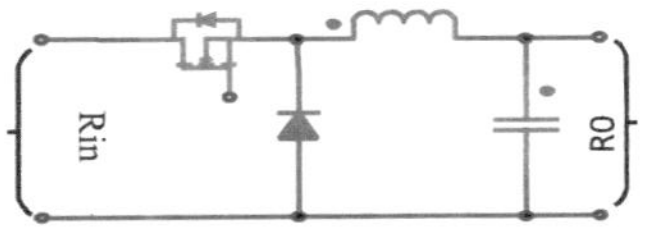 BUCK CONVERTER	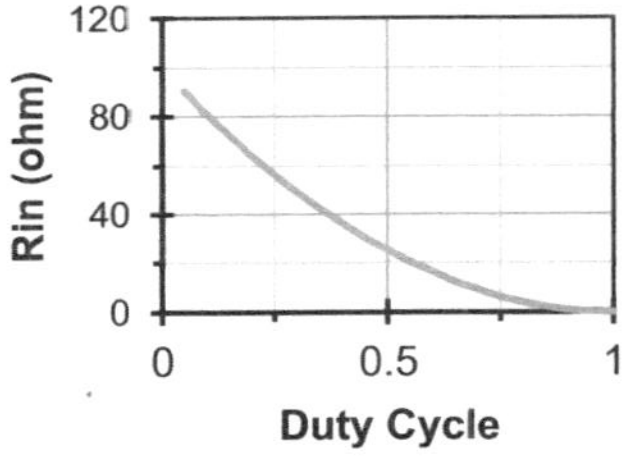	$R_{in} = \frac{R_o}{D^2}$	• Provide a better option with fewer, smaller external components. • Both step-ups and step-downs are possible [20].
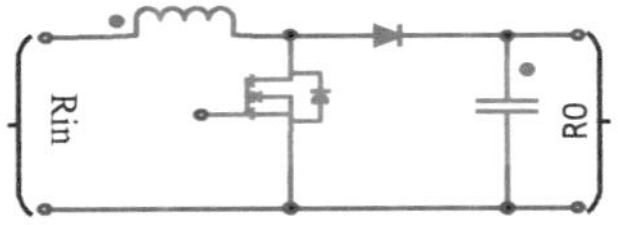 BOOST CONVERTER	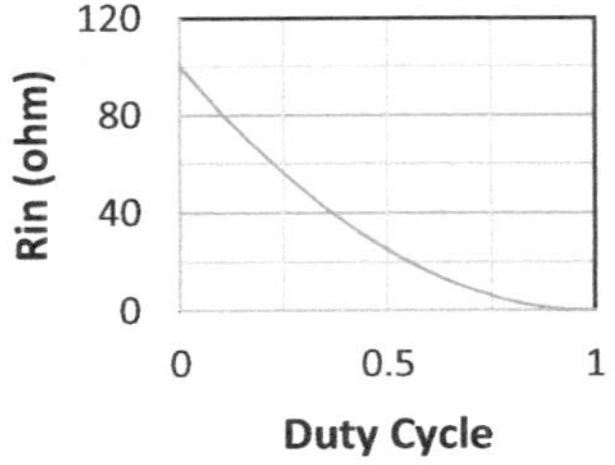	$R_{in} = R_o(1 - D)^2$	• It gives a high voltage output. • It has low operating duty cycle. • It has lower voltage on MOSFET.
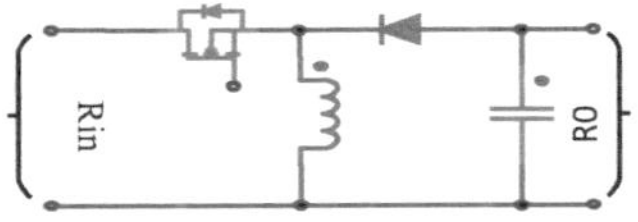 BUCK-BOOST CONVERTER	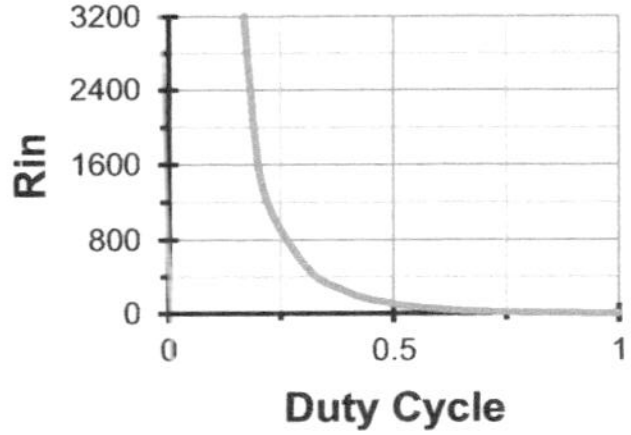	$R_{in} = (1-D)^2 \times \frac{R_o}{D^2}$	• It has a voltage inversion capability. • It can provide reverse polarity protection. • It can minimize losses and maximize energy transfer efficiency [21].

(Continued)

Table 4.1 (Continued)

Circuit topology	*Duty cycle and characteristic equation*	*Input and output relationship*	*Advantages in smart city*
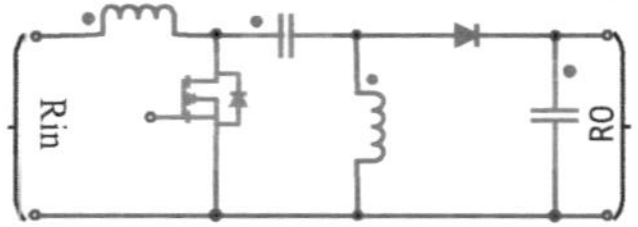 SEPIC CONVERTER	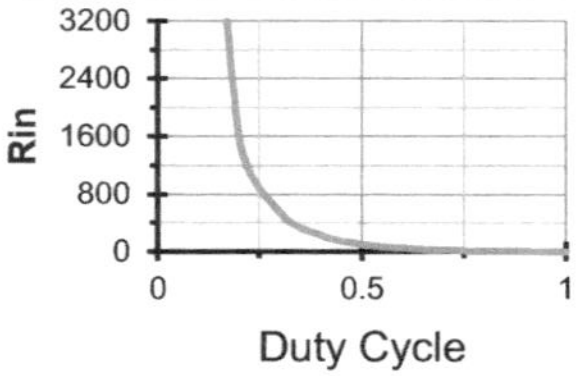	$R_{in} = (1-D)^2 \times \frac{R_o}{D^2}$	• The orientation of the output voltage will match that of the input voltage. • Reduced input current ripple. • High efficiency and stable operation.
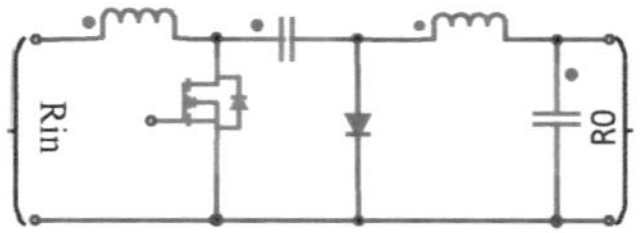 Cuk Converter	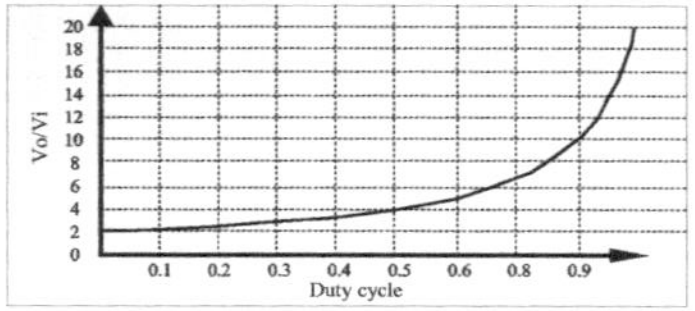	$R_{in} = R_o \frac{(1-D)}{D}$	• High efficiency. • It is used for high-frequency operation. • It is easy to stabilize regulator loop [22].
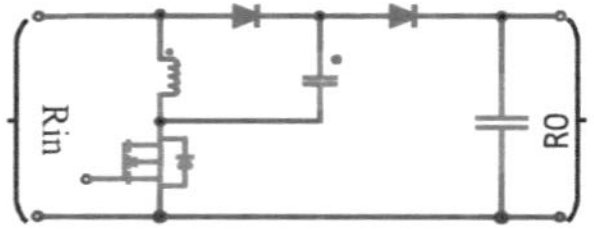 POSLL Converter	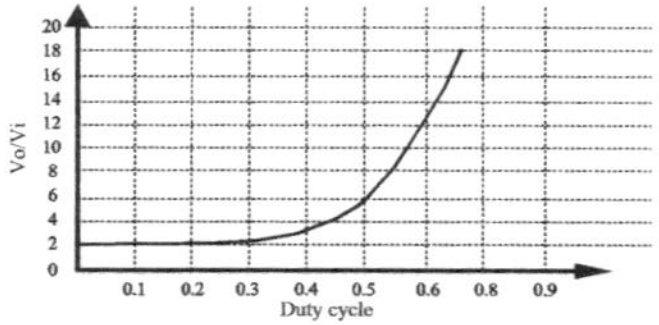	$R_{in} = R_o \frac{(1-D)}{(2-D)}$	• It gives high voltage at low ripple. • It is a cheap topology in simple configuration. • Its performance is good for a transient varying load [23, 24].

4.2.1.2 Reliability in smart grids

As technology advances quickly, the nation and its citizens have new expectations for the electricity quality of the SGs. When the SGs' electricity grid quality deteriorates, both the nation and its citizens will suffer significant economic losses. Therefore, the utilization of cutting-edge PE in SGs is crucial to the consistent output of these systems. Additionally, it will meet the everyday electrical demands of regular people [1, 14].

4.2.1.3 Optimization of resources in smart grids

The government has given the development of renewable energy more attention because countries have a significant demand for energy, and its existing energy use efficiency and per capita share. The SGs can effectively address China's current energy shortage situation by making extensive use of cutting-edge PE technology and building a scientific and logical power system, ensuring long-distance and massively safe power transportation and, ultimately, energy efficiency. Long-term development objectives include the following [15, 16]:

A. **Maximum utilization of energy resources:** The new type of distributed energy sources like renewables, and so on, may soon share the major power generation share in the grid as compared to conventional sources without significantly affecting the infrastructure of the transmission network [17]. The phenomena of energy distribution are not balanced in terms of science and reason. The SGs may also be used as a carrier by all pertinent power departments to optimize the distribution of power resources.
B. **Increasing the power grid's safety performance:** The present flexible AC and DC transmission technologies offer quicker reaction times and more potent control capabilities than classic transmission technologies, which may further increase the real efficiency of SGs transmission construction [18]. A few of the more sophisticated PE systems may also offer tailored services to various power users. Its operation has the potential to increase the overall grade and degree of service offered by the electrical networks while reducing energy consumption, efficiently reducing losses, and consuming less energy.
C. **Optimal planning and building design in smart cities:** Modern cities have stricter regulations on the use of power. The previous high-altitude power generators (PGs) are currently unable to meet the needs of normal users on a daily basis, and this situation will only become worse as losses increase and airborne exposure increases the frequency of mishaps. DC and AC gearbox cables that are flexible, however, are buried well beneath the surface of the earth. It is suitable to enhance the long-distance transmission of power as this will not only have no

adverse effects on the urban environment but will also not change the electromagnetic influence of the cable. The intensity can offer more precise and efficient design techniques for the city's upcoming development.

D. **Applications for storing energy from a combination of solar and wind sources:** Solar panels and windmills are a few of examples of clean, sustainable energy. They are becoming more and more popular in the use of green and environmentally friendly energy, and they represent the general direction of the future growth of high-quality power. They are able to fill as many of the expanding societal demands as they can. It is vital to focus on the coordination of power generation and electricity demand in the process of actual application, though, due to the unstable elements in these types of energy sources [19]. The power system should be connected using high-power converters, which efficiently convert energy and aid in establishing a suitable link between high-power voltage sources and storage devices. This will increase the power energy as well as provide a significant quantity of energy storage.
E. **Advantageous for the effective implementation of smart grid distribution:** The power distribution system for the SGs must satisfy the high standards for electrical energy set by the users. Along with sophisticated PE and automation, it painstakingly studies each step of the distribution process and gives customers supply solutions that are specifically catered to their unique needs. To ensure that the electrical energy being distributed is of high quality, the information received by sensory technology for measurement should be incorporated over the network, and control system coordination should be enhanced.

The primary functions of DC-to-DC converters in solar PV enabled plants are illustrated in Figure 4.2.

4.2.2 Role of renewable energy

Many major cities throughout the world are benefiting from the adoption of smart city technology as the urban environment grows more complex and linked. To optimize the provision of municipal services, boost safety and security, and consume less energy and resources, these technologies rely

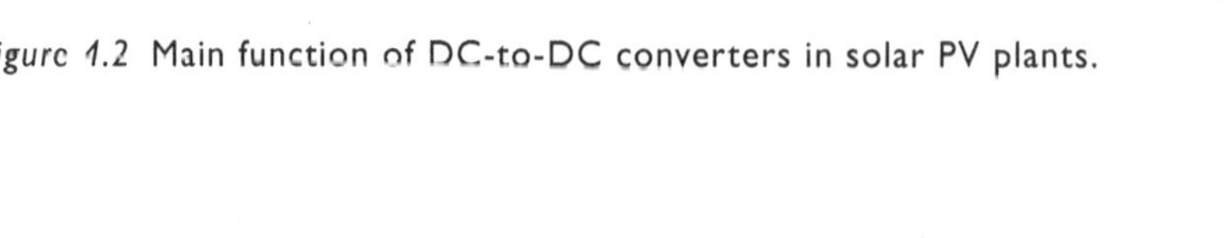

Figure 4.2 Main function of DC-to-DC converters in solar PV plants.

on the use of data [25], analytics, and enhanced automation. In this setting, renewable energy is essential for establishing resilient smart cities that are sustainable for future generations. Here are some of the major functions that renewable energy plays in smart cities.

A. **Reducing carbon footprint:** Without emitting any greenhouse gases, renewable energy sources like sun, wind, and hydroelectricity provide electricity. Smart communities may dramatically lower their carbon footprint and lessen the consequences of adopting climate change by these energy sources. Furthermore, technical improvements have made the usage of renewable energy more accessible and economical, making it a viable alternative for both large and small communities [26].
B. **Solar power:** One of the most often used renewable energy sources in SCs is solar electricity. Cities may generate power locally and reduce their dependency on fossil fuels by placing solar panels on buildings and other structures. Advanced sensors and controls are also being used by smart cities to optimize the use of solar energy [27, 28], ensuring that any extra power is saved for later use or returned to the grid. Figure 4.3 shows the block diagram of maximum power point tracking (MPPT)-based solar PV connected to grid and battery storage system.
C. **Wind power:** Another attractive renewable energy source for smart cities is wind power, particularly in regions with constant winds. To produce power and lessen their dependency on fossil fuels, cities are mounting wind turbines on roofs and in other places. In addition, they

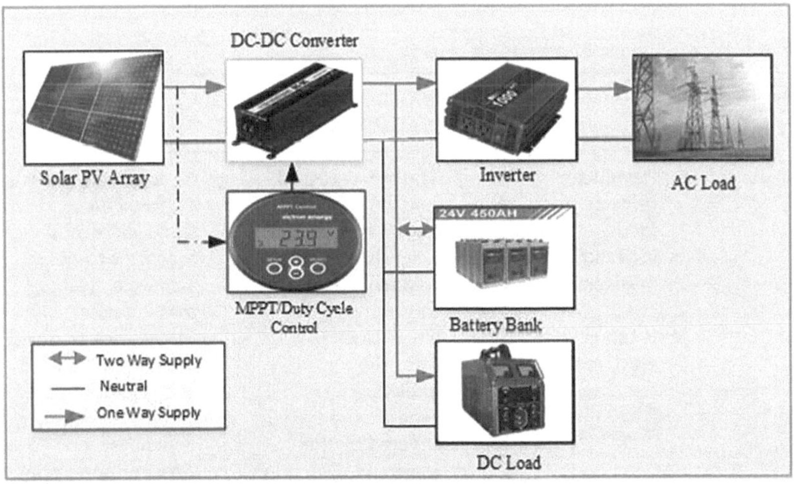

Figure 4.3 Block diagram of solar PV based grid-connected system with MPPT control and battery storage.

are optimizing the usage of wind energy and forecasting future energy requirements using sophisticated data analytics and modeling [29].

D. **Hydroelectric power:** The force of flowing water rotates turbines and produces electricity, thereby producing hydroelectric power. In order to produce electricity for homes and businesses, smart cities are examining ways to harness the power of water from rivers and other sources. Additionally, they are managing the water flow and maximizing the production of hydropower by employing cutting-edge sensors and controls [30].

4.2.2.1 Improving resilience and reliability of energy systems

The resilience and dependability of energy systems in smart cities can be increased through the use of renewable energy [31]. During severe weather or other situations, communities can prevent blackouts and power outages by decentralizing energy generation and storage. In addition, using renewable energy helps ease the load on energy systems, making them less susceptible to outages or other threats. Table 4.2 shows the advantages, disadvantages, and role of renewable energy in smart cities.

A. **Decentralized energy system:** Decentralized energy systems, which might incorporate energy storage options like batteries or other energy storage devices, rely on local renewable energy sources to provide power. Smart cities can guarantee that energy is accessible in times of need and that homes and businesses can continue to run even when the grid is down by establishing a decentralized energy system [32].

Table 4.2 Pros and cons of renewable energy

Renewable energy	*Advantages*	*Disadvantages*	*Role in smart city*
Solar power	• It reduces reliance on fossil fuels. • It has low maintenance cost. • It can reduce electricity bills.	• The initial cost of installing PV cells can be high. • It is reliant on weather conditions to generate electricity. • The amount of electricity generated is affected by the angle and direction of the panels.	• Reduced dependence on fossil fuel. • Reduces carbon footprint to help achieve a clean environment. • The choice of power source for isolated and grid-connected systems is heavily dependent on weather conditions.

(Continued)

Table 4.2 (Continued)

Renewable energy	*Advantages*	*Disadvantages*	*Role in smart city*
Wind power	• The energy it produces is clean. • The carbon footprint is small. • It is a cost-effective alternative to fossil fuels.	• It is reliant on weather conditions to generate electricity. • Wind turbines can be noisy and unappealing visually. • Bird and bat deaths have been associated with wind turbines.	• By using wind power, the atmosphere is clean. Utilizing wind power contributes to a cleaner atmosphere and helps mitigate climate change. • It is installed in a wind area where air always blows and it generates power [34].
Hydro power	• It is a reliable source of energy. • It has low operating costs. • It helps to manage water supply and improve water quality.	• Hydroelectric power plants can have a negative impact on fish and wildlife habitats. • The construction of dams can lead to the displacement of people and animals. • Hydro energy production requires substantial capital investment.	• It stores energy as we need it. • It produces a sustainable energy which makes the smart cities' energy reliable. • It is used for water management, irrigation, and flood control.
Geothermal energy	• It is a reliable source of energy. • It has low carbon emissions. • It provides constant energy.	• It is only available in certain locations. • The construction of geothermal power plants can be expensive. • It can lead to land subsidence.	• By using this energy smart cities reduce carbon footprint and contribute to sustainable energy. • It can provide a consistent and base load power. • It also minimizes the failure of the grid [35].
Biomass energy	• It reduces waste disposal in landfills. • It is an abundant source of energy. • It has a low carbon footprint.	• Land use changes required for biomass production can lead to deforestation and higher greenhouse gas emissions. • Biofuels can compete with food production. • Biomass combustion can release air pollutants [36].	• Biomass energy can contribute to waste management strategies. • It is used for storing thermal energy, making for flexibility in supply. • It contributes to reducing greenhouse gases and also improving air quality [37].

B. **Microgrid:** Small-scale energy systems called microgrids can function without the help of the national power grid. Microgrids are being used by smart cities to guarantee that vital infrastructures, like hospitals and emergency response centers, have access to dependable and resilient energy sources in times of need [33].

4.2.3 Role of IoT

Figure 4.4 shows the resolved and unresolved challenges of IoT in SCs. The role of IoT in SCs are as follows [10]:

A. **Smart meter:** Smart meters provide real-time data on energy consumption, enabling residents and utility providers to more effectively monitor and control usage. It is a computerized device that measures and monitors the usage of energy, gas, or water in homes. They enhance billing accuracy, promote energy efficiency, and support demand-response programs.
B. **Smart home:** An IoT-enabled smart home uses sensors and IoT devices to automate and regulate numerous parts of the home environment. These products, which include smart thermostats, lighting systems, security systems, and appliances, may be managed and linked remotely. Through features such as remote monitoring, automatic routines, and

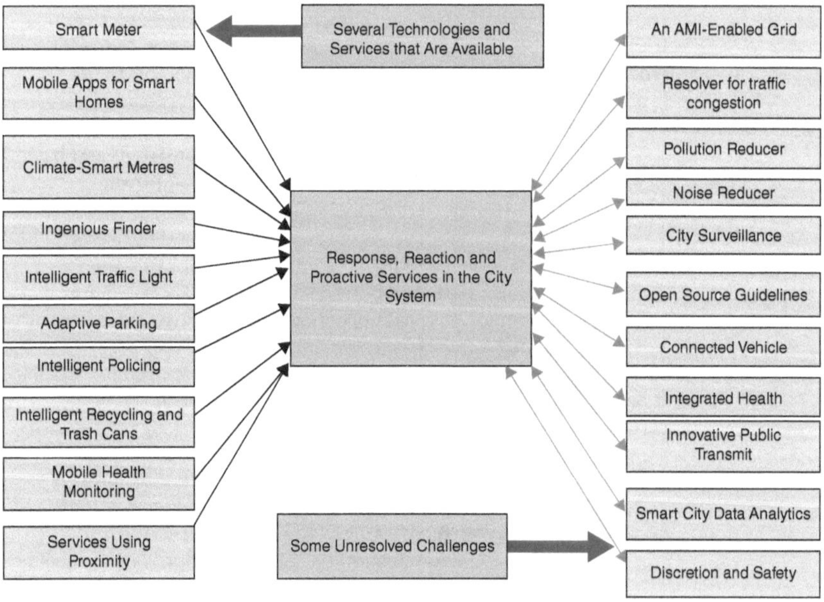

Figure 4.4 IoT in smart cities: available and unresolved technology.

connection with voice assistants, smart homes provide convenience, energy efficiency, and better security.

C. **Smart environment meters:** IoT sensors are used in smart environment meters to measure and monitor environmental factors like air quality, temperature, humidity, and noise levels. This information assists authorities in identifying pollution sources, assessing the impact on public health, and taking required steps to enhance the environment and residents' well-being.
D. **Smart traffic light:** Smart traffic lights use IoT technology to improve traffic flow and decrease congestion. They are linked to sensors and traffic management systems, and their signal timings are constantly adjusted based on real-time traffic circumstances. This improves traffic efficiency, reduces travel times, and increases overall road safety.
E. **Smart parking:** Smart parking systems monitor parking spot availability in real-time using IoT sensors and data analytics. This information is then relayed to vehicles via mobile applications or digital signs, making it easier for them to discover and reserve parking places. Smart parking decreases traffic congestion, increases parking utilization, and reduces the amount of time spent looking for parking.
F. **Smart waste management (SWM):** SWM solutions based on IoT enable effective garbage collection and disposal. Sensors in smart bins monitor fill levels, allowing garbage collection routes and timetables to be optimized. Furthermore, these devices may deliver real-time bin status notifications and optimize waste segregation procedures. Smart trash management lowers costs, increases cleanliness, and encourages environmentally friendly waste practices.
G. **Smart policing:** Smart policing uses IoT devices, surveillance cameras, and data analytics to enhance public safety and law enforcement. IoT sensors can identify suspicious activity, monitor crowd movements, and offer law enforcement authorities with real-time situational awareness. Smart policing shortens reaction times, increases crime prevention, and aids with traffic management.
H. **Remote health monitors:** IoT-based remote health monitoring devices offer continuous monitoring of patients' health problems outside of traditional healthcare facilities. Wearable gadgets capture vital signs and relay information to healthcare practitioners, such as fitness trackers or medical sensors. Remote health monitors enhance patient care, allow for early intervention, and save healthcare costs by reducing the number of hospital visits.

4.3 ADVANCED APPLICATIONS IN SMART CITIES

Smart cities are urban areas that use cutting-edge technology to improve the quality of life for their citizens. As a result of several technological, financial, and regulatory obstacles that limit the widespread deployment of smart cities, the concept has continued to evolve. When the IoT and big data technologies are combined, it may be possible to create smart cities in previously unimagined ways.

Some of the cutting-edge applications used in smart cities include intelligent applications, process digitization, the use of UAV resources in conjunction with information and communication technologies (ICT), real-time data processing and assessment, and artificial intelligence.

4.3.1 Applications of power electronics

Power electronics is a branch of electrical engineering, and it deals with handling high voltages and currents to provide electricity that may meet a variety of demands. It is used in a wide variety of items, from household appliances to spacecraft equipment. Electrical energy is converted using power transistors like the MOSFET and IGBT2, as well as other semiconductor switching components like diodes and thyristors. Power electronics devices are needed by many items, including battery chargers, computers, televisions, and variable-speed drives (VSDs).

4.3.1.1 FACTS smart grid technology

FACTS technology strives to enhance the reliability and dependability of the electrical grid's distribution and transmission networks. It makes use of current power electronic equipment while incorporating contemporary control.

Technology reacts quickly to the features and network topologies of the present AC transmission systems regulation. The static overcompensating and history-controlled series compensating FACTS technology have grown with the advancement and continuous creation of PE technology. The device uses a unified power flow controller and static synchronous compressor technologies. FACTS technology's compatibility and quick regulation may help to balance the present energy supply and demand, increase the capacity of power transmission and distribution lines, and address other problems.

4.3.1.2 SVC technology in smart grid

In the 1980s, China looked abroad for solutions to its power transmission challenges. During that period, six sets of static overcompensate equipment PGs were imported. Following this, China made further discoveries and advancements, leading to significant improvements in power transmission. This approach has the potential to quickly increase resistance while simultaneously altering and preserving system voltage stability, PGs system stability, and gearbox capacity. For line assembly, flat steel or copper wire is often utilized; to connect with subterranean leads, grounded current box shells are utilized. We must create voltage protection guidelines because substations are vulnerable to voltage changes. In order to ensure the security of critical equipment, we must construct a reliable and sufficient grounding system.

4.3.1.3 Smart grid power generation

How can we successfully enforce environmentally friendly energy usage during this time of rapid development? We are currently dealing with a serious issue. Due to implementation of SGs there are a lot of PE technologies used [38]. This can aid in preserving energy, reducing energy shortages, and preserving the environment. In Figure 4.5 it is shown that the utilization of cutting-edge technologies as well as some sustainable resources, such as wind power, hydro, and geo energy, are part of the use of current PE technology in the process of power production in the SGs. Modern technology must be used to interface the electric machine group's electronics with the electric grid in order to better manage variable-speed operation, which is essential for producing power from wind energy.

4.3.2 Applications of renewable energy

Renewable energy finds numerous applications in smart cities. Advanced technologies and innovative solutions as shown in Figure 4.6 are employed for the super efficiency of energy and an eco-friendly environment. Some of the applications of renewable energy in smart cities are as follows:

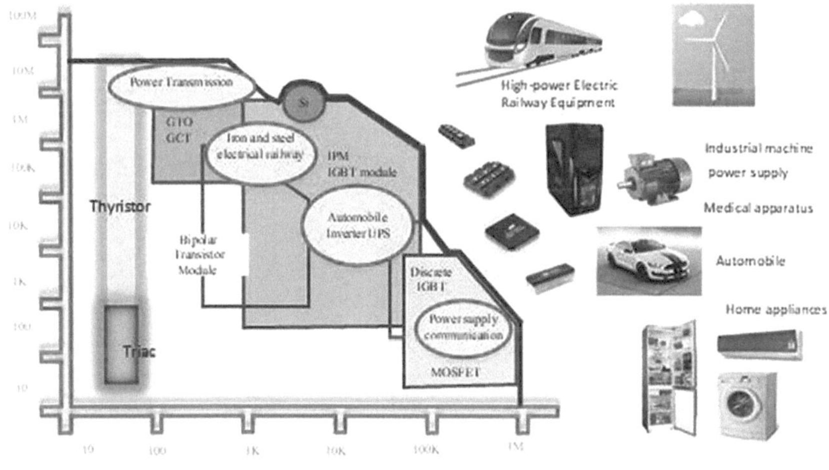

Figure 4.5 Application of PE in smart cities.

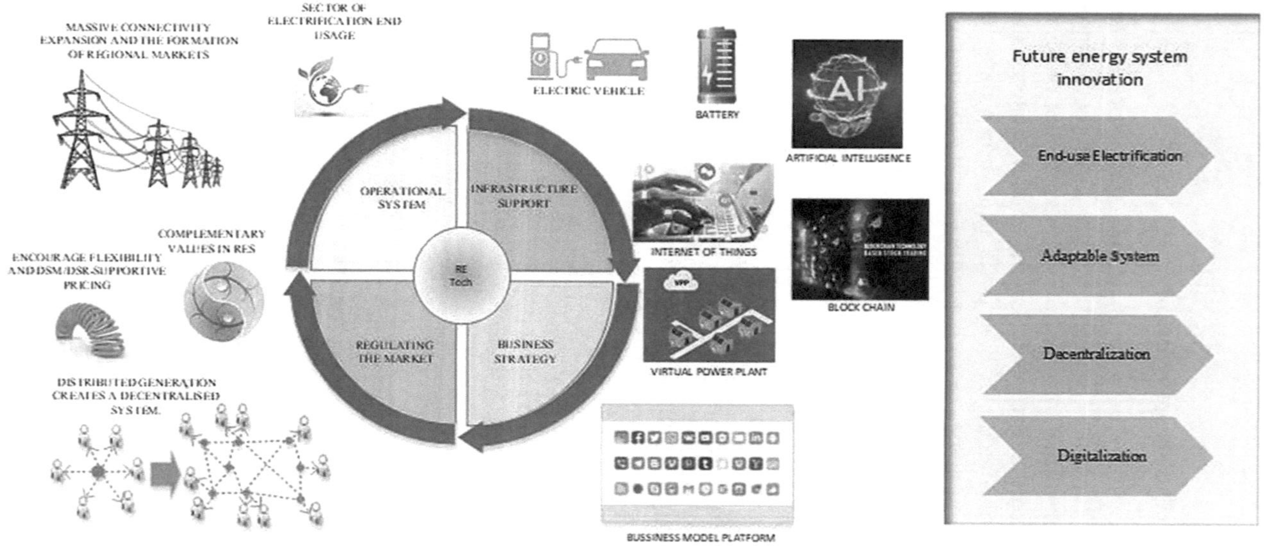

Figure 4.6 Renewable based innovative power systems infrastructure for future prospective.

4.3.2.1 Power generation

Solar panels utilize the photovoltaic effect to convert sunlight into electricity. They can be mounted on rooftops, facades, or in smart city solar parks. Solar energy provides a decentralized method of energy generation, allowing individual buildings or communities to generate their own power. By utilizing clean and abundant solar energy, it minimizes dependency on traditional power systems and reduces carbon emissions [39]. Wind turbines transform the conversion of wind's motion-induced energy to electricity. To maximize energy output, SCs can strategically deploy wind turbines in favorable locations such as coastal areas, plains, or high spots. Wind power is a scalable alternative that smart cities may use to diversify their energy mix and minimize their dependency on fossil fuels [40]. Small-scale hydroelectric systems may be included in smart cities, particularly in places with rivers or other water sources. These systems use turbines to produce power from the flow or fall of water. Small-scale hydroelectric systems, such as canals or water distribution networks, can be incorporated into metropolitan settings to provide a sustainable and consistent power source [30, 41].

Smart cities may produce power in a decentralized fashion by combining solar panels, wind turbines, and small-scale hydroelectric systems, lowering the burden on existing energy networks and fostering energy independence. This decentralized strategy contributes to the city's overall sustainability by ensuring a stable power supply, reducing transmission losses, and lowering carbon emissions [42, 43].

4.3.2.2 Smart grid integration

As a result of climate change, fuel security, and other issues, renewable energy technologies will supply an increasing amount of our power in the future. The use of renewable energy carries both a chance and an issue. It goes without saying that as more sustainable energy sources are connected into the grid, it will be harder to adequately control and regulate the system's electrical characteristics. Additionally to significant inefficiencies that lead to grid energy losses, the threats include frequency and voltage changes and outages [44].

The grid needs to employ more renewable energy sources and manage demand and supply of energy more efficiently due to the urban environment. Even if it costs more in the near term, consumers in households, businesses, and factories will continue to use sustainable energy sources and storage. In addition to saving money, they are driven by independence, sustainability, dependability, security, high-quality power, and excellent customer service. Before SCs become a reality, major infrastructural changes are required, particularly in the area of energy distribution. The intermittent usage of sustainable energy in SCs, as well as the necessity for ways of decreasing the community's peak load when demand for power is high, are driving much of the need for updated infrastructure. Energy storage has previously been employed effectively by utilities to address comparable distribution system concerns. Furthermore, distributed energy storage technologies may be able to give comparable assistance to smart cities across the country [45].

4.3.2.3 Microgrid

A microgrid is described as a collection of linked loads and DERs contained inside clearly defined limits that operate as a single controlled entity in relation to the grid. A microgrid-capable system can operate in both grid-connected and island modes. As a result, microgrid safety in both grid-connected and island modes is critical for safe and dependable operation. Intermittent DERs, bidirectional power flow, and dynamic microgrid configuration are some influencing elements for safe and reliable microgrid operation utilizing standard protection techniques [33, 46].

A microgrid is described by the Department of Energy's Microgrid Exchange set as "a group of interconnected loads and distributed energy resources within clearly defined electrical boundaries that act as a single controllable entity with respect to a grid." Microgrids are therefore made up of low-voltage (LV) distribution systems with distributed energy resources (DERs) (micro-turbines, fuel cells, solar production, and so on) and storage devices (fly wheels, energy capacitors, and batteries). If linked to a grid, such a system can run on a nonautonomous or autonomous basis if separated from the main grid. If controlled and coordinated properly, the functioning of micro sources in the network can give significant improvements to overall system performance [47]. This includes generation, storage, and load control. The generating will be a few kilowatts, often less than a megawatt (MW), and can serve part or full loads as needed. The microgrid can operate in two modes: grid linked mode most of the time and isolation mode, delivering partial or full load depending on the capacity of the DER. There are also isolated microgrids that function without grid connections in distant regions when grid power is unavailable.

4.3.2.4 Energy storage

Transferring electrical energy from a power network into a form that can be held until it is changed back into electrical energy is known as energy storage [48]. Electrical energy has the unique property that the generated energy must meet the load demand, making it a particularly dynamic system. Furthermore, load centers are often positioned a long distance away from the bulk producing units. This needs huge gearbox systems, and line failure causes significant power shortages. During high demand hours, a few power lines get congested. However, there is significant potential if the electrical energy produced by base load plants during off-peak hours (particularly nights) can be stored in some form and used to meet peak loads at a later time. Peak load is typically met by operating expensive generating units like oil and gas power plants [49]. Household energy management systems need energy storage to smooth out problems like poor power quality, grid outages, and plug-in hybrid automobile integration. Energy storage may be effectively employed to help utilities and consumers deal with the difficulties that come with ensuring a consistent supply of high-quality energy.

4.3.2.5 *Electrical vehicle charging infrastructure*

The electric car battery has a large amount of energy storage capacity, and the charging requirements will put a strain on the power system. PHEVs will have a significant impact on the design and functionality of smart distribution networks since electric vehicles (EV) manufacturers plan to introduce a significant number of vehicles to the market in the near future. Energy use will significantly increase, demand patterns will shift, substation transformer loads will increase, and there will be a host of additional problems. The EV's battery may collect and store energy before releasing it to the grid or your house when the car is parked, making it an additional source of energy. This idea of "vehicle to grid" (V2G) gives rise to a number of ancillary services including spinning reserves and charging stations [50–52].

The EVs must have electronic interfaces for connecting to the grid at any time and from any location, as well as unique IP addresses that permit controlled energy exchange, metering capabilities, and a bidirectional communication interface to communicate with an aggregator entity that will manage the numerous EVs [53, 54].

4.3.2.6 *Smart building*

Many parts of our existence require that the built environment be digitized, controlled, and understood, employing sensors that capture and analyze data to communicate in digital language. Data about a building's energy efficiency, lighting and temperature, air quality, asset locations, mobility, and space usage, to name a few, are collected indoors and outside by the intelligent sensor [55]. The system operator responsible for managing a smart building oversees its operations, ensuring it runs smoothly to improve residents' quality of life. It demonstrates how the building is operated to provide comfort and enhance the quality of life for its occupants. Heating, ventilation, and air conditioning (HVAC) systems account for a sizable portion of energy consumption (35% in commercial buildings and 40% in residential), and numerous studies have been done to examine various aspects of the built environment, including how to improve energy efficiency. It is based on a smart sensor that is used to optimize energy use on device use and alter comfort. The effects of various sensor types on the built environment and occupant productivity have been verified [56].

The BIM-based approach is used for preconstruction design and verification, as well as postconstruction facility management, of the smartly designed environment in the age of smart networks. It is possible (and has been successfully proven) to use Revit and the BIM toolbox in a simple but functional prototype for a smart home energy management system. The development of BIM software tools is an outgrowth of current research into the integration of data and decision systems in future intelligent networked applications for powered buildings, as well as efforts to autonomously control HVACR (Heating, Ventilation, Air Conditioning, and Refrigeration) systems.

4.3.2.7 Energy management system

By controlling and overseeing DER, an energy management system (EMS) installed at a consumer location offers auxiliary services for the power system along with energy savings and cost cuts for consumers. Through DER, consumers can provide supplementary services such as load frequency management, demand response, and voltage adjustment of distribution feeders. A building's energy systems are used more effectively and at a lesser cost thanks to integrated management of DER by EMS [57].

Based on the Integrated-Distributed Energy Management System (IDEMS) concept of demand forecasting, ideal planning, and 24/7 operation, an energy management model was developed. Using operational scenarios that included metrics like energy cost, CO_2 mitigation, and energy savings, the models were assessed.

This chapter proposed an IDEMS for residential buildings. The system works with installed systems in other residences and controls equipment in a residential home for more effective functioning. It is anticipated that the IDEMS would result in cost savings, lower energy use, and/or lower CO_2 emissions. On the basis of the anticipated energy consumption from the previous day, IDEMS develops an ideal operational plan (schedule) in advance. The plan is then put into action and modified in accordance with the situation [49].

4.3.2.8 Environmental and conservation

Before renewables are recognized as an environmentally responsible, dependable, and economically advantageous technology, a number of challenges must be overcome. Both the creation of norms (green norms) for recycled materials and the control of equipment disposal would be necessary for industry. Governments' increased use of financial incentives and environmental sanctions would also increase supply-side pressure, pushing businesses to use green energy more responsibly. Companies building new renewable energy plants must think about the effects on the surrounding environment [58].

Both an evaluation of the consequences of climate change and operational adaptations, such as quicker and more frequent ramping, are required for renewable energy. A sizable increase in the share of renewable energy in total energy output requires a sizable investment in green energy technologies. Due to lengthy planning processes for new manufacturing, businesses must spend. FDI inflows, which frequently provide developing countries with not just the necessary financial resources but also management expertise, a more productive work environment, a more diversified talent pool, and distribution networks, may effectively assist these technologies [59]. The global economy might become much more unstable as a result of climate change. The negative effects of pollution on health are also becoming more significant. These cover a wide range of issues, including how people use renewable energy, the cost and quality of energy, the impact of energy security on economic issues, information sources, and general ecological concerns.

4.3.3 Applications of IoT technologies

IoT and other information and communication technologies are critical for the development of SCs. They enable smart cities to be smart by providing connection and data collecting capabilities. Almost every facet of a smart city's operation is reliant on communication architecture, advantages, and disadvantages as shown in Table 4.3. The most essential IoT applications in smart cities are as follows:

4.3.3.1 IoT in healthcare

Smart healthcare is necessary for SCs. It is crucial to include technology, especially information and communication technologies, to support healthcare services. For improving hospital operations, home-based healthcare, telemedicine, senior care, and other sectors, IoT has immense promise. Lives can be saved by keeping an eye on patients receiving care at home, regularly reviewing their condition with medical professionals, identifying potentially dangerous situations, and alerting healthcare organizations to seek immediate assistance. These technologies make it easier to keep track of organ donors, transplant candidates, and timely matches. A reliable communication environment, like the one provided by IoT, is essential under these circumstances.

The senior population and patients can be monitored using IoT-based communication infrastructure to manage their healthcare needs at home. It makes sure that the patients take the right medications as directed. Along with their vital signs and other activities, it keeps an eye on their eating and sleeping habits. Many IoT devices are integrated into clothes as IoT device sizes continue to shrink, and these garments might track the health of those who wear them. Additionally, the collected data may be combined to

Table 4.3 Application, advantages, disadvantages, and communication protocol used for IoT services

IoT services	*Advantages*	*Disadvantages*	*Communication protocols*	*Potential applications in smart city*
AWS IoT	Security of data transfers	Exclusive platform	MQQT, HTTPS, LoRaWAN, CoAP, Bluetooth	Data visualization, monitoring, and analysis from wired or wireless sensors.
Oracle IoT cloud	Millions of device endpoints supported heterogeneous connection	There is no support for devices based on open source	ODBC/JDBC, WebSocket, REST	M2M platform, real-time data capturing [70].
Microsoft research Lab of Things	(Platform, drivers, and applications) Is accessible for academic research	Accessible only for Microsoft-based products	MQTT, CoAP, and HTTP	Services for smart homes.
Open remote	There are a number of protocols that are supported. Cloud computing	High expenses	MQTT, CoAP, KNX, HTTP, WebSocket	Services for buildings, home automation, healthcare, and smart cities.
KAA	IoT cloud platform that is open. Big data assistance	Supported hardware modules are restricted	MQQT	Open-source platform with applications in agriculture, healthcare, industrial IoT, consumer electronics, and smart homes.
Things Board	Data security and device authentication	Platform is quite new and has not been tried on a big scale	MQTT, CoAP, and HTTP	Data collection, processing, visualization, and device administration are all possible with an open-source IoT platform.

analyze healthcare within a SC, which might result in wise resource allocation to deal with developing healthcare challenges. The IoT might be an ideal platform for collecting healthcare data from across cities, analyzing trends, and taking decisive steps to address any expanding healthcare issue inside SCs [10, 60].

4.3.3.2 IoT in transportation

The transportation industry is rapidly developing. Many highly novel and popular elements are being added to the transportation experience by IoT and other communication technologies. This is true for all forms of transportation, including land, air, and sea. Travel activities are expanding as a result of a growing worldwide population, more global awareness, and looser travel laws. In order to better manage all modes of transport and provide better services to residents in smart cities, there is a tremendous demand and necessity for integrating technology. This is because the capacity limits of current transport techniques are being reached [61, 62].

The mobility experience both within and outside of SCs is optimized in large part thanks to the Internet of Things. It is anticipated that the use of IoT in transportation would transform status queries and traffic management. Smart transportation results from the use of IoT to gather and analyze transportation data in real-time, identify difficulties in transportation, make wise decisions, find solutions, and put those answers into action. The technical foundation for smart transportation, which will benefit passengers, transportation businesses, and local government administration, is in place. This is not only a necessary and timely tactic, but it is also a quick and affordable solution. Additionally, such a strategy permits and encourages the emerging trend of autonomous vehicles [63].

Vehicle-to-vehicle (VTV) and vehicle-to-infrastructure (VTI) connections must be widely available for vehicles (land, air, and sea) in order for an IoT-based smart transportation network to function. They collect, handle, and disperse data. Real-time monitoring of traffic conditions in a smart city allows for the application of appropriate measures, such as traffic rerouting and changing the timing of traffic lights. It is possible to monitor parking lots and lead drivers to open spots in real-time; this functionality is expected to be especially popular in urban settings. Based on these circumstances, it is anticipated that the application of IoT in smart city transportation management would significantly impact the collection of real-time traffic data, its analysis, and its free exchange with all parties concerned in order to improve traffic conditions [64].

The anticipated decline in vehicle ownership is another rising mobility issue for which smart cities must prepare and plan. Instead of owning a car, people will use public transport as a service. The younger generation is already becoming more and more enamored with this. A fleet of vehicles is dispersed around a smart city using transport as a service. Anybody who requires transportation will use a smartwatch or other wearable device to unlock a car, drive it to the necessary destination, and leave it there. IoT-enabled ubiquitous connectivity is necessary to provide transport as a service.

4.3.3.3 IoT in grid energy

With the world's population expanding and urbanization on the increase, energy consumption is predicted to skyrocket. As a result, another critical topic to solve in order to satisfy the energy demands of SCs is the efficient use of energy. Emerging technologies like the Internet of Things can be beneficial in this sector. The integration of sensing IoT gadgets in a smart energy grid offers efficient, adaptive, and dependable energy transmission and distribution. The deeper deployment of communication and information technology is predicted to revolutionize the present electric power distribution system [65].

A smart grid that uses widespread connection of things and devices via IoT helps utility companies and users to keep track of energy use, predict energy demand, and efficiently and effectively manage the energy system. While the usage of IoT enhances the energy infrastructure, it also introduces new issues. The vulnerability of the smart grid owing to its cyberinfrastructure is one of the primary issues. The magnitude of the smart grid and the related entry sites complicate the security problems even further. The context-aware nature of IoT operations benefits the electricity grid system's stability and resilience [66]. This is especially important when it comes to meeting the needs of the energy system's increasingly diversified energy source supply. This includes widely scattered energy sources including wind turbines, solar farms, and solar panels. For these localized energy sources, more fault tolerance, greater precision, a larger coverage area, and extraction of localized features are needed. Additionally, a diversified energy system like this calls for the management of energy flow to and from smaller sources, such as solar-powered homes. IoT might be a great tool for managing and controlling operations by monitoring the energy grid and collecting valuable data [67].

4.3.3.4 IoT in agriculture

Analyzing modern agricultural history reveals that the agriculture sector has been significantly impacted by the usage of developing technology. Improved crop yield has been one of the most significant effects of technological integration in agriculture. This has benefited the expansion of the population. With over 10 billion people on the earth by 2050, it will be crucial to deploy cutting-edge technology to substantially enhance food production. We must also use any area that is suitable for agriculture. Rooftops, vertical areas, and even moving cargo containers are shown. Modern agricultural technology, such as greenhouses, successfully boosts food production while protecting valuable resources, such as water, to feed the world's expanding population [68].

Using SCs may encourage this. The IoT in agriculture has great promise for managing all land uses, including greenhouses. IoT gadgets support the monitoring of crop and plant health as well as the agricultural climate (temperature, humidity, and irrigation). To make wiser decisions, enhance crop management, and boost crop productivity through optimal control decision. In order to do this, the farm must be irrigated, fertilized, and occasionally sprayed with pesticides. As a result, the use of IoT in agriculture has a significant impact on the health and yield of crops. For instance, water may be conserved since the data collected can be used to accurately calculate the amount and timing of irrigation. This approach is expected to save a significant number of resources [69].

4.4 SUMMARY

The necessity of power electronics, renewable energy, and the IoT in the creation and operation of SCs has been examined in this chapter for the Role & Applications of Power Electronics, Renewable Energy, and IoT in smart cities. It explored the numerous benefits of these technologies and how they help metropolitan areas achieve sustainability, efficiency, and higher living standards. The chapter began by giving a general overview of smart cities and emphasizing their goal in using cutting-edge technology to improve the overall metropolitan infrastructure and services. Also, it highlighted the necessity of effective energy management systems in smart cities, considering the importance of protecting the environment and the expanding requirement for power. The discussion of PE, a crucial subject, then concentrated on how well it can transform and regulate electrical power. The use of PE in sectors including energy production, distribution, and consumption is examined in this chapter. It describes how equipment used in power electronics, such as inverters, converters, and controllers, enables the integration of renewable energy sources like solar and wind into the power grid, permitting a cleaner and more sustainable energy mix. In the effort to create sustainable SCs, renewable energy is crucial. The chapter discusses the advantages, drawbacks, and integration of various renewable energy sources into the urban energy infrastructure, including solar, wind, hydro, and geothermal energy. It emphasizes the importance of energy storage devices in offsetting intermittent renewable energy sources and guaranteeing a steady supply of electricity. The chapter also looks at how IoT technologies are used to link and communicate among various devices, sensors, and infrastructure components in smart cities. It highlights how real-time data gathering, analysis, and decision-making are made easier by IoT, which results in better resource management and urban services. In-depth discussion is had on the use of

IoT in smart grids, intelligent transportation systems, waste management, and public safety. The chapter also emphasized how PEs, renewable energy, and IoT may work together to create smart cities. It detailed how various technologies interact to build a more productive and sustainable urban ecology. It also demonstrated the effective use of advanced technologies and their key beneficial effects on energy efficiency, decreased carbon emissions, and improved energy infrastructure across the world.

REFERENCES

[1] Z. Zhang, Y. Tang, Y. Qi, and J. Ren, "Application of Advanced Power Electronic Technology in Smart Grid Application of Advanced Power Electronic Technology in Smart Grid," 2018, doi: 10.1088/1757-899X/394/4/042017

[2] N. Haegel and S. Kurtz, "Global Progress toward Renewable Electricity: Tracking the Role of Solar," *IEEE J. Photovoltaics*, vol. 11, no. 6, pp. 1335–1342, 2021, doi: 10.1109/JPHOTOV.2021.3104149

[3] Y. A. Medvedkina and A. V. Khodochenko, "Renewable Energy and Their Impact on Environmental Pollution in the Context of Globalization," in *2020 International Multi-Conference on Industrial Engineering and Modern Technologies (FarEastCon)*, Oct. 2020, pp. 1–4, doi: 10.1109/FarEastCon50210.2020.9271508

[4] R. Srivastava, M. Amir, F. Ahmad, S. K. Agrawal, A. Dwivedi, and A. K. Yadav, "Performance Evaluation of Grid Connected Solar Powered Microgrid: A Case Study," *Front. Energy Res.*, vol. 10, 2022, doi: 10.3389/fenrg.2022.1044651

[5] Z. Tang, Y. Yang, and F. Blaabjerg, "Power Electronics: The Enabling Technology for Renewable Energy Integration," *CSEE J. Power Energy Syst.*, vol. 8, no. 1, pp. 39–52, 2022, doi: 10.17775/CSEEJPES.2021.02850

[6] M. Amir, A. K. Prajapati, and S. S. Refaat, "Dynamic Performance Evaluation of Grid-Connected Hybrid Renewable Energy-Based Power Generation for Stability and Power Quality Enhancement in Smart Grid," *Front. Energy Res.*, vol. 10, 2022, doi: 10.3389/fenrg.2022.861282

[7] M. Ali, A. Iqbal, and M. R. Khan, "AC-AC Converters," in *Power Electronics Handbook*, Elsevier, 2018, pp. 417–456. doi: 10.1016/B978-0-12-811407-0.00014-3

[8] N. L. Panwar, S. C. Kaushik, and S. Kothari, "Role of Renewable Energy Sources in Environmental Protection: A Review," *Renewable and Sustainable Energy Reviews*, vol. 15, no. 3. pp. 1513–1524, 2011, doi: 10.1016/j.rser.2010.11.037

[9] A. Iqbal, M. Amir, V. Kumar, A. Alam, and M. Umair, "Integration of Next Generation IIoT with Blockchain for the Development of Smart Industries," *Emerg. Sci. J.*, vol. 4, pp. 1–17, 2020, doi: 10.28991/esj-2020-SP1-01

[10] M. Ilyas, "IoT Applications in Smart Cities," in *2021 IEEE International Conference on Electronic Communications, Internet of Things and Big Data, ICEIB 2021*, 2021, pp. 44–47, doi: 10.1109/ICEIB53692.2021.9686400

[11] S. K. Ram, S. Chourasia, B. B. Das, A. K. Swain, K. Mahapatra, and S. Mohanty, "A Solar Based Power Module for Battery-Less IoT Sensors towards Sustainable Smart Cities," in *2020 IEEE Computer Society Annual Symposium on VLSI (ISVLSI)*, Jul. 2020, pp. 458–463, doi: 10.1109/ISVLSI49217.2020.00-14

[12] B. Hammi, A. Fayad, R. Khatoun, S. Zeadally, and Y. Begriche, "A Lightweight ECC-Based Authentication Scheme for Internet of Things (IoT)," *IEEE Syst. J.*, vol. 14, no. 3, pp. 3440–3450, 2020, doi: 10.1109/JSYST.2020.2970167
[13] R. Khatoun and S. Zeadally, "Cybersecurity and Privacy Solutions in Smart Cities," *IEEE Commun. Mag.*, vol. 55, no. 3, pp. 51–59, 2017, doi: 10.1109/MCOM.2017.1600297CM
[14] K. Singh, M. Amir, F. Ahmad, and S. S. Refaat, "Enhancement of Frequency Control for Stand-Alone Multi-Microgrids," *IEEE Access*, vol. 9, pp. 79128–79142, 2021, doi: 10.1109/ACCESS.2021.3083960
[15] W. Li and X. Zhang, "Simulation of the Smart Grid Communications: Challenges, Techniques, and Future Trends," *Computers and Electrical Engineering*, vol. 40, no. 1. pp. 270–288, 2014, doi: 10.1016/j.compeleceng.2013.11.022
[16] Zaheeruddin, K. Singh, and M. Amir, "Intelligent Fuzzy TIDF-II Controller for Load Frequency Control in Hybrid Energy System," *IETE Tech. Rev.*, vol. 39, no. 6, pp. 1355–1371, 2022, doi: 10.1080/02564602.2021.1994476
[17] S. Li, W. Chen, X. Yin, D. Chen, and Y. Teng, "A Novel Integrated Protection for VSC-HVDC Transmission Line Based on Current Limiting Reactor Power," *IEEE Transactions on Power Delivery*, vol. 35, no. 1. pp. 226–233, 2020, doi: 10.1109/TPWRD.2019.2945412
[18] K. Singh, M. Dahiya, A. Grover, R. Adlakha, and M. Amir, "An Effective Cascade Control Strategy for Frequency Regulation of Renewable Energy Based Hybrid Power System with Energy Storage System," *J. Energy Storage*, vol. 68, p. 107804, 2023, doi: 10.1016/j.est.2023.107804
[19] K. Singh, M. Amir, and Y. Arya, "Optimal Dynamic Frequency Regulation of Renewable Energy Based Hybrid Power System Utilizing a Novel TDF-TIDF Controller," *Energy Sources, Part A Recover. Util. Environ. Eff.*, vol. 44, no. 4, pp. 10733–10754, 2022, doi: 10.1080/15567036.2022.2158251
[20] H. Sharma, M. Sharma, C. Sharma, A. Haque, and Z. A. Jaffery, "Performance Analysis of Solar Powered DC-DC Buck Converter for Energy Harvesting IoT Nodes," in *2018 3rd International Innovative Applications of Computational Intelligence on Power, Energy and Controls with their Impact on Humanity (CIPECH)*, Nov. 2018, pp. 26–29, doi: 10.1109/CIPECH.2018.8724183
[21] S. Palanidoss and T. V. S. Vishnu, "Experimental Analysis of Conventional Buck and Boost Converter with Integrated Dual Output Converter," in *2017 International Conference on Electrical, Electronics, Communication, Computer, and Optimization Techniques (ICEECCOT)*, Dec. 2017, pp. 323–329, doi: 10.1109/ICEECCOT.2017.8284521
[22] B. B. Tuvar and M. H. Ayalani, "Analysis of a Modified Interleaved Non-Isolated Cuk Converter with Wide Range of Load Variation and Reduced Ripple Content," in *2019 3rd International Conference on Trends in Electronics and Informatics (ICOEI)*, Apr. 2019, pp. 406–411, doi: 10.1109/ICOEI.2019.8862665
[23] K. Sarasvathi and K. Divya, "Analysis and Design of Superlift Luo Boost Converter," in *Proc. 4th Int. Conf. Electr. Energy Syst. ICEES 2018*, pp. 591–595, 2018, doi: 10.1109/ICEES.2018.8442339
[24] F. L. Luo and H. Ye, "Positive Output Super-Lift Converters," *IEEE Trans. Power Electron.*, vol. 18, no. 1, pp. 105–113, 2003, doi: 10.1109/TPEL.2002.807198
[25] M. Amir Zaheeruddin, and A. Haque, "Intelligent Based Hybrid Renewable Energy Resources Forecasting and Real Time Power Demand Management

System for Resilient Energy Systems," *Sci. Prog.*, vol. 105, no. 4, p. 00368504 2211321, 2022, doi: 10.1177/00368504221132144

[26] P. Asopa, P. Purohit, R. R. Nadikattu, and P. Whig, "Reducing Carbon Footprint for Sustainable Development of Smart Cities using IoT," in *2021 Third International Conference on Intelligent Communication Technologies and Virtual Mobile Networks (ICICV)*, Feb. 2021, pp. 361–367, doi: 10.1109/ICICV50876.2021.9388466

[27] P. Chamoso, A. González-Briones, S. Rodríguez, and J. M. Corchado, "Tendencies of Technologies and Platforms in Smart Cities: A State-of-the-Art Review," *Wirel. Commun. Mob. Comput.*, vol. 2018, 2018, doi: 10.1155/2018/3086854

[28] M. Amir and S. Z. Khan, "Assessment of Renewable Energy: Status, Challenges, COVID-19 Impacts, Opportunities, and Sustainable Energy Solutions in Africa," *Energy Built Environ.*, 2021, doi: 10.1016/j.enbenv.2021.03.002

[29] Z. Wang et al., "Research on the Active Power Coordination Control System for Wind/Photovoltaic/Energy Storage," in *2017 IEEE Conference on Energy Internet and Energy System Integration (EI2)*, Nov. 2017, pp. 1–5, doi: 10.1109/EI2.2017.8245403

[30] R. J. R. Kumar and A. Jain, "Hydro/Wind Driven Induction Generator Model for Load Flow Solution of Smart Electricity Infrastucture," in *2016 3rd MEC International Conference on Big Data and Smart City (ICBDSC)*, Mar. 2016, pp. 1–6, doi: 10.1109/ICBDSC.2016.7460376

[31] K. S. Mumbere, Y. Sasaki, N. Yorino, Y. Zoka, A. Bedawy, and Y. Tanioka, "An Interconnected Prosumer Energy Management System Model for Improved Outage Resilience," in *2022 IEEE PES/IAS PowerAfrica*, Aug. 2022, pp. 1–5, doi: 10.1109/PowerAfrica53997.2022.9905352

[32] U. Sangpanich, "A Novel Method of Decentralized Battery Energy Management for Stand-Alone PV-Battery Systems," in *2014 IEEE PES Asia-Pacific Power and Energy Engineering Conference (APPEEC)*, Dec. 2014, pp. 1–5, doi: 10.1109/APPEEC.2014.7066077

[33] M. N. Mojdehi and J. Viglione, "Microgrid Protection: A Planning Perspective," *Proc. IEEE Power Eng. Soc. Transm. Distrib. Conf.*, vol. 2018, pp. 1–9, 2018, doi: 10.1109/TDC.2018.8440389

[34] S. Xuewei et al., "Research on Energy Storage Configuration Method Based on Wind and Solar Volatility," in *2020 10th International Conference on Power and Energy Systems (ICPES)*, Dec. 2020, pp. 464–468, doi: 10.1109/ICPES51309.2020.9349645

[35] J. L. Casteleiro-Roca, L. A. Fernandez-Serantes, J. L. Calvo-Rolle, I. Machon-Gonzalez, M. J. Crespo-Ramos, and H. Lopez-Garcia, "Study of the Effect of a Geothermal Heat Exchanger over the Ground," in *2013 International Conference on New Concepts in Smart Cities: Fostering Public and Private Alliances (SmartMILE)*, Dec. 2013, pp. 1–6, doi: 10.1109/SmartMILE.2013.6708197

[36] J. P. Paredes-Sanchez, A. J. Gutierrez-Trashorras, and J. M. Gonzalez-Caballin, "Bio-Smartcity: Biomass Supply to a Smartcity. A Case Study," in *2013 International Conference on New Concepts in Smart Cities: Fostering Public and Private Alliances (SmartMILE)*, Dec. 2013, pp. 1–4, doi: 10.1109/SmartMILE.2013.6708171

[37] M. Ataei, S. Malekshah, M. Ghanbarnejad, and A. J. Irani, "Implementation, Operation and Economical Assessment of the First 3MW Biomass Distributed Energy Resource: A Case Study of Iran," in *2017 Smart Grid Conference (SGC)*, Dec. 2017, pp. 1–8, doi: 10.1109/SGC.2017.8308862

[38] S. N. V. B. Rao et al., "Day-Ahead Load Demand Forecasting in Urban Community Cluster Microgrids Using Machine Learning Methods," *Energies*, vol. 15, no. 17, p. 6124, 2022, doi: 10.3390/en15176124

[39] S. Lu, B. Zhou, L. Chen, and W. Yao, "Discuss of the future Power Generation Structure in China Southern Power Grid with High Penetration of PV Generation," in *2018 International Conference on Power System Technology (POWERCON)*, Nov. 2018, pp. 1196–1201, doi: 10.1109/POWERCON.2018.8602066

[40] W. Niu, F. Yang, and Y. Yuan, "Power Generation Scheduling for Long Distance Consumption of Wind-Solar-Thermal Power Based on Game-Theory," in *2022 7th Asia Conference on Power and Electrical Engineering (ACPEE)*, Apr. 2022, pp. 773–777, doi: 10.1109/ACPEE53904.2022.9783976

[41] Y. Susilowati, P. Irasari, and A. Susatyo, "Study of Hydroelectric Power Plant Potential of Mahakam River Basin East Kalimantan Indonesia," in *2019 International Conference on Sustainable Energy Engineering and Application (ICSEEA)*, Oct. 2019, pp. 1–7, doi: 10.1109/ICSEEA47812.2019.8938641

[42] K. Chaithanya Bandla and N. P. Padhy, "Decentralized Control for Coordinated Power Management Among Multiple HESS in DC Microgrid," in *2020 IEEE International Conference on Power Electronics, Smart Grid and Renewable Energy (PESGRE2020)*, Jan. 2020, pp. 1–6, doi: 10.1109/PESGRE45664.2020.9070769

[43] N. Shah, A. Haque, M. Amir, and A. Kumar, "Investigation of Renewable Energy Integration Challenges and Condition Monitoring Using Optimized Tree in Three Phase Grid System," in *2023 7th International Conference on Computing Methodologies and Communication (ICCMC)*, Feb. 2023, pp. 1582–1588, doi: 10.1109/ICCMC56507.2023.10083636

[44] I. El-Dessouki and N. Saeed, "Smart Grid Integration into Smart Cities," in *2021 IEEE International Smart Cities Conference (ISC2)*, Sep. 2021, pp. 1–4, doi: 10.1109/ISC253183.2021.9562769

[45] "Advancements That Will Facilitate the Integration of Smart Grid Technologies," in *2012 IEEE PES Innovative Smart Grid Technologies (ISGT)*, Jan. 2012, pp. 1–1, doi: 10.1109/ISGT.2012.6175694

[46] X. Chen and I. Khan, "A Tutorial on Current Controlled DC-DC Converter used in Microgrid System," in *2019 IEEE 16th International Conference on Smart Cities: Improving Quality of Life Using ICT & IoT and AI (HONET-ICT)*, Oct. 2019, pp. 232–234, doi: 10.1109/HONET.2019.8908027

[47] C. Dapeng, J. Zhaoxia, and T. Huijuan, "Optimal Bidding Strategy for Microgrids in the Day-Ahead Energy Market Considering the Price Elasticity of Load Demand," in *2018 International Conference on Power System Technology (POWERCON)*, Nov. 2018, pp. 794–799, doi: 10.1109/POWERCON.2018.8601985

[48] Z. Shi, W. Wang, and Y. Huang, "Simultaneous Optimization of Renewable Energy and Energy Storage Capacity with Hierarchical Control," *CSEE J. Power Energy Syst.*, 2020, doi: 10.17775/CSEEJPES.2019.01470

[49] H. Aki, T. Wakui, and R. Yokoyama, "Optimal Management of Fuel Cells in a Residential Area by Integrated-Distributed Energy Management System (IDEMS)," in *2016 IEEE Power & Energy Society Innovative Smart Grid Technologies Conference (ISGT)*, Sep. 2016, pp. 1–5, doi: 10.1109/ISGT.2016.7781156

[50] I. Aizpuru, A. Arruti, J. Anzola, U. Iraola, M. Mazuela, and A. Rujas, "Universal Electric Vehicle Charging Infrastructure Analysis Tool," in *2020 IEEE Vehicle Power and Propulsion Conference (VPPC)*, Nov. 2020, pp. 1–5, doi: 10.1109/VPPC49601.2020.9330990

[51] M. Amir, Zaheeruddin, A. Haque, F. I. Bakhsh, V. S. B. Kurukuru, and M. Sedighizadeh, "Intelligent Energy Management Scheme-Based Coordinated Control for Reducing Peak Load in Grid-Connected Photovoltaic-Powered Electric Vehicle Charging Stations," *IET Gener. Transm. Distrib.*, 2023, doi: 10.1049/gtd2.12772

[52] P. Xue et al., "Robust Joint Planning of Electric Vehicle Charging Infrastructures and Distribution Networks," in *2022 7th Asia Conference on Power and Electrical Engineering (ACPEE)*, Apr. 2022, pp. 135–139, doi: 10.1109/ACPEE53904.2022.9783920

[53] D. Atkar, P. Chaturvedi, H. M. Suryawanshi, P. Nachankar, D. Yadeo, and S. Krishna, "Solid State Transformer for Electric Vehicle Charging Infrastructure," in *2020 IEEE International Conference on Power Electronics, Smart Grid and Renewable Energy (PESGRE2020)*, Jan. 2020, pp. 1–6, doi: 10.1109/PESGRE45664.2020.9070447

[54] M. Amir, Zaheeruddin, and A. Haque, "Optimal Scheduling of Charging/Discharging Power and EVs Pattern Using Stochastic Techniques in V2G System," in *2021 IEEE Transportation Electrification Conference (ITEC-India)*, Dec. 2021, pp. 1–6, doi: 10.1109/ITEC-India53713.2021.9932455

[55] A. Latifah, S. H. Supangkat, and A. Ramelan, "Smart Building: A Literature Review," in *2020 International Conference on ICT for Smart Society (ICISS)*, Nov. 2020, pp. 1–6, doi: 10.1109/ICISS50791.2020.9307552

[56] V. Matsouliadis, K. Siountri, and D. D. Vergados, "Analysis and Design of Services in Building Units in the Context of a Smart City," in *2020 11th International Conference on Information, Intelligence, Systems and Applications (IISA)*, Jul. 2020, pp. 1–6, doi: 10.1109/IISA50023.2020.9284377

[57] M. P. De Novais, V. De Almeida Xavier, L. H. Xavier, and J. Hwang, "Modelling E-Waste Management Data in Smart Cities," in *2021 2nd Sustainable Cities Latin America Conference (SCLA)*, Aug. 2021, pp. 1–4, doi: 10.1109/SCLA53004.2021.9540096

[58] M. D. Ibrahim, Y. Al Amoudy, and F. Hameed, "Integrated Analysis for Renewable Energy-Economic Development-Environmental Sustainability (RE-ED-ES) Nexus: Analysis of MENA Countries," in *2020 Advances in Science and Engineering Technology International Conferences (ASET)*, Feb. 2020, pp. 1–6, doi: 10.1109/ASET48392.2020.9118296

[59] I. C. Hoarca, N. Bizon, and F. M. Enescu, "Using the Potential of Renewable Energy Sources in Romania to Reduce Environmental Pollution," in *2021 13th International Conference on Electronics, Computers and Artificial Intelligence (ECAI)*, Jul. 2021, pp. 1–6, doi: 10.1109/ECAI52376.2021.9515074

[60] B. Hammi, R. Khatoun, S. Zeadally, A. Fayad, and L. Khoukhi, "IoT Technologies for Smart Cities," *IET Networks*, vol. 7, no. 1. Institution of Engineering and Technology, pp. 1–13, Jan. 2018, doi: 10.1049/iet-net.2017.0163

[61] D. Cheng, C. Li, and N. Qiu, "The Application Prospects of NB-IoT in Intelligent Transportation," in *2021 4th International Conference on Advanced Electronic Materials, Computers and Software Engineering (AEMCSE)*, Mar. 2021, pp. 1176–1179, doi: 10.1109/AEMCSE51986.2021.00240

[62] M. Amir, A. Haque, V. S. B. Kurukuru, F. Bakhsh, and A. Ahmad, "Agent Based Online Learning Approach for Power Flow Control of Electric Vehicle Fast Charging Station Integrated with Smart Microgrid," *IET Renew. Power Gener.*, 2022, doi: 10.1049/rpg2.12508

[63] M. Derawi, Y. Dalveren, and F. A. Cheikh, "Internet-of-Things-Based Smart Transportation Systems for Safer Roads," in *2020 IEEE 6th World Forum on Internet of Things (WF-IoT)*, Jun. 2020, pp. 1–4, doi: 10.1109/WF-IoT48130.2020.9221208

[64] M. Akhtar, M. Raffeh, F. ul Zaman, A. Ramzan, S. Aslam, and F. Usman, "Development of Congestion Level Based Dynamic Traffic Management System Using IoT," in *2020 International Conference on Electrical, Communication, and Computer Engineering (ICECCE)*, Jun. 2020, pp. 1–6, doi: 10.1109/ICECCE49384.2020.9179375

[65] M. Kumar, A. F. Minai, A. A. Khan, and S. Kumar, "IoT Based Energy Management System for Smart Grid," in *2020 International Conference on Advances in Computing, Communication & Materials (ICACCM)*, Aug. 2020, pp. 121–125, doi: 10.1109/ICACCM50413.2020.9213061

[66] M. Amir and Zaheeruddin, "ANN Based Approach for the Estimation and Enhancement of Power Transfer Capability," in *2019 International Conference on Power Electronics, Control and Automation (ICPECA)*, Nov. 2019, pp. 1–6, doi: 10.1109/ICPECA47973.2019.8975665

[67] N. Muleta and A. Q. H. Badar, "Study of Energy Management System and IoT Integration in Smart Grid," in *2021 1st International Conference on Power Electronics and Energy (ICPEE)*, Jan. 2021, pp. 1–5, doi: 10.1109/ICPEE50452.2021.9358769

[68] S. Heble, A. Kumar, K. V. V. D. Prasad, S. Samirana, P. Rajalakshmi, and U. B. Desai, "A Low Power IoT Network for Smart Agriculture," in *2018 IEEE 4th World Forum on Internet of Things (WF-IoT)*, Feb. 2018, pp. 609–614, doi: 10.1109/WF-IoT.2018.8355152

[69] A. Anand, N. K. Trivedi, V. Gautam, R. G. Tiwari, D. Witarsyah, and A. Misra, "Applications of Internet of Things(IoT) in Agriculture: The Need and Implementation," in *2022 International Conference Advancement in Data Science, E-learning and Information Systems (ICADEIS)*, Nov. 2022, pp. 01–05, doi: 10.1109/ICADEIS56544.2022.10037505

[70] Y. Li, X. Cheng, Y. Cao, D. Wang, and L. Yang, "Smart Choice for the Smart Grid: Narrowband Internet of Things (NB-IoT)," *IEEE Internet Things J.*, vol. 5, no. 3, pp. 1505–1515, 2018, doi: 10.1109/JIOT.2017.2781251

Chapter 5

Smart grid concept and technologies for smart cities

Ahteshamul Haque
Jamia Millia Islamia, New Delhi, India

K. V. S. Bharath
Silicon Austria Labs, Graz, Austria

Suwaiba Mateen
Jamia Millia Islamia, New Delhi, India

5.1 INTRODUCTION

The need for energy has increased because of urbanization, rising living standards, and technological advancements. As a result, electricity consumption has increased to levels that, if left unchecked, may become unmanageable. This is a worrying condition for both the preservation of the environment and providing sustainable energy. Cities only cover 3% of the earth's land, but consume more than 80% of the world's energy and produce 75% greenhouse gases [1]. The traditional power grid is a centralized model that consists of a complex network of power lines. There is a loss of energy in these grids due to the system inefficiencies. Issues like unreliability, poor power quality, blackouts, and no integration of distributed energy sources in the grid constitute some of the major problems with the power grid. Lack of monitoring and real-time control in the power grid has led to the development of smart grids to act as a real-time solution.

The concept of a smart grid has emerged as a transformative solution for the modernization of power systems, particularly in the context of smart cities. A smart grid represents a paradigm shift in the way we generate, distribute, and consume electricity, incorporating advanced technologies and intelligent systems to optimize energy management and enhance grid reliability. In this chapter, we delve into the fundamental concepts and technologies that underpin smart grids and explore their crucial role in the development of smart cities.

A smart grid is defined as an integrated energy network that seamlessly integrates communication, automation, and advanced analytics into traditional power grids. Smart grids provide real-time monitoring, control, and optimization of energy flows by utilizing the potential of digitalization, permitting efficient energy generation, distribution, and consumption [2].

 DOI: 10.1201/9781032669809-5

This holistic approach to grid management empowers cities to address pressing challenges, such as the integration of distributed energy sources, grid resiliency, and demand-side management. The smart grid, as defined by the US National Institute of Standards and Technologies (NIST) [3], is a modified grid that uses two-way communication and allows bidirectional flow of energy. The UK's Department of Energy and Climate Change underlines that a smarter grid increases system operators' awareness of data about the supply-demand balance, enabling them to manage the system wisely and shift demand from peak time to off-peak hours [4].

As urban centers face increasing energy demands and environmental concerns, the need for sustainable and efficient energy systems has become imperative. Smart grids offer a range of benefits that go beyond traditional grid infrastructure, ensuring a more sustainable future [5]. These benefits include improved grid stability and reliability, cost savings, enhanced flexibility, and environmental sustainability. Table 5.1 shows the key differences between the traditional power grid and the smart grid.

In the realm of smart cities, smart grids serve as a critical foundation for integrated urban systems. They enable the seamless integration of various smart city domains. The interoperability of these systems allows for

Table 5.1 Comparison between traditional power grid and smart grid

Characteristics	*Traditional power grid*	*Smart grid*
Technology	Electromechanical: The electromechanical technology has no means of communication.	Digital: The digital technology allows for an increase in communication between devices.
Distribution	One-way distribution: The power flow is unidirectional.	Two-way distribution: Power flow is bidirectional. Any customer with access to alternative energy sources can provide energy back to the grid.
Generation	Centralized: All the power is generated from a main plant. Integration with distributed energy sources is not possible.	Distributed: Power can be distributed from distributed energy sources. This reduces the peak demand load and reduces power outages.
Sensors	Few sensors: Number of sensors in the system are less, making it difficult to pinpoint the location of a fault.	Sensors throughout: The infrastructure has multiple sensors placed on the lines and helps in pinpointing the location of a fault.
Monitoring	Manual: In traditional power grids, energy distribution must be monitored manually.	Self: It is possible to balance power loads and troubleshoot outages without the help of a technician.

(*Continued*)

Table 5.1 (Continued)

Characteristics	*Traditional power grid*	*Smart grid*
Restoration	Manual: In case of repairing faults, technicians must travel physically to the site of the fault. Because of this the duration of outage increases.	Self-healing: For simple troubleshooting, sensors detect the faults and repairs without the intervention of technicians. In case of severe faults, the technicians are notified by the grid to begin the necessary repairs.
Equipment	Failure & Blackout: The traditional power grid is prone to failures. This leads to outages.	Adaptive & Islanding: In a smart grid system, it is possible to have power flow around any faulty area. This limits the area affected by power outages.
Control	Limited: Once the power leaves the power plant or substation, companies have no control over the energy distribution.	Pervasive: In smart grids, energy companies have more control over power distribution. It is possible to monitor energy and its consumption from the power plant to the end consumer.
Customer choices	Fewer: The customer choices in traditional power grid system infrastructure are less.	Many: For smart grids, infrastructure can be shared. This means consumers have more choice in how they want to receive the energy.

optimized energy consumption patterns, streamlined operations, and enhanced urban livability. Moreover, smart grids synergize with other key smart city domains such as transportation and water waste management, fostering integrated and sustainable urban development.

As cities worldwide recognize the immense potential of smart grids, significant efforts are underway to deploy and integrate these technologies. Numerous global initiatives, standards, and policies have been established to accelerate the adoption of smart grids and propel the transformation of cities into smart cities. However, challenges such as regulatory frameworks, financial considerations, and public acceptance must be addressed to enable widespread implementation.

By exploring the intricacies of smart grid concepts and technologies, this chapter aims to provide a comprehensive understanding of their role in the context of smart cities. It serves as a guide in harnessing the potential of smart grids to create sustainable, efficient, and resilient urban energy systems.

5.2 FUNDAMENTALS OF SMART GRID

A conceptual model of a smart grid is presented by NIST [6]. The model is divided into seven domains viz., markets, operations, service providers, generation, transmission, distribution, and customer, as shown in Figure 5.1. Each domain of the model includes smart grid actors and applications. The devices and/or programs that make decisions and help in exchanging information are known as actors, while the tasks performed by these actors in a domain are known as applications.

To better identify the actors and possible communications path in smart grids, a high-level conceptual model with an overarching perspective is shown in Figure 5.2. The shown model acts as a useful tool for analyzing interactions within and outside the domain.

In the Figure 5.2, RTO/ISO, transmission, and distribution operations are the sub-domain of the operations domain. Similarly domain service providers, utility provider and third-party providers are the sub-domains. Actors, such as meter, energy services interface, and so on, interact with actors from other domains or networks. They are known as the gateway actor. A collection of interconnected computers and other information and communication technologies are known as the information network. The network includes the enterprise bus, wide area networks, field area networks, premises networks, and so on. These networks may be implemented using public and non-public networks in combination. The logical exchange of data between domains is shown by the communication path.

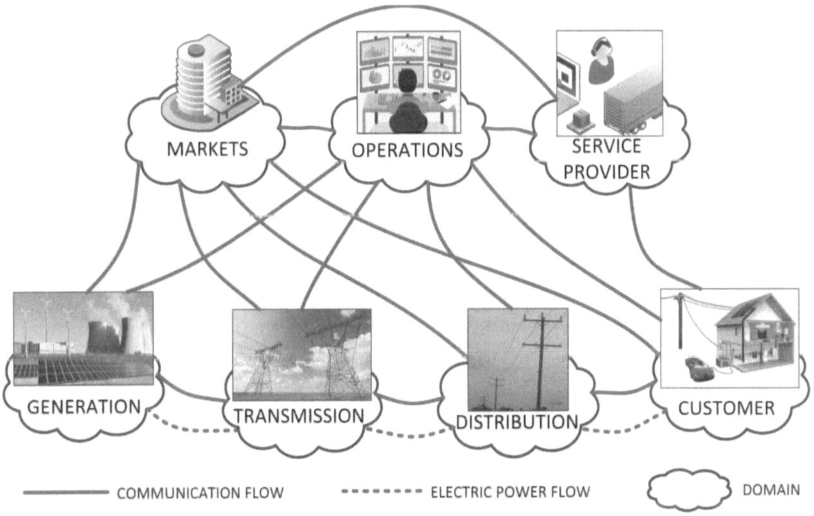

Figure 5.1 Interaction between domains of smart grid.

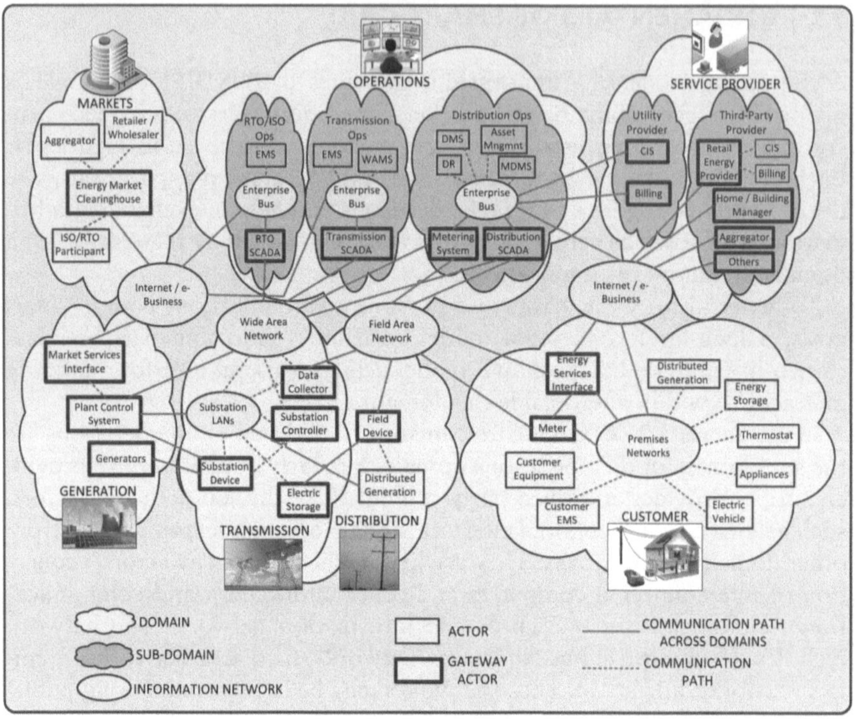

Figure 5.2 Conceptual model of smart grid.

5.2.1 Characteristics

Smart grids possess several characteristics that distinguish them from traditional power grids. These characteristics enable them to enhance grid operations, improve efficiency, and support the integration of renewable energy sources. Some of the key characteristics discussed by the authors in [7, 8] are:

1. Advanced metering and monitoring: Smart grids incorporate advanced metering infrastructure (AMI) that provides real-time data on energy consumption, generation, and grid conditions. This enables accurate monitoring of energy flows, load profiles, and system performance.
2. Intelligent control and automation: Smart grids utilize automation and intelligent control systems to optimize grid operations. These systems enable rapid fault detection, isolation, and restoration, enhancing grid reliability and reducing outage durations.
3. Distributed energy resources (DER) Integration: Smart grids are designed to seamlessly integrate distributed energy resources. This integration enables efficient management and utilization of DERs while ensuring grid stability and power quality.

4. Demand-side response (DSR): Smart grids facilitate demand response programs, allowing consumers to actively manage their energy usage.
5. Grid resilience and self-healing: Smart grids employ self-healing mechanisms that automatically isolate faults and re-route power to minimize disruptions. These features improve grid resilience, reducing outage durations and improving overall reliability.
6. Grid optimization and planning: Smart grids utilize advanced analytics, optimization algorithms, and forecasting techniques to optimize grid operations and planning.
7. Enhanced grid security: Smart grids incorporate robust cybersecurity measures to protect against cyber-attacks. Security protocols and encryption techniques are implemented to safeguard critical grid infrastructure.

These characteristics collectively enable smart grids to optimize energy management, improve system efficiency, enhance reliability, and facilitate the integration of renewable energy sources.

5.2.2 Benefits

Several countries around the world offer incentives to promote the adoption and conversion to smart grids. The United States has implemented incentives at the federal, state, and local levels. The US Department of Energy (DOE) offers funding programs such as the Smart Grid Investment Grants and the Grid Modernization Initiative. Similar funding programs are implemented in Germany, China, Australia, and Japan. South Korea has prioritized smart grid development as part of its national energy strategy. The country offers financial incentives and subsidies to utilities and consumers for implementing smart grid technologies. In European countries, a policy known as "Cap and Trade" is designed to reduce greenhouse gas emissions from industrial sectors and power generation [9]. It involves setting a cap on greenhouse gas emissions for covered sectors, allocating emission allowances to participating entities, and creating a market for trading these allowances. Entities with lower emissions can sell their surplus allowances to those with higher emissions, providing an economic incentive for emission reductions. Denmark, a European country, known for its leadership in renewable energy and smart grid integration, offers feed-in tariffs, favorable pricing structures, and tax incentives for renewable energy producers. As the traditional grid transforms to a smart grid, many benefits of modernized network are discussed in [10] and are as follows:

1. Overall improvement in power quality and reliability of the grid.
2. Improvement in system resiliency and efficiency.
3. Optimized operation of power plants.
4. Increased opportunities for enhancing system security.
5. Integration of the DERs.

5.3 TECHNOLOGIES IN SMART GRID

To control the smart grid, various technologies are utilized. These technologies play a critical role in modernizing and optimizing power systems, enhancing operational efficiency, grid reliability, and the integration of renewable energy sources, while also enabling more effective energy management and customer engagement. The electric power system is divided into three sub-systems: generation, distribution, and transmission system. Different smart grid technologies utilized for each sub-system are shown by the flowchart in Figure 5.3.

Automatic Generation Control (AGC) and Economic Dispatch Calculation (EDC) are used in the generating sub-system. AGC is a control system used to maintain the balance between electricity generation and demand. It continuously adjusts the output of power generation sources based on load changes, ensuring stable grid operation and frequency control [11]. In the transmission system, Energy Management System (EMS) and Wide Area Management System (WAMS) are utilized. EMS is a software-based technology that optimizes the transmission of electrical energy. It integrates data from various sources, such as power plants, substations, and load centers, to enable real-time monitoring, control, and optimization of energy flows. WAMS employs phasor measurement units (PMUs) and communication infrastructure to provide real-time monitoring and control of the power grid over wide geographic areas [12]. Some of these technologies discussed by the authors are tabulated in Table 5.2.

In the distribution power system level, Automatic Voltage Regulation (AVR) technology is used to regulate and stabilize voltage levels. It ensures

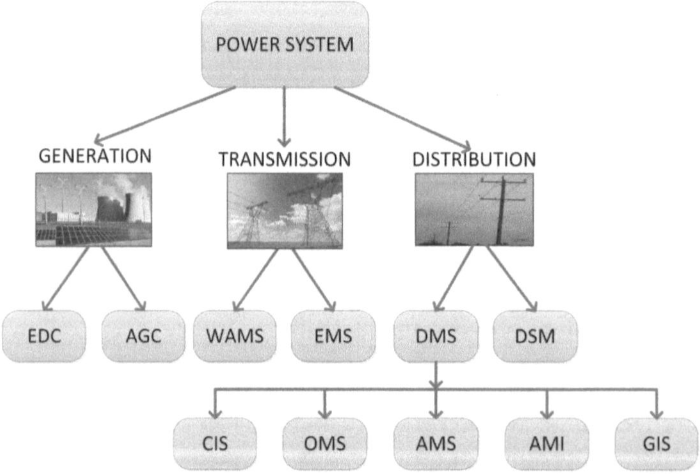

Figure 5.3 Classification of power system.

Table 5.2 Technologies of smart grid

Technique	*Power system level*	*Ref.*	*Objectives*
Automatic Generation Control (AGC)	Generation Level	[13]	• Impact of data • Attack detection
		[14]	• Accommodate intermittency of DERs • Mitigate high-frequency load fluctuations
Wide Area Management System (WAMS)	Transmission Level	[15]	• Identify best phasor data communication technique in WAMS • Multicast authentication for WAMS
		[16]	• Sectionalizing method for the power system restoration • Utilizing WAMS to fully observe each island
Energy Management System (EMS)	Transmission Level	[17]	• Analysis of DER behavior in EMS • Comparative analysis of optimization techniques used to achieve different EMS objectives
		[18]	• Balance between supply and demand • Reduce peak load demand in all conditions
Automatic Voltage Regulation (AVR)	Distribution Level	[19]	• Recurrent radial basis function network (RRBFN) as a driving control for system • Power control using Fuzzy Sliding mode
		[20]	• Global Neighborhood Algorithm (GNA) used for AVR • Comparison with Particle Swarm Optimization (PSO)
Outage Management System (OMS)	Distribution Level	[21]	• OMS analysis in different operating conditions • Integrate data from AMI, SCADA, and DMS
		[22]	• Model-driven view-based approach to build smart grid • Reduce outage time

(*Continued*)

Table 5.2 (Continued)

Technique	*Power system level*	*Ref.*	*Objectives*
Advanced Metering Infrastructure (AMI)	Distribution Level	[23]	• Bidirectional, self-configurational and auto-update capabilities in AMI • Cost reduction and improve scalability
		[24]	• Providing security to AMI • Key management system (KMS) utilized for safeguarding AMI
Geographical Information System (GIS)	Distribution Level	[25]	• Spatial representation of the electricity network • Trackable real-time value of power flow of DERs
		[26]	GIS integrated automation of a holomorphic embedded power flow (HEPF)

that voltage remains within specified limits, mitigating voltage fluctuations and maintaining optimal operating conditions for connected devices and equipment. Distribution Management System (DMS) is a software platform that facilitates real-time monitoring, control, and optimization of distribution networks. It integrates data from various sources, including sensors, smart meters, and substation equipment, enabling utilities to manage grid operations efficiently, detect faults, and optimize load distribution. Another software system utilized in the distribution system is the Outage Management System (OMS) which detects, analyzes, and manages power outages. It integrates real-time data from various sources to identify outage locations, estimate restoration times, and optimize outage response, helping utilities restore power more efficiently and improve customer communication. The advanced metering infrastructure (AMI) consists of smart meters, communication networks, and data management systems, and Meter Data Management (MDM) collects, stores, and manages the vast amount of data generated by smart meters. They ensure the accuracy, security, and integrity of meter data, facilitate billing processes, and support data analytics for grid optimization and customer engagement. Demand-Side Management (DSM) refers to various strategies and technologies aimed at influencing consumer electricity consumption patterns. It includes time-based pricing, load shifting, and demand response programs. The geospatial data is combined with the Geographical Information System (GIS) to create visual representations and analyses of the grid infrastructure [27].

5.4 DEMAND RESPONSE

Demand response (DR) is a strategy or program implemented in the electricity sector that aims to adjust consumption patterns in response to changing conditions in the power grid. It involves the voluntary reduction or shifting of electricity usage by consumers during times of high demand, grid instability, or in response to price signals. The goal of DR is to balance electricity supply and demand, optimize grid operation, and enhance overall system efficiency. Through demand response, consumers can actively participate in managing their electricity usage and contribute to grid stability. They may receive signals from utilities or grid operators indicating periods of high demand or grid constraints, such as during heatwaves or system emergencies. In response, consumers can adjust their energy consumption by reducing non-essential electricity usage, shifting activities to off-peak hours, or modifying their energy consumption patterns. Authors in [28, 29] have provided a full analysis of the functions of DR in smart cities. Figure 5.4 shows the categorization of various forms of DR.

Demand response programs can take various forms, from incentive-based programs to price-based programs. These programs motivate consumers to reduce or shift their electricity usage through financial incentives, lower electricity rates, or other benefits. The incentive-based programs are divided into direct and indirect control loads. Under the direct control line programs, the energy consumption for controllable devices is allowed to alter with prior notification. The indirect control line program is meant for appliances that can be disrupted over a short period of time [30].

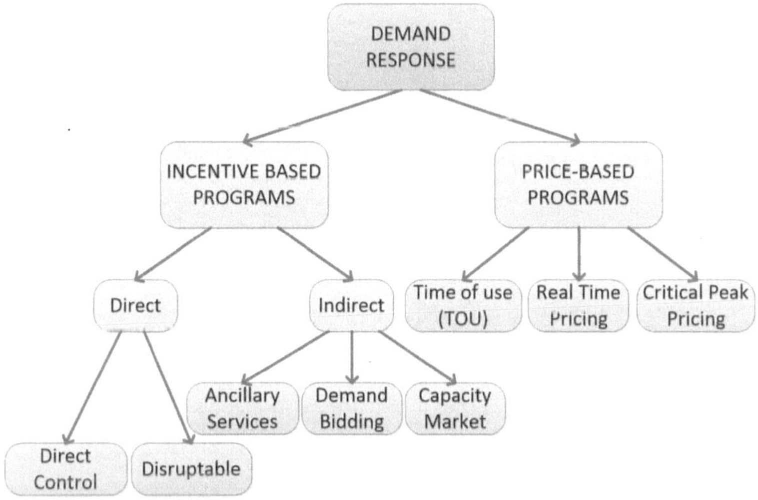

Figure 5.4 Categorization of demand response (DR).

Furthermore, demand response plays a crucial role in supporting the integration of renewable energy sources into the grid [31]. As the renewable generation is variable and dependent on weather conditions, demand response can help manage fluctuations by aligning electricity consumption with periods of high renewable generation.

5.4.1 Reliability

The reliability of DR programs depends on several factors, such as the design and implementation of the program, the participation and engagement of consumers, and the coordination with grid operators and utilities. While performing traditional reliability assessments, it is crucial to differentiate between demand response action and natural interruptions [32]. The DR is planned in advance, but the latter is unplanned and consumers will not be able to take any preventative measures for it. As discussed by authors in [32] compared to conventional reliability tests, the impact of DR on reliability requires additional and complex measurements.

Figure 5.5 shows the chronological steps for a DR event as defined by The North American Energy Standards Board (NAESB) [33]. In demand response programs, particularly those targeting residential or commercial customers, there can be a delay or ramp period before consumer demand is successfully reduced to the requested levels. This period is influenced by the dynamics of

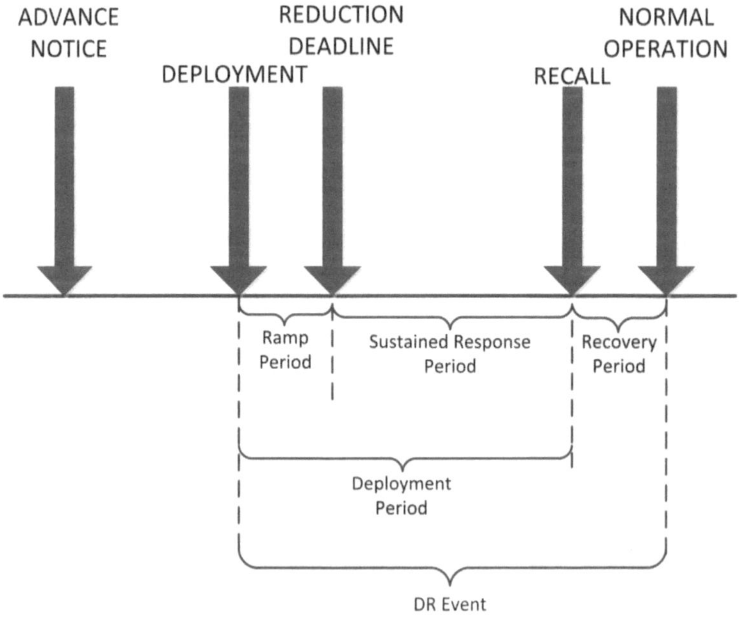

Figure 5.5 Chronological order for DR event.

the loads and the response capabilities of the consumers. However, for specific customers, such as large-scale industrial plants, this process can be expedited. The ramp period may depend on the response capabilities of the consumers participating in the DR program. Loads in the HVAC system category or lighting can be adjusted relatively quickly, allowing for a shorter ramp period. However, other loads, such as industrial processes or certain equipment, may require more time to reduce their electricity consumption. Depending on the load factors and the consumers' responsiveness, the ramp time often lasts a few seconds to a few minutes. It is important for demand response program administrators and grid operators to consider the expected ramp period when designing and implementing demand response events.

Several techniques are available in the literature that deal with the reliability of DRs [34]. These techniques are summarized in Table 5.3. Sequential Monte Carlo Simulation (SMCS) uses computational techniques that can be applied to analyze and optimize demand response techniques [35, 36].

The DR programs can be divided into three categories based on the party in charge [40]: reliability-based programs, rate-based programs, and demand reduction bids. Reliability-based programs are also known as incentive-based programs. These DR programs are primarily focused on maintaining grid reliability and stability. Grid operators or utilities provide incentives to consumers to reduce their electricity demand during peak

Table 5.3 Techniques for reliability assessment of DR

Ref.	*Techniques*	*Mode*	*Network*	*DR Instrument*	*Criteria*	*ICT Impact*
[35]	Analytical SMCS	Grid connected operation	Smart distribution network	Incentive payments	• Minimum interruption cost • Payback incentives	Yes
[36]	SMCS	Islanded operation	Smart distribution system (SDS)	Incentive payments	Minimum interruption cost	Yes
[37]	Analytical	Grid connected operation	Residential distribution network	Incentive payments	Minimum interruption cost	No
[38]	SMCS	Grid connected operation	Distributed generation and storage (DG&S) network	Incentive payments	Disconnect or shift load	No
[39]	SMCS	Islanded operation	Distribution generation (DG) network	Time-of-use pricing	• Maximum income of supplier • Minimum payments to customers	No

hours. The incentives can take various forms, such as financial rewards, bill credits, or other benefits. Participants in these programs voluntarily reduce their electricity consumption to support grid reliability and may receive compensation or incentives for their participation.

Rate-based demand response programs use pricing structures to motivate consumers to adjust their electricity usage. These programs involve the implementation of time-varying rates. By aligning their usage with off-peak hours, participants can take advantage of lower prices and potentially reduce their overall electricity costs. For demand reduction bid programs consumers bid to curtail their electricity demand in response to specific grid needs. In these programs, consumers submit bids indicating the amount of load reduction and price. Grid operators or utilities evaluate the bids and select the most cost-effective options to meet grid requirements.

5.4.2 Application

Aggregated demand response programs or controllable demands have the potential to serve as Virtual Energy Storage Systems (VESS). By coordinating the energy requirement of multiple participants, aggregated DR enables load shifting and demand reduction during peak periods. This load flexibility and demand response capability contribute to grid balancing and voltage support, similar to the functions of energy storage [41]. Additionally, by aligning electricity consumption with periods of high renewable energy generation, aggregated DR facilitates the integration of renewables without the need for physical storage, reducing reliance on fossil fuel-based generation during peak demand. Overall, aggregated demand response acts as a virtual energy storage system, providing grid operators with valuable flexibility, load management, and grid stability services. A detailed comparative analysis of the benefits of using VESS is provided by Cheng, Sami, and Wu [42]. In the paper, the authors have proved that demand response is capable of replacing 50% of the energy storage system (ESS) market share.

Demand response plays a vital role in frequency regulation services within the electricity grid. By leveraging the flexibility of consumer electricity usage, DR enables load adjustments in real-time to help maintain grid frequency stability. Wu et al. [43] have proposed a control algorithm that provides frequency control services for refrigerators. Short, Infield, and Freris [44] have examined how dynamically regulated loads keep grid frequency within a certain range even after a sudden loss of generation. The study found decreased reliance on quickly deployable backup generators and a significant delay in frequency fall. Cheng et al. [45] have proposed an aggregated control for bitumen tanks that provides reserve capacity for frequency regulations. Comparing the model to frequency-sensitive generators, it was proved to be more dependable and quicker to react. For frequency regulation support, a comprehensive central DR is proposed [46]. The proposed model is simulated on a 13-bus IEEE benchmark.

Demand response plays a crucial role in providing ancillary services within the smart grid. DR resources are utilized for various ancillary services, such as frequency regulation, voltage support, operating reserves, and contingency reserves. By adjusting their electricity consumption in real-time, DR participants contribute to grid stability, balance supply and demand, and support reliable grid operation. They can respond to grid signals, market prices, or emergency situations to provide the necessary flexibility and responsiveness required for ancillary services. Shoreh et al. [47] have presented a detailed review on the role of ancillary services in industries. Through their active participation, demand response resources help optimize grid performance, reduce reliance on traditional ancillary service providers, and enhance the overall reliability and efficiency of the electricity grid.

DR applications extend beyond operational smart grid management and also play a significant role in generation, transmission, and distribution expansion planning. By considering DR as a resource in the planning process, utilities and system operators can optimize infrastructure investments, improve resource allocation, and enhance grid reliability. Hashemi and Shayeghi [48] have proposed a comprehensive model to assess DR technique and its role in generation, transmission, and distribution. DR can provide valuable insights into demand patterns, load flexibility, and customer preferences, helping to identify areas where demand reduction or load-shifting strategies can offset the need for new generation or transmission infrastructure. Integrating DR into expansion planning enables a more holistic and cost-effective approach by factoring in the potential of load management, peak shaving, and demand-side measures as alternatives or supplements to traditional infrastructure upgrades. This approach allows for a more efficient and resilient grid, reducing the overall costs and negative environmental impacts because of expanding generation, transmission, and distribution systems.

5.5 DATA MANAGEMENT

5.5.1 Energy data management in smart cities

The lack of resources is a major obstacle to the development of a functioning smart grid [49]. To efficiently run a dependable network, a smart city integrates engineering solutions and informatics skills. It includes a broad spectrum of intelligent services, including smart grids, transportation, and communication. The smart grid serves as the foundation of the smart city's hierarchical structure. Intelligent data systems should be a part of smart grids for proper operation. Such a vast volume of information needs to be carefully captured and processed. Capturing and processing the massive amount of information in smart grids requires careful attention. The intelligent data systems in smart grids play a crucial role in ensuring their proper functioning. These systems are responsible for collecting, analyzing, and managing data within the grid. One of the key challenges in implementing intelligent data systems in smart grids is the scalability and interoperability of data. Smart grids generate a vast amount of data in real-time, and the data systems need to handle this volume efficiently. They should be capable of processing and analyzing the data in a timely manner to provide actionable insights for grid operators and stakeholders. Potdar et al. [50] analyzed energy data management for smart cities in detail.

Another challenge is ensuring the security and privacy of the data. Smart grids deal with sensitive information, such as energy consumption patterns and customer data. Robust security measures and data governance protocols are necessary to protect the integrity and confidentiality of the data. Furthermore, data integration and interoperability are crucial for the effective functioning of smart grids. Data systems should be able to integrate data from various sources, enabling seamless communication and coordination between different grid elements. This allows for efficient monitoring, control, and optimization of the grid's operations.

To address these challenges, advanced data management techniques and technologies, such as big data analytics, machine learning, and artificial intelligence, can be employed. These technologies enable real-time data processing, predictive analytics, and intelligent decision-making, leading to improved grid performance, energy efficiency, and reliability. Cooperation between utilities, grid operators, technology providers, and regulators is necessary to establish standards, protocols, and data exchange frameworks that facilitate interoperability and data sharing.

Energy data management in smart cities involves data collection, pre-processing and integration, data storage, analysis and insights, data visualization, decision-making, and billing and tariff management [51]. The key aspects of energy data management in smart cities are shown in Figure 5.6 by the sequentially organized flow of data.

5.5.1.1 Data collection

In smart cities, data collection involves the systematic gathering of information from various sources. Some key aspects of data collection in smart cities include sensor networks, internet of things (IoT) devices, open data platforms, mobile applications and Geographic Information Systems (GIS). Smart cities deploy a wide range of sensors throughout the urban environment to collect real-time data on various parameters [52]. Sensor networks capture data at regular intervals, providing continuous and detailed information about the city's environment and infrastructure. IoT devices, such as smart meters that are included in the AMI, smart appliances, and connected vehicles, generate valuable data that can be collected and analyzed. These devices are equipped with sensors and communication capabilities, allowing them to capture and transmit data related to energy consumption and other

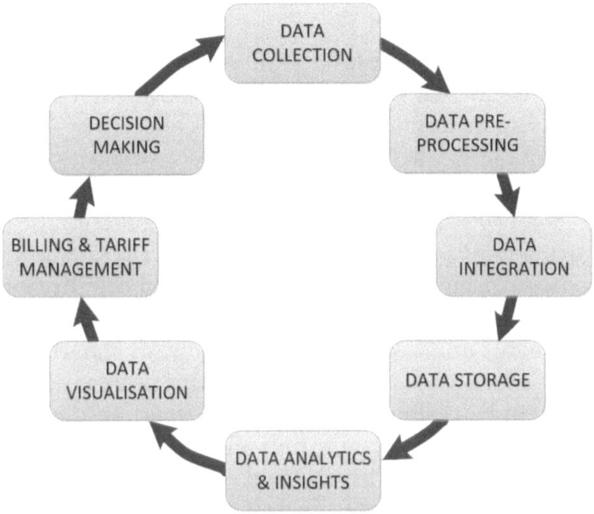

Figure 5.6 Flow of data in smart cities.

relevant metrics. GIS technology is also used to collect and manage spatial data, such as maps, satellite imagery, and geospatial information. GIS platforms provide a comprehensive view of the city's infrastructure, land use, visualization of generation, distribution facilities and other spatial features [53]. This data supports urban planning, transportation management, and emergency response.

To ensure effective data collection in smart cities, data privacy and security should be considered [49]. Clear data governance policies, consent mechanisms, and anonymization techniques are implemented to protect citizens' privacy while still enabling data-driven initiatives. A novel KMS scheme for smart meters is proposed by Keping et al. [54]. This data collection procedure ensures confidentiality, integrity, and authentication. Taking privacy into account, a lightweight message authentication system is proposed by Abbasinezhad-Mood and Nikooghadam [55] whereas Kim et al.'s [56] focus of the study is on reducing the storage requirement for data collection. Saputro and Akkaya [57] have proposed a total of three novel data collection mechanisms. An overall improvement in TCP performance in IEEE 802.11s-based wireless mesh AMI networks is observed. The authors have proposed a novel control mechanism based on the p-center facility problem to reduce the latency in data delivery at gateway location.

Data collection in smart cities is an ongoing process, with continuous efforts to expand the scope of data collection, improve data quality, and explore emerging technologies for data capture. The collected data serves as a foundation for evidence-based decision-making, urban planning, and the development of innovative solutions that contribute to the overall well-being and sustainability of the city.

5.5.1.2 Data pre-processing

Data pre-processing handles the large and diverse datasets collected in smart cities. It involves a series of steps to clean, transform, and prepare the data for further analysis and application [58]. Data collected in smart cities may contain errors, outliers, missing values, or inconsistencies. Data cleaning involves identifying and addressing these issues to ensure data quality and reliability. Techniques such as outlier detection, imputation of missing values, and removing duplicates are applied to cleanse the data [59]. Smart city data often consists of high-resolution data collected at different time intervals or spatial resolutions. Aggregation involves summarizing or grouping data into meaningful units, such as hourly, daily, or monthly averages, or aggregating data at the district or city level. Data aggregation helps reduce the data volume and simplifies analysis while retaining the essential information. Zhu et al.'s [60] proposal for an authentication and data aggregation mechanism for a fog-based smart grid is put forward.

By performing the data pre-processing steps, smart cities can ensure that the collected data is standardized and ready for analysis. Proper data pre-processing enhances the accuracy, reliability, and effectiveness of subsequent data analysis, modeling, and decision-making processes in smart city applications.

5.5.1.3 Data integration

Smart cities generate data from multiple sources and in various formats. The process of data integration entails merging information from several sources and harmonizing it into a single format. This allows for a comprehensive view and analysis of the city's data. Integration can be achieved through data transformation, normalization, or by using common identifiers or keys to link datasets. A data integration for smart grid ecosystems is proposed by Guerrero et al. [61]. The heterogeneous data source integration and metadata mining are applied. At times when the data is missing, data integration can help to mitigate the problems related to missing data. Jurado et al. [62] have studied the problem of forecasting applications with missing data. Because of missing data, forecasting of data might be delayed for next few hours or even days.

5.5.1.4 Data storage

In the data storage step, a vast amount of data generated in smart cities is managed. With the proliferation of sensors, IoT devices, and other data sources, efficient and scalable data storage solutions are necessary. Smart cities produce large volumes of data that grow exponentially over time. Scalable storage solutions, such as cloud-based storage, are needed to accommodate the increasing data volume and provide flexibility for future expansion [63].

To achieve high availability and data redundancy techniques such as data replication, fault tolerance mechanisms, and backup and recovery procedures can be utilized. Some smart city applications require real-time data processing and analysis. In such cases, data storage systems need to support low-latency storage and retrieval of data, enabling timely decision-making and responsiveness. Keeping this in mind, Pinheiro et al. [64] combined the Kappa and Lambda for data storage by using the Simplified Mandatory Access Control Kernel (SMACK) framework. A different storage technique is proposed in [65] that utilizes a graph storage engine to store data effectively.

5.5.1.5 Data analytics and insights

Advanced analytics techniques are applied to energy data to extract valuable insights. Data analytics helps identify consumption patterns, energy waste, anomalies, and opportunities for energy-efficiency improvements. It enables smart cities to optimize energy consumption, predict demand, detect faults, and proactively manage energy resources. In order to reduce time delays and process local data to save network bandwidths, Ahsan and Bais [66] have proposed a machine learning algorithm and reduced the overall cost of system.

5.5.1.6 Data visualization

Energy data management systems provide visualization tools to track energy consumption, generation, and performance metrics. Interactive dashboards and visualization tools enable stakeholders to understand energy usage patterns, identify areas for improvement, and make informed decisions regarding energy-efficiency measures. Real-time monitoring helps detect abnormal energy usage and supports proactive energy management. There are several visualization tools available in the market, such as, Tableau, Power BI, QlikView/QlikSense, ArcGIS, MapInfo, Grafana, Maptitude, and so on. GIS software can be utilized for visualizing smart grid data on maps [67].

5.5.1.7 Billing and tariff management

Energy data management systems support accurate billing and tariff management for residential, commercial, and industrial customers. By integrating consumption data with appropriate billing algorithms and tariff structures, cities can generate accurate energy bills and incentivize energy conservation behaviors through dynamic pricing or time-of-use tariffs. A combined billing and customer care system is proposed by Jindal [68]. The proposed system will enable utilities to manage payments, consumer accounts, meters, and so on, on one interface.

5.5.1.8 Decision-making

Smart grids offer significant advancements in real-time and automated decision-making compared to traditional grids. Decision-making relies on the continuous monitoring and analysis of data to identify energy consumption patterns, detect anomalies, and make informed decisions regarding energy management and optimization. Decision-making involves analyzing energy consumption patterns, identifying energy-intensive areas, and implementing energy-saving measures such as optimizing building operations, upgrading to energy-efficient appliances, and implementing smart lighting systems. Data-driven decisions can lead to significant energy and cost savings [69]. Overall, energy data management in smart cities empowers decision-makers with timely and accurate insights to optimize energy usage, enhance efficiency, and support sustainable energy practices. By leveraging the wealth of energy data available, smart cities can make data-driven decisions that drive positive environmental, economic, and social outcomes.

5.5.2 Cyber security of smart cities

Cyber security is a critical aspect of ensuring the resilience and trustworthiness of smart cities. As smart cities heavily rely on interconnected devices, networks, and data, they become more vulnerable to cyber threats [70]. Due to the significant improvements in automation over the past 10 years, cybersecurity challenges, and responses have been intensively researched in the literature [71, 72]. Cyber-attacks are classified into Denial of Service (DoS) attacks, replay attacks, and deception attacks [73]. In a DoS attack, the access to a particular service or resource is denied by overwhelming it with a flood of illegitimate requests or by exploiting vulnerabilities. This results in legitimate users being unable to access the targeted service or resource. DoS attacks can be executed through various means, such as sending a high volume of network traffic, exploiting resource exhaustion, or leveraging application vulnerabilities. Queuing models, Bernoulli models, and Markov models are commonly used in performance analysis to quantify the degradation of systems. Queuing models are mathematical models that analyze systems with queues. Queuing models consider factors like arrival rate, service rate, and number of servers to predict system performance and measure the impact of different parameters on system degradation. In the work of Pang, Liu, and Dong [74], the adverse impacts from weak DoS attack are mitigated by utilizing a round-trip-based predictive control. The Bernoulli models are probabilistic models that can be used to estimate the probability of system degradation or failure [75]. And Markov models are stochastic models that analyze systems with states and transitions between states. They capture the dynamic behavior of systems and provide insights into system degradation over time [76].

The replay attacks involve the interception and malicious replay of legitimate data packets or messages. The attacker captures valid data transmissions and later replays them to gain unauthorized access or perform malicious actions [77]. On the other hand, deception attacks involve the manipulation or alteration of data, systems, or communication to mislead or deceive the target. This can include masquerading as a legitimate entity, forging or modifying data packets, or manipulating system behavior to trick users or systems into taking unintended actions [73]. Deception attacks undermine the integrity, confidentiality, or availability of information or systems.

There are several detection schemes used in cyber systems to identify and respond to potential threats such as Bayesian detection, Weighted Least Square (WLS), Kalman Filters-based-2-Detector, and Quasi-FDI (Fault Detection and Isolation) Techniques. In Han et al. [78], a detection method based on Bayesian hypothesis testing is proposed to improve practical advantages of the system overall. WLS assigns weights to different observations based on their reliability or importance. It is commonly used in fault detection applications to identify deviations or outliers in observed data by minimizing the weighted sum of squared residuals [79]. Kalman filters can be used as a detection scheme to monitor and track system states, predict future states, and detect anomalies or deviations from expected behavior. Cai et al. [80] have proposed a novel Kalman filter design method with low conservativeness. Quasi-FDI techniques are fault detection and isolation methods used to identify and localize faults in a system [81]. These detection schemes leverage statistical analysis, probabilistic modeling, and estimation techniques to identify abnormal behavior or deviations from expected patterns in cyber systems. They play a crucial role in detecting potential cyber threats and mitigation to protect the integrity, availability, and confidentiality of the systems and their data. Many researchers have conducted a thorough survey to explore the development of detection schemes for cyber-attacks. Table 5.4 shows the detection techniques utilized in literature.

One of the severe cyber-attacks in history is the devastating attack against the Ukrainian electricity infrastructure in June 2017. It resulted in a six-hour outage. Because it was the first significant cyberattack to result in a significant collapse affecting 80,000 people, this incident was recognized as a turning point in cybersecurity. Some of the other examples of notable cyber-attacks include the WannaCry ransomware attack, in 2017, that impacted numerous organizations globally, exploiting a vulnerability in Microsoft Windows systems. It disrupted operations in healthcare, logistics, government agencies, and other sectors. In December 2020, the SolarWinds cyberattack was discovered, targeting various organizations, including government agencies and technology companies. The attackers compromised the SolarWinds Orion software, which was widely used for network monitoring, to distribute a supply chain attack and gain unauthorized access to networks.

Table 5.4 Intrusion detection system (IDS) techniques in smart grids

Ref.	*Detection system*	*Protection range*	*Domain*
[82]	Multi-attribute SCADA-Specific IDS	SCADA	Network-based
[83]	Model-based techniques	SCADA	Host-based
[84]	Model-based Intrusion Detection for Smart grid (MINDS)	Substation	Network-based
[85]	Anomaly inference algorithm	Substation	Host-based
[86]	Multidimensional IDS	Substation	Integrated
[87]	Real-time anomaly detection (ReTAD) algorithm	WAMS	Host-based
[88]	Global Positioning System (GPS) carrier-to-noise ratio (C/No) based spoofing detection	GPS (PMU)	Host-based
[89]	Behavior-rule based intrusion detection system (BRIDS)	Distribution system	Host-based
[90]	Support vector regression and impact difference	AMI	Host-based

5.6 COMMUNICATION AND NETWORK INFRASTRUCTURE

Communication and network infrastructure play a vital role in the functioning of smart cities. They enable the seamless exchange of information, data, and services among various components and systems within the city. Some key aspects of communication and network infrastructure in smart cities include connectivity, sensor networks, data communication, cloud computing and edge computing, and command and control centers [91]. Smart cities require robust and reliable connectivity to support the flow of data. This includes both wired and wireless technologies such as fiber optic networks, 4G/5G cellular networks, Wi-Fi, and IoT (internet of things) protocols. High-speed and low-latency connectivity is essential to enable real-time communication and data transmission between sensors, devices, and systems.

5.6.1 Communication protocols and technologies in smart cities

Communication protocols and technologies facilitate the exchange of information between various grid components, including power generation sources, smart meters, and energy management systems. Figure 5.7 shows few smart city applications and the communication protocols utilized.

As depicted in the Figure 5.7, smart buildings use protocol IEEE 801.15.1 (or Bluetooth). It is designed for short-range communication within a limited area, typically up to 100 meters. Therefore, it is used to connect devices like smartphones, laptops, headphones, and smart building [92]. Also, protocols like Zigbee (IEEE 802.15.4) are used for low-power, low-data-rate

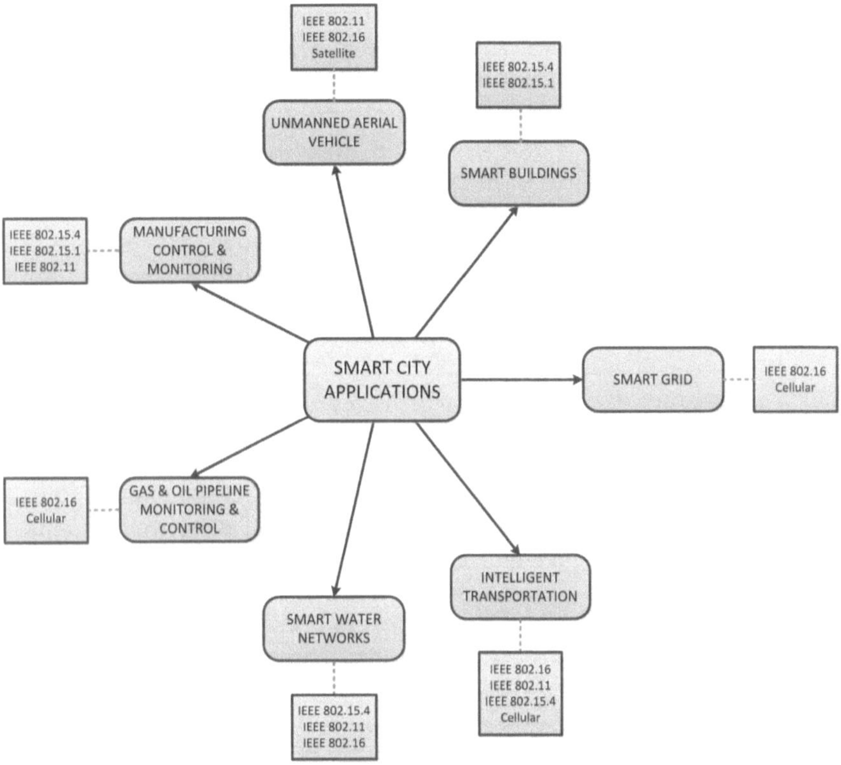

Figure 5.7 Communication protocols in smart cities.

wireless communication in applications such as home automation, smart water networks [93], and manufacturing control and monitoring [94].

For applications requiring longer ranges and higher data rates, protocols in the Local Area Network (LAN) class are used. IEEE 802.11 (WiFi) is a widely used LAN protocol for wireless communication such as intelligent transportation [95].

When it comes to wide-range communication, protocols in the Wide Area Network (WAN) class are utilized. Examples include IEEE 802.16 (WiMAX) for wireless metropolitan area networks, cellular networks (e.g., 4G/5G), and satellite communication. These protocols provide extended coverage and higher data transmission capabilities suitable for applications like unmanned aerial vehicles (UAVs) [96] and smart grid systems [97].

5.6.2 Role of internet of things (IoT) in communication

IoT plays a significant role in enabling smart grid communication by connecting various devices, sensors, and systems. IoT devices, such as smart meters, collect real-time data on energy consumption, grid performance, and environmental conditions. This data is transmitted over communication networks to central monitoring systems, enabling utilities to monitor and manage the grid more effectively. These devices allow for remote control and automation of grid components [98]. For example, smart grid technologies enable utilities to remotely control the operation of distributed energy resources (DERs) or adjust load settings based on grid conditions. This enhances grid flexibility and enables more efficient energy management. Devices such as smart thermostats or appliances, enable demand response programs by providing real-time information about energy usage. This allows utilities to adjust load patterns and manage peak demand more effectively, contributing to grid stability and reliability. The IoT devices and sensors provide valuable data for grid optimization and predictive maintenance. By analyzing real-time data, utilities can identify potential issues, optimize grid operations, and schedule maintenance activities more efficiently, reducing downtime and improving overall grid performance. Valuable insights into energy consumption patterns can help consumers and utilities identify areas for energy-efficiency improvements. By analyzing these patterns, individuals and organizations can make informed decisions, leading to reduced energy waste, cost savings, and a more sustainable energy ecosystem.

Communication protocols enable devices to exchange data and messages in a standardized and interoperable manner. There are a large number of IoT protocols. Some of the widely used protocol include Message Queuing Telemetry Transport (MQTT), Data Distribution Service (DDS), Advance Message Queuing Protocol (AMQP), Constrained Application Protocol (CoAP), Hypertext Transfer Protocol (HTTP), WebSocket, and others [99]. MQTT and CoAP are lightweight protocol designed for constrained devices and low-bandwidth networks. In the context of IoT, HTTP is often used for device management, configuration, and data exchange with cloud platforms or web services. In addition to the aforementioned protocols, there are also coexistence protocols such as Zigbee and Bluetooth Low Energy (BLE) that are specifically designed for short-range communication within IoT networks. These protocols are commonly used in smart home automation, industrial monitoring, and other IoT applications. These protocols facilitate efficient data transmission, device discovery, and secure communication between devices and the cloud.

Figure 5.8 shows a schematic of a home automation system to flip lights on or off remotely [100]. A battery-powered IoT switch is used that communicates directly with lamps using Zigbee protocol. The gateway is utilized to translate the Zigbee messages to MQTT protocol and transfer it to the network and cloud server. Using the mobile application, the data on the cloud can be accessed easily and the switch in the home can be controlled remotely.

The combination of communication protocols and IoT technologies within the smart grid enables efficient data exchange, real-time monitoring, control, and optimization of the grid infrastructure. This contributes to improved grid reliability and enhanced energy management.

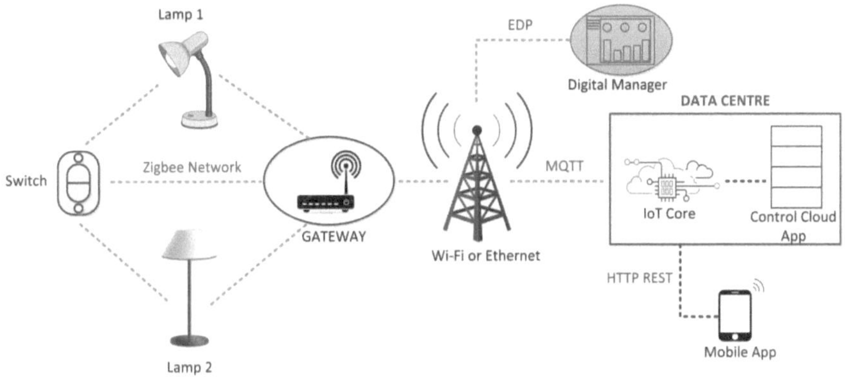

Figure 5.8 Home automation system.

5.7 CASE STUDIES AND BEST PRACTICES

Case studies of smart grid installation may be found all around the world. Here are a few examples of typical Supervisory Control and Data Acquisition (SCADA) system deployment and expansion cases for the smart grid [101, 102].

1. Pacific Gas and Electric (PG&E) is an American utility public company that utilizes PMUs to enhance its distribution system. PMUs are advanced monitoring devices that provide high-resolution measurements of voltage, current, and frequency. By deploying PMUs across their distribution network, PG&E aims to improve the situational awareness and overall performance of their system. The real-time monitoring and analysis capabilities of PMUs enable PG&E to proactively address issues and deliver a more efficient and reliable power supply to their customers.
2. Omaha Public Power District (OPPD) is an electric utility serving customers in the state of Nebraska, United States. OPPD is committed to providing reliable and affordable electricity to its customers while also embracing innovative technologies and practices to enhance the efficiency and sustainability of its operations.
3. Singapore has made significant strides in adopting smart grid technologies and building a resilient and sustainable energy infrastructure. The city-state has deployed advanced metering infrastructure, distribution automation, and demand response programs. These initiatives have helped optimize energy consumption, reduce carbon emissions, and enhance grid reliability. In 2014, Singapore launched its Smart Nation initiative had introduced many smart technologies. Also in 2021, it announced the development of a vehicle-free eco-smart city with five residential districts.

4. Copenhagen in Denmark is often cited as a leading example of a smart city with an advanced smart grid infrastructure. The city has implemented a district heating system that integrates renewable energy sources and uses smart meters to optimize energy distribution and consumption. The smart grid in Copenhagen enables better management of energy resources, reduced emissions, and increased energy efficiency.
5. Barcelona in Spain has implemented a successful smart grid initiative called "Smart City Barcelona." It incorporates advanced metering infrastructure, smart grid sensors, and a centralized data platform for real-time monitoring and control. The smart grid implementation has led to improved energy efficiency, reduced power outages, and better integration of renewable energy sources.

Many cities are joining the movement in investing in smart technologies. Some of the notable cities include Masdar city in UAE, Tokyo in Japan, Oslo in Norway, Amsterdam in the Netherlands, New York in USA, Seoul in South Korea, and so on.

Smart grid deployments have taught many lessons. The lessons learned and best practices for implementing smart grids include stakeholder collaboration and engagement, clear regulatory support and policy frameworks, robust cybersecurity measures, leveraging data analytics and visualization, conducting pilot projects for testing and scalability, and consumer engagement and education [103].

Successful smart grid implementations require collaboration and coordination among utilities, government agencies, technology providers, and consumers. Engaging stakeholders and fostering partnerships ensure a holistic approach to planning, implementation, and operation of the smart grid infrastructure. Clear regulatory support and well-defined policy frameworks help in facilitating the adoption of smart grids. Governments should provide incentives, mandates, and regulations that encourage utilities and consumers to invest in smart grid technologies [104]. Also, for smart grids, ensuring robust cybersecurity measures is essential. Implementing encryption, authentication protocols, and continuous monitoring helps protect the smart grid infrastructure from potential cyber-attacks. Leveraging data analytics and visualization tools helps utilities make informed decisions, identify energy consumption patterns, and optimize grid operations [105]. Real-time data analysis enables proactive maintenance, load forecasting, and demand response management. Conducting pilot projects in specific areas allows utilities to test and evaluate the feasibility and effectiveness of smart grid technologies before full-scale deployment. Scalability should be a key consideration from the early stages to ensure that the smart grid infrastructure can accommodate future growth and evolving needs. And most importantly it is essential to engage consumers and provide them with information and tools to actively participate in demand response programs and manage their energy consumption. Consumer awareness campaigns, user-friendly interfaces, and feedback mechanisms help drive behavioral changes and optimize energy usage.

By fostering collaboration among stakeholders, establishing supportive policies, ensuring cybersecurity, utilizing data analytics, conducting pilot projects, and educating and engaging consumers, cities can enhance the success of smart grid implementations.

5.8 SUMMARY

This chapter aims to provide an understanding of the smart grid and the technologies for smart cities. The smart grid concept and technologies form a critical infrastructure for smart cities, enabling efficient, reliable, and sustainable energy management. Smart grids leverage advanced technologies to optimize energy generation, distribution, and consumption. These technologies are analyzed and compared in detail. Demand response (DR) and its reliability and application with respect to smart cities are discussed. The communication and networking infrastructure, including protocols like Bluetooth and IoT, to enable real-time data exchange and control are analyzed. Smart grids facilitate efficient energy data management, cybersecurity measures, and advanced analytics for informed decision-making. Lessons learned and best practices include stakeholder collaboration, clear regulatory support, robust cybersecurity, data analytics, pilot projects, and consumer engagement. Successful smart grid implementations in cities worldwide, such as Singapore, Copenhagen, Barcelona, and so on, have demonstrated the benefits of these technologies, including improved energy efficiency, integration with renewable resources, reduced carbon emissions, and enhanced grid reliability.

The future trends of smart grids in smart cities are shaped by emerging technologies, evolving energy landscapes, and changing consumer demands. The use of advanced data analytics and AI algorithms will enable more sophisticated analysis of energy data, leading to better demand forecasting, load balancing, and grid optimization. AI-powered applications can help utilities and consumers make data-driven decisions and automate processes for improved efficiency. Blockchain technology holds the potential to revolutionize energy transactions by enabling peer-to-peer energy trading, transparent billing, and secure data exchange. Smart grids can leverage blockchain to facilitate direct energy transactions between consumers and prosumers, promoting renewable energy usage and decentralization. With the availability of real-time energy data and smart grid technologies, consumers will have more opportunities to actively participate in energy management. This includes personalized energy usage insights, demand response programs, and energy-efficiency recommendations, empowering consumers to make informed choices and reduce their carbon footprint.

REFERENCES

[1] "United Nations: Goal 11: Make cities inclusive, safe, resilient and sustainable." https://www.un.org/sustainabledevelopment/cities/ (accessed May 06, 2023).

[2] G. Betis, C. G. Cassandras, and C. A. Nucci, "Smart cities [scanning the issue]," *Proc. IEEE*, vol. 106, no. 4, pp. 513–517, Apr. 2018, doi: 10.1109/JPROC.2018.2812998

[3] A. Gopstein, C. Nguyen, C. O'Fallon, N. Hastings, and D. Wollman, "NIST framework and roadmap for smart grid interoperability standards, release 4.0," Gaithersburg, MD, Feb. 2021. doi: 10.6028/NIST.SP.1108r4

[4] Department for Business, Energy & Industrial Strategy, Ofgem, The Rt Hon Anne-Marie Trevelyan MP, and Lord Callanan, "Smart technologies and data to future-proof UK energy," Jul. 20, 2021. [Online]. Available: https://www.gov.uk/government/news/smart-technologies-and-data-to-future-proof-uk-energy

[5] Y. Cunjiang, Z. Huaxun, and Z. Lei, "Architecture design for smart grid," *Energy Procedia*, vol. 17, pp. 1524–1528, 2012, doi: 10.1016/j.egypro.2012.02.276

[6] Office. of the National Cordinator for Smart Grid, "NIST Framework and Roadmap for Smart Grid Interoperability Standards, Release 1.0," 2010.

[7] O. Majeed Butt, M. Zulqarnain, and T. Majeed Butt, "Recent advancement in smart grid technology: Future prospects in the electrical power network," *Ain Shams Eng. J.*, vol. 12, no. 1, pp. 687–695, Mar. 2021, doi: 10.1016/j.asej.2020.05.004

[8] A. A. Abdullah and T. M. Hassan, "Smart grid (SG) properties and challenges: an overview," *Discov. Energy*, vol. 2, no. 1, p. 8, Nov. 2022, doi: 10.1007/s43937-022-00013-x

[9] J. Carl and D. Fedor, "Tracking global carbon revenues: A survey of carbon taxes versus cap-and-trade in the real world," *Energy Policy*, vol. 96, pp. 50–77, Sep. 2016, doi: 10.1016/j.enpol.2016.05.023

[10] Z. Mahmood, Ed., *Smart Cities*. Cham: Springer International Publishing, 2018. doi: 10.1007/978-3-319-76669-0

[11] M. L. Tuballa and M. L. Abundo, "A review of the development of smart grid technologies," *Renew. Sustain. Energy Rev.*, vol. 59, pp. 710–725, Jun. 2016, doi: 10.1016/j.rser.2016.01.011

[12] Y. Shi, H. D. Tuan, T. Q. Duong, H. V. Poor, and A. V. Savkin, "PMU placement optimization for efficient state estimation in smart grid," *IEEE J. Sel. Areas Commun.*, vol. 38, no. 1, pp. 71–83, Jan. 2020, doi: 10.1109/JSAC.2019.2951969

[13] S. Sridhar and M. Govindarasu, "Model-based attack detection and mitigation for automatic generation control," *IEEE Trans. Smart Grid*, vol. 5, no. 2, pp. 580–591, Mar. 2014, doi: 10.1109/TSG.2014.2298195

[14] A. Keyhani and A. Chatterjee, "Automatic generation control structure for smart power grids," *IEEE Trans. Smart Grid*, vol. 3, no. 3, pp. 1310–1316, Sep. 2012, doi: 10.1109/TSG.2012.2194794

[15] T. T. Tesfay and J.-Y. Le Boudec, "Experimental comparison of multicast authentication for wide area monitoring systems," *IEEE Trans. Smart Grid*, vol. 9, no. 5, pp. 4394–4404, Sep. 2018, doi: 10.1109/TSG.2017.2656067

[16] S. A. Nezam Sarmadi, A. S. Dobakhshari, S. Azizi, and A. M. Ranjbar, "A sectionalizing method in power system restoration based on WAMS," *IEEE Trans. Smart Grid*, vol. 2, no. 1, pp. 190–197, Mar. 2011, doi: 10.1109/TSG.2011.2105510
[17] S. K. Rathor and D. Saxena, "Energy management system for smart grid: An overview and key issues," *Int. J. Energy Res.*, vol. 44, no. 6, pp. 4067–4109, May 2020, doi: 10.1002/er.4883
[18] M. Meliani, A. El Barkany, I. El Abbassi, A. M. Darcherif, and M. Mahmoudi, "Energy management in the smart grid: State-of-the-art and future trends," *Int. J. Eng. Bus. Manag.*, vol. 13, p. 184797902110329, Jan. 2021, doi: 10.1177/18479790211032920
[19] O. Ting-Chia, T.-P. Tsao, C.-M. Hong, and C.-H. Chen, "Hybrid control system for automatic voltage regulator in smart grid," in *2013 International Conference on Machine Learning and Cybernetics*, Jul. 2013, pp. 1103–1108. doi: 10.1109/ICMLC.2013.6890757
[20] H. Gozde, M. C. Taplamacioglu, and M. Ari, "Simulation study for global neighborhood algorithm based optimal automatic voltage regulator (AVR) system," in *2017 5th International Istanbul Smart Grid and Cities Congress and Fair (ICSG)*, Apr. 2017, pp. 46–50. doi: 10.1109/SGCF.2017.7947634
[21] J. Dirkman, "Best practices for creating your smart grid network model," Schneider Electric Inc.: Fort Collins, CO, USA, 2013.
[22] E. Burger, V. Mittelbach, and A. Koziolek, "View-based and model-driven outage management for the smart grid," *CEUR Workshop Proc.*, vol. 1742, pp. 1–8, 2016.
[23] M. Orlando et al., "A smart meter infrastructure for smart grid IoT applications," *IEEE Internet Things J.*, vol. 9, no. 14, pp. 12529–12541, Jul. 2022, doi: 10.1109/JIOT.2021.3137596
[24] A. Ghosal and M. Conti, "Key management systems for smart grid advanced metering infrastructure: A survey," *IEEE Commun. Surv. Tutorials*, vol. 21, no. 3, pp. 2831–2848, 2019, doi: 10.1109/COMST.2019.2907650
[25] A. D. Ashkezari, N. Hosseinzadeh, A. Chebli, and M. Albadi, "Development of an enterprise Geographic Information System (GIS) integrated with smart grid," *Sustain. Energy, Grids Networks*, vol. 14, pp. 25–34, Jun. 2018, doi: 10.1016/j.segan.2018.02.001
[26] K. Kaippilly Radhakrishnan, J. Moirangthem, S. K. Panda, and G. Amaratunga, "GIS Integrated automation of a near real-time power-flow service for electrical grids," *IEEE Trans. Ind. Appl.*, vol. 54, no. 6, pp. 5661–5670, Nov. 2018, doi: 10.1109/TIA.2018.2855645
[27] I. Alotaibi, M. A. Abido, M. Khalid, and A. V. Savkin, "A comprehensive review of recent advances in smart grids: A sustainable future with renewable energy resources," *Energies*, vol. 13, no. 23, pp. 1–41, 2020, doi: 10.3390/en13236269
[28] X. Yan, Y. Ozturk, Z. Hu, and Y. Song, "A review on price-driven residential demand response," *Renew. Sustain. Energy Rev.*, vol. 96, pp. 411–419, Nov. 2018, doi: 10.1016/j.rser.2018.08.003
[29] P. Warren, "A review of demand-side management policy in the UK," *Renew. Sustain. Energy Rev.*, vol. 29, pp. 941–951, Jan. 2014, doi: 10.1016/j.rser.2013.09.009

[30] Y.-K. Wu and K.-T. Tang, "Frequency support by demand response – Review and analysis," *Energy Procedia*, vol. 156, pp. 327–331, Jan. 2019, doi: 10.1016/j.egypro.2018.11.150
[31] S. Kakran and S. Chanana, "Smart operations of smart grids integrated with distributed generation: A review," *Renew. Sustain. Energy Rev.*, vol. 81, pp. 524–535, Jan. 2018, doi: 10.1016/j.rser.2017.07.045
[32] S. Mohagheghi, F. Yang, and B. Falahati, "Impact of demand response on distribution system reliability," in *2011 IEEE Power and Energy Society General Meeting*, Jul. 2011, pp. 1–7. doi: 10.1109/PES.2011.6039365
[33] S. Coe, A. Ott, and D. Pratt, "Demanding standards," *IEEE Power Energy Mag.*, vol. 8, no. 3, pp. 55–59, May 2010, doi: 10.1109/MPE.2010.936350
[34] A. Escalera, B. Hayes, and M. Prodanović, "A survey of reliability assessment techniques for modern distribution networks," *Renew. Sustain. Energy Rev.*, vol. 91, pp. 344–357, Aug. 2018, doi: 10.1016/j.rser.2018.02.031
[35] A. L. A. Syrri and P. Mancarella, "Reliability and risk assessment of post-contingency demand response in smart distribution networks," *Sustain. Energy, Grids Networks*, vol. 7, pp. 1–12, Sep. 2016, doi: 10.1016/j.segan.2016.04.002
[36] B. Zeng, G. Wu, J. Wang, J. Zhang, and M. Zeng, "Impact of behavior-driven demand response on supply adequacy in smart distribution systems," *Appl. Energy*, vol. 202, pp. 125–137, Sep. 2017, doi: 10.1016/j.apenergy.2017.05.098
[37] A. Safdarian, M. Z. Degefa, M. Lehtonen, and M. Fotuhi-Firuzabad, "Distribution network reliability improvements in presence of demand response," *IET Gener. Transm. Distrib.*, vol. 8, no. 12, pp. 2027–2035, Dec. 2014, doi: 10.1049/iet-gtd.2013.0815
[38] I. Hernando-Gil, I.-S. Ilie, and S. Z. Djokic, "Reliability performance of smart grids with demand-side management and distributed generation/storage technologies," in *2012 3rd IEEE PES Innovative Smart Grid Technologies Europe (ISGT Europe)*, Oct. 2012, pp. 1–8. doi: 10.1109/ISGTEurope.2012.6465883
[39] H. Xie, Z. Bie, B. Hua, and G. Li, "Reliability assessment of distribution power systems considering the TOU pricing," in *2013 IEEE International Conference of IEEE Region 10 (TENCON 2013)*, Oct. 2013, pp. 1–4. doi: 10.1109/TENCON.2013.6719049
[40] I.-S. Ilie, I. Hernando-Gil, A. J. Collin, J. L. Acosta, and S. Z. Djokic, "Reliability performance assessment in smart grids with demand-side management," in *2011 2nd IEEE PES International Conference and Exhibition on Innovative Smart Grid Technologies*, Dec. 2011, pp. 1–7. doi: 10.1109/ISGTEurope.2011.6162650
[41] M. Cheng, S. S. Sami, and J. Wu, "Virtual energy storage system for smart grids," *Energy Procedia*, vol. 88, pp. 436–442, Jun. 2016, doi: 10.1016/j.egypro.2016.06.021
[42] M. Cheng, S. S. Sami, and J. Wu, "Benefits of using virtual energy storage system for power system frequency response," *Appl. Energy*, vol. 194, pp. 376–385, May 2017, doi: 10.1016/j.apenergy.2016.06.113
[43] J. Wu, W. Hung, J. Ekanayake, N. Jenkins, T. Coleman, and M. Cheng, "Primary frequency response in the great britain power system from dynamically controlled refrigerators," in *22nd International Conference and Exhibition on Electricity Distribution (CIRED 2013)*, 2013, pp. 0507–0507. doi: 10.1049/cp.2013.0772

[44] J. A. Short, D. G. Infield, and L. L. Freris, "Stabilization of grid frequency through dynamic demand control," *IEEE Trans. Power Syst.*, vol. 22, no. 3, pp. 1284–1293, Aug. 2007, doi: 10.1109/TPWRS.2007.901489
[45] M. Cheng et al., "Power system frequency response from the control of bitumen tanks," *IEEE Trans. Power Syst.*, vol. 31, no. 3, pp. 1769–1778, May 2016, doi: 10.1109/TPWRS.2015.2440336
[46] S. A. Pourmousavi and M. H. Nehrir, "Real-time central demand response for primary frequency regulation in microgrids," *IEEE Trans. Smart Grid*, vol. 3, no. 4, pp. 1988–1996, Dec. 2012, doi: 10.1109/TSG.2012.2201964
[47] M. H. Shoreh, P. Siano, M. Shafie-Khah, V. Loia, and J. P. S. Catalão, "A survey of industrial applications of demand response," *Electr. Power Syst. Res.*, vol. 141, pp. 31–49, Dec. 2016, doi: 10.1016/j.epsr.2016.07.008
[48] Y. Hashemi and H. Shayeghi, "Demand response application in generation, transmission, and distribution expansion planning," in *Demand Response Application in Smart Grids*, Cham: Springer International Publishing, 2020, pp. 163–191. doi: 10.1007/978-3-030-31399-9_7
[49] V. Potdar, A. Chandan, S. Batool, and N. Patel, "Big energy data management for smart grids—Issues, challenges and recent developments," in *Smart Cities: Development and Governance Frameworks*, Mahmood, Z., Ed, 2018, pp. 177–205. doi: 10.1007/978-3-319-76669-0_8
[50] A. Gharaibeh et al., "Smart cities: A survey on data management, security, and enabling technologies," *IEEE Commun. Surv. Tutorials*, vol. 19, no. 4, pp. 2456–2501, 2017, doi: 10.1109/COMST.2017.2736886
[51] H. Daki, A. El Hannani, A. Aqqal, A. Haidine, and A. Dahbi, "Big Data management in smart grid: Concepts, requirements and implementation," *J. Big Data*, vol. 4, no. 1, p. 13, Dec. 2017, doi: 10.1186/s40537-017-0070-y
[52] S. Uludag, K.-S. Lui, W. Ren, and K. Nahrstedt, "Secure and scalable data collection with time minimization in the smart grid," *IEEE Trans. Smart Grid*, vol. 7, no. 1, pp. 43–54, Jan. 2016, doi: 10.1109/TSG.2015.2404534
[53] X. Li, Z. Lv, J. Hu, B. Zhang, L. Shi, and S. Feng, "XEarth: A 3D GIS platform for managing massive city information," in *2015 IEEE International Conference on Computational Intelligence and Virtual Environments for Measurement Systems and Applications (CIVEMSA)*, Jun. 2015, pp. 1–6. doi: 10.1109/CIVEMSA.2015.7158625
[54] Y. Keping, M. Arifuzzaman, Z. Wen, D. Zhang, and T. Sato, "A key management scheme for secure communications of information centric advanced metering infrastructure in smart grid," *IEEE Trans. Instrum. Meas.*, vol. 64, no. 8, pp. 2072–2085, Aug. 2015, doi: 10.1109/TIM.2015.2444238
[55] D. Abbasinezhad-Mood and M. Nikooghadam, "An ultra-lightweight and secure scheme for communications of smart meters and neighborhood gateways by utilization of an ARM Cortex-M microcontroller," *IEEE Trans. Smart Grid*, vol. 9, no. 6, pp. 6194–6205, Nov. 2018, doi: 10.1109/TSG.2017.2705763
[56] Y.-J. Kim, V. Kolesnikov, H. Kim, and M. Thottan, "SSTP: A scalable and secure transport protocol for smart grid data collection," in *2011 IEEE International Conference on Smart Grid Communications (SmartGridComm)*, Oct. 2011, pp. 161–166. doi: 10.1109/SmartGridComm.2011.6102310
[57] N. Saputro and K. Akkaya, "Investigation of smart meter data reporting strategies for optimized performance in smart grid AMI networks," *IEEE*

Internet Things J., vol. 4, no. 4, pp. 894–904, Aug. 2017, doi: 10.1109/JIOT.2017.2701205

[58] R. Cody, *Cody's Data Cleaning Techniques Using SAS*, Third edition, Cary: NC, USA: SAS Institute Inc. 2017.

[59] X. Chu and I. F. Ilyas, "Qualitative data cleaning," *Proc. VLDB Endow.*, vol. 9, no. 13, pp. 1605–1608, Sep. 2016, doi: 10.14778/3007263.3007320

[60] L. Zhu et al., "Privacy-preserving authentication and data aggregation for fog-based smart grid," *IEEE Commun. Mag.*, vol. 57, no. 6, pp. 80–85, Jun. 2019, doi: 10.1109/MCOM.2019.1700859

[61] J. I. Guerrero, A. García, E. Personal, J. Luque, and C. León, "Heterogeneous data source integration for smart grid ecosystems based on metadata mining," *Expert Syst. Appl.*, vol. 79, pp. 254–268, Aug. 2017, doi: 10.1016/j.eswa.2017.03.007

[62] S. Jurado, À. Nebot, F. Mugica, and M. Mihaylov, "Fuzzy inductive reasoning forecasting strategies able to cope with missing data: A smart grid application," *Appl. Soft Comput.*, vol. 51, pp. 225–238, Feb. 2017, doi: 10.1016/j.asoc.2016.11.040

[63] "Big Data for smart grid," in *Smart Grid Using Big Data Analytics*, Chichester, UK: John Wiley & Sons, Ltd, 2017, pp. 485–491. doi: 10.1002/9781118716779.ch11

[64] G. Pinheiro, E. Vinagre, I. Praça, Z. Vale, and C. Ramos, "Smart grids data management: A case for cassandra," 2018, pp. 87–95. doi: 10.1007/978-3-319-62410-5_11

[65] R. De Virgilio, "Smart RDF data storage in graph databases," in *2017 17th IEEE/ACM International Symposium on Cluster, Cloud and Grid Computing (CCGRID)*, May 2017, pp. 872–881. doi: 10.1109/CCGRID.2017.108

[66] U. Ahsan and A. Bais, "Distributed big data management in smart grid," in *2017 26th Wireless and Optical Communication Conference (WOCC)*, Apr. 2017, pp. 1–6. doi: 10.1109/WOCC.2017.7928971

[67] M. Stefan, J. G. Lopez, M. H. Andreasen, and R. L. Olsen, "Visualization techniques for electrical grid smart metering data: A survey," in *2017 IEEE Third International Conference on Big Data Computing Service and Applications (BigDataService)*, Apr. 2017, pp. 165–171. doi: 10.1109/BigDataService.2017.26

[68] A. Jindal, "Combined billing and customer care systems for all utilities in a smart city," 2022, pp. 21–31. doi: 10.1007/978-981-16-8727-3_3

[69] M. Yu and S. H. Hong, "A real-time demand-response algorithm for smart grids: A stackelberg game approach," *IEEE Trans. Smart Grid*, pp. 1–1, 2015, doi: 10.1109/TSG.2015.2413813

[70] F. Aloul, A. R. Al-Ali, R. Al-Dalky, M. Al-Mardini, and W. El-Hajj, "Smart grid security: Threats, vulnerabilities and solutions," *Int. J. Smart Grid Clean Energy*, pp. 1–6, 2012, doi: 10.12720/sgce.1.1.1-6

[71] A. Gupta, A. Anpalagan, G. H. S. Carvalho, A. S. Khwaja, L. Guan, and I. Woungang, "RETRACTED: Prevailing and emerging cyber threats and security practices in IoT-enabled smart grids: A survey," *J. Netw. Comput. Appl.*, vol. 132, pp. 118–148, Apr. 2019, doi: 10.1016/j.jnca.2019.01.012

[72] C. Peng, H. Sun, M. Yang, and Y.-L. Wang, "A survey on security communication and control for smart grids under malicious cyber attacks," *IEEE*

Trans. Syst. Man, Cybern. Syst., vol. 49, no. 8, pp. 1554–1569, Aug. 2019, doi: 10.1109/TSMC.2018.2884952
[73] D. Ding, Z. Wang, Q.-L. Han, and G. Wei, "Security control for discrete-time stochastic nonlinear systems subject to deception attacks," *IEEE Trans. Syst. Man, Cybern. Syst.*, vol. 48, no. 5, pp. 779–789, May 2018, doi: 10.1109/TSMC.2016.2616544
[74] Z. H. Pang, G. P. Liu, and Z. Dong, "Secure networked control systems under denial of service attacks*," *IFAC Proc. Vol.*, vol. 44, no. 1, pp. 8908–8913, Jan. 2011, doi: 10.3182/20110828-6-IT-1002.02862
[75] S. Amin, G. A. Schwartz, and S. Shankar Sastry, "Security of interdependent and identical networked control systems," *Automatica*, vol. 49, no. 1, pp. 186–192, Jan. 2013, doi: 10.1016/j.automatica.2012.09.007
[76] G. K. Befekadu, V. Gupta, and P. J. Antsaklis, "Risk-sensitive control under Markov modulated denial-of-service (DoS) attack strategies," *IEEE Trans. Automat. Contr.*, vol. 60, no. 12, pp. 3299–3304, Dec. 2015, doi: 10.1109/TAC.2015.2416926
[77] P. Lee, A. Clark, L. Bushnell, and R. Poovendran, "A passivity framework for modeling and mitigating wormhole attacks on networked control systems," *IEEE Trans. Automat. Contr.*, vol. 59, no. 12, pp. 3224–3237, Dec. 2014, doi: 10.1109/TAC.2014.2351871
[78] K. Han et al., "Attack detection method based on bayesian hypothesis testing principle in CPS," *Procedia Comput. Sci.*, vol. 187, pp. 474–480, 2021, doi: 10.1016/j.procs.2021.04.086
[79] R. Deng, G. Xiao, and R. Lu, "Defending against false data injection attacks on power system state estimation," *IEEE Trans. Ind. Informatics*, vol. 13, no. 1, pp. 198–207, Feb. 2017, doi: 10.1109/TII.2015.2470218
[80] F. Cai, S. Liao, Y. Chen, and W. Wang, "Kalman filter of switching system under hybrid cyber attack," *IEEE Trans. Autom. Sci. Eng.*, pp. 1–9, 2023, doi: 10.1109/TASE.2023.3277960
[81] W. Ao, Y. Song, and C. Wen, "Adaptive cyber-physical system attack detection and reconstruction with application to power systems," *IET Control Theory Appl.*, vol. 10, no. 12, pp. 1458–1468, Aug. 2016, doi: 10.1049/iet-cta.2015.1147
[82] Y. Yang et al., "Multiattribute SCADA-specific intrusion detection system for power networks," *IEEE Trans. Power Deliv.*, vol. 29, no. 3, pp. 1092–1102, Jun. 2014, doi: 10.1109/TPWRD.2014.2300099
[83] "Detecting integrity attacks on SCADA systems," *IEEE Trans. Control Syst. Technol.*, vol. 22, no. 4, pp. 1396–1407, Jul. 2014, doi: 10.1109/TCST.2013.2280899
[84] A. Hahn and M. Govindarasu, "Model-based intrustion detection for the smart grid (MINDS)," in *Proceedings of the Eighth Annual Cyber Security and Information Intelligence Research Workshop*, Jan. 2013, pp. 1–4. doi: 10.1145/2459976.2460007
[85] C.-W. Ten, J. Hong, and C.-C. Liu, "Anomaly detection for cybersecurity of the substations," *IEEE Trans. Smart Grid*, vol. 2, no. 4, pp. 865–873, Dec. 2011, doi: 10.1109/TSG.2011.2159406
[86] Y. Yang, H.-Q. Xu, L. Gao, Y.-B. Yuan, K. McLaughlin, and S. Sezer, "Multidimensional intrusion detection system for IEC 61850-based SCADA

networks," *IEEE Trans. Power Deliv.*, vol. 32, no. 2, pp. 1068–1078, Apr. 2017, doi: 10.1109/TPWRD.2016.2603339

[87] J. Wu, J. Xiong, P. Shil, and Y. Shi, "Real time anomaly detection in wide area monitoring of smart grids," in *2014 IEEE/ACM International Conference on Computer-Aided Design (ICCAD)*, Nov. 2014, pp. 197–204. doi: 10.1109/ICCAD.2014.7001352

[88] Y. Fan, Z. Zhang, M. Trinkle, A. D. Dimitrovski, J. Bin Song, and H. Li, "A cross-layer defense mechanism against GPS spoofing attacks on PMUs in smart grids," *IEEE Trans. Smart Grid*, vol. 6, no. 6, pp. 2659–2668, Nov. 2015, doi: 10.1109/TSG.2014.2346088

[89] R. Mitchell and I.-R. Chen, "Behavior-rule based intrusion detection systems for safety critical smart grid applications," *IEEE Trans. Smart Grid*, vol. 4, no. 3, pp. 1254–1263, Sep. 2013, doi: 10.1109/TSG.2013.2258948

[90] Y. Liu, S. Hu, and T.-Y. Ho, "Leveraging strategic detection techniques for smart home pricing cyberattacks," *IEEE Trans. Dependable Secur. Comput.*, vol. 13, no. 2, pp. 220–235, Mar. 2016, doi: 10.1109/TDSC.2015.2427841

[91] L. Belli et al., "IoT-enabled smart sustainable cities: Challenges and approaches," *Smart Cities*, vol. 3, no. 3, pp. 1039–1071, Sep. 2020, doi: 10.3390/smartcities3030052

[92] L. Gurgen, O. Gunalp, Y. Benazzouz, and M. Galissot, "Self-aware cyber-physical systems and applications in smart buildings and cities," in *Design, Automation & Test in Europe Conference & Exhibition (DATE), 2013*, 2013, pp. 1149–1154. doi: 10.7873/DATE.2013.240

[93] S. Kartakis, E. Abraham, and J. A. McCann, "WaterBox," in *Proceedings of the 1st ACM International Workshop on Cyber-Physical Systems for Smart Water Networks*, Apr. 2015, pp. 1–6. doi: 10.1145/2738935.2738939

[94] J. Lee, B. Bagheri, and H.-A. Kao, "A cyber-physical systems architecture for Industry 4.0-based manufacturing systems," *Manuf. Lett.*, vol. 3, pp. 18–23, Jan. 2015, doi: 10.1016/j.mfglet.2014.12.001

[95] Y. P. Fallah, C. Huang, R. Sengupta, and H. Krishnan, "Design of cooperative vehicle safety systems based on tight coupling of communication, computing and physical vehicle dynamics," in *Proceedings of the 1st ACM/IEEE International Conference on Cyber-Physical Systems*, Apr. 2010, pp. 159–167. doi: 10.1145/1795194.1795217

[96] F. Mohammed, A. Idries, N. Mohamed, J. Al-Jaroodi, and I. Jawhar, "UAVs for smart cities: Opportunities and challenges," in *2014 International Conference on Unmanned Aircraft Systems (ICUAS)*, May 2014, pp. 267–273. doi: 10.1109/ICUAS.2014.6842265

[97] S. Karnouskos, "Cyber-physical systems in the SmartGrid," in *2011 9th IEEE International Conference on Industrial Informatics*, Jul. 2011, pp. 20–23. doi: 10.1109/INDIN.2011.6034829

[98] S. Ashraf, "A proactive role of IoT devices in building smart cities," *Internet Things Cyber-Physical Syst.*, vol. 1, pp. 8–13, 2021, doi: 10.1016/j.iotcps.2021.08.001

[99] R. P. Janani, K. Renuka, A. Aruna, and K. Lakshmi Narayanan, "IoT in smart cities: A contemporary survey," *Glob. Transitions Proc.*, vol. 2, no. 2, pp. 187–193, Nov. 2021, doi: 10.1016/j.gltp.2021.08.069

[100] R. Faludi, "How do IoT devices communicate?," 2021. [Online]. Available: https://www.digi.com/blog/post/how-do-iot-devices-communicate
[101] M. S. Thomas and J. D. McDonald, *Power System SCADA and Smart Grids*, 1st ed., Boca Raton: CRC Press, Taylor & Francis Group, 2015.
[102] O. Lai, "Top 7 smart cities in the world in 2023," *ENERGY,FUTURE, SOLUTIONS*, Mar. 2023. [Online]. Available: https://earth.org/top-7-smart-cities-in-the-world/
[103] P. Acharjee, "Strategy and implementation of smart grids in India," *Energy Strateg. Rev.*, vol. 1, no. 3, pp. 193–204, Mar. 2013, doi: 10.1016/j.esr.2012.05.003
[104] J. Wang, M. A. Biviji, and W. M. Wang, "Lessons learned from smart grid enabled pricing programs," in *2011 IEEE Power and Energy Conference at Illinois*, Feb. 2011, pp. 1–7. doi: 10.1109/PECI.2011.5740488
[105] C. W. Gellings, "Lessons learned from the world's smart grid demonstrations," in *Smart Grid Planning and Implementation*, River Publishers, 2020, pp. 191–228. doi: 10.1201/9781003151968-5

Chapter 6

Smart agriculture for smart cities

Akhtar Kalam and Neirat Mohamad Fayez Mustafa
Victoria University, Melbourne, Australia

6.1 INTRODUCTION

Currently, the construction of Eco-Cities is "forefront of national and global agendas" [1]. Various technologies integrate to achieve eco-smart cities. Sensing and tracking methods are applied through the internet connection, information gathering, and cloud computing. Diverse systems are incorporated including smart energy metering, water efficiency, water recycling, reducing emissions, enhancing efficiency in food supply chains, and waste management [2] to achieve eco-intelligent cities. An additional concern raised is whether the new systems can merge with messy, complex cities. Another opinion concerns resident privacy and city owner surveillance [2]. The world population is predicted to reach 9.7 billion in 2050 and around 11 billion in 2100 [3]. Two-thirds of the population live in urban areas. This will raise food supply and security challenges [4]. To feed this huge number of people, it is necessary to improve crop production. Smart city citizens are gradually increasing. As a result, smart farming is in demand. Smart agriculture has extended, ensuring that food is available to all smart city citizens.

In the 1700s, environmental farming appeared. Controlled environment farming includes various types such as indoor growing, indoor agriculture, vertical farming, and so on. [5]. In the last decade, greenhouses have played an essential role in the food supply. However, a conservatory is unlike indoor controlled farming [5]. In 2011, an earthquake and tsunami hit Japan and caused the Fukushima Daiichi nuclear disaster. As a result, 5% of agriculture was destroyed. The Japanese government searched for a solution. Japanese food scientists suggested a solution where crops could grow safely away from contaminated water or soil. This solution was the introduction to the vertical farming model. The Japanese government assisted vertical farm businesses [5]. As a result, vertical farming numbers grew significantly in Japan. Various countries, such as South Korea, Scandinavian countries, and the USA use vertical farming technologies. Statistics show that over the last ten years, vertical farming has increased sharply [5].

Conventional agriculture faces difficulties such as global warming, plant disease, insect attacks, soil degradation, decrease in soil quality, shortage of global resources like water, land, work labor, and insufficient knowledge to deal with various types of harvests. Conventional farming used old

 DOI: 10.1201/9781032669809-6

processes without integrating supply chains, market prices, and weather prediction [6]. Traditional agriculture demolishes forests, which increases the carbon footprint that leads to climate warming. Vertical farming assisted in saving 60–70% of world forests by restoring around two trillion trees [5]. Furthermore, farmland decreases globally due to climate change, global warming, industrialization, and residential buildings. Researchers have concerns and consider that vertical farming's initial cost will be expensive. However, operating prices are lower than conventional farming [7]. Conventional agriculture is inappropriate for harsh environments such as hot and frozen atmospheres and in crowded cities. For this reason, the indoor vertical farming solution is a suitable solution [4]. Smart farming integrated with appropriate technologies is the solution to overcome these difficulties [8].

Smart farming integrates various technologies and applications, internet connection, cloud, IoT equipment, and platforms [8]. This integration supports farmers to be conscious of all farming elements such as crop condition, water quality, soil quality, and so on. Furthermore, the collected extensive data is used for artificial intelligence analysis and automatic action [8]. Fruits, vegetables, and crops are produced through smart farming. Agriculture trays are arranged vertically in an array supported with robotic devices and IoT sensors (Figure 6.1). IoT sensors and actuators are constantly measuring farming environments, viz. light, humidity, temperature,

Figure 6.1 Smart vertical farming.

(Source: Aerofarms)

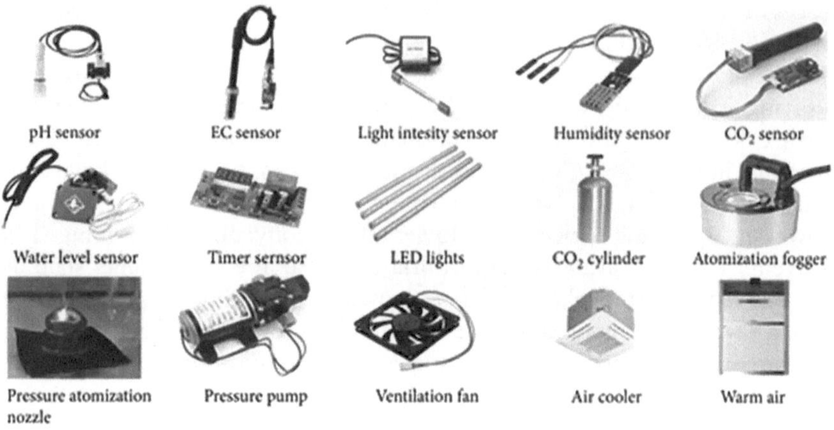

Figure 6.2 Sensors and actuators used in smart farming.

(Source: Ref. [9])

CO_2, pH, and so on. The digital environment provides full monitoring and control of the agriculture process.

Similar to other IoT systems globally, smart farming contains several IoT sensors, as shown in Figure 6.2, to operate the farming business, viz. humidity sensor, air temperature sensor, soil pH sensor, water volume sensor, soil moisture sensor, and so on [6, 9].

Developed technologies are integrated with vertical farming, such as analytics, robotics, artificial intelligence, IoT, and so on, which help grow crops indoors [5, 10]. Vertical farming allows a chance to improve crop product quality and boost agricultural production. Another advantage of vertical farming is enabling the crops to grow in the cities, providing fresh food to the customer, and minimizing transportation environmental impacts [11]. Vertical farming is a method where the plants are stacked vertically in towers [10]. Vertical farming in urban areas can produce around 60% of residents' food consumption [5], and it consumes fewer resources and recycles natural waste [5].

Substantial factors are being used to develop and improve vertical farming. Those factors help reduce the initial cost, decrease the operation cost, varies crop production, and utilizes energy-efficient methods [10]. Vertical farming costs are split into the initial cost to construct the farm and the operating costs to operate the farm [11]. Studies show that large vertical farms have high operating costs due to high electrical consumption, water management, and skilled labor costs [11]. Information technology helps to improve vertical farming. Thus, it makes agricultural production more competitive and beneficial [10]. Various companies such as Nuvege, AeroFarms, Mirai, plant lab, GE, and so on aim to enhance vertical farming procedures.

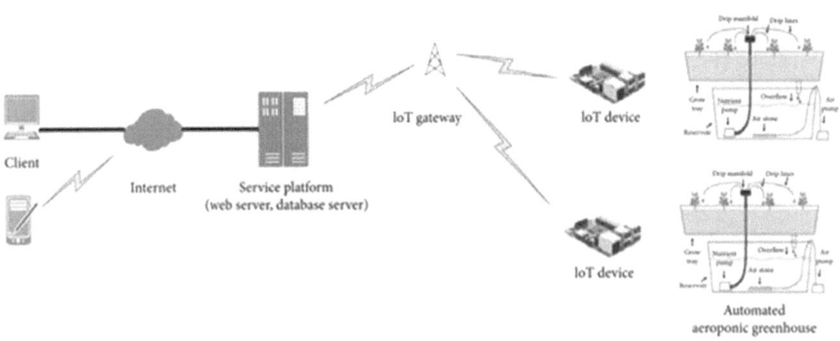

Figure 6.3 Automated smart farming system using IoT devices.
(Source: Ref. [9])

Smart farming ensures proper interaction and interconnection among system components. In smart farming quality is assured through reliable connection and data security [12]. IoT is a technology that shapes the future of communication and information technology [8]. In this chapter, IoT technologies are integrated into intelligent farming (Figure 6.3). IoT technologies develop crop quality and quantity, decrease costs by managing the resources, and ensures crop management through crop monitoring and field monitoring [8, 9].

6.2 VERTICAL FARMING AGRICULTURE TECHNOLOGIES

6.2.1 Hydroponics

The hydroponics farming method is commonly used in vertical indoor farms. This style may include soil or related materials (Figure 6.4): for example, rock material, coconut coir, vermiculite, perlite, sand, and so on. Moreover, some plants may be grown soil-free in nutrient materials [13]. Plants irrigate in liquid and may be assisted with nutrients and minerals. However, plant roots might be grown in a liquid solution. A hydroponics farming system saves around 60–70% of water consumption over conventional outdoor agriculture [10]. Various parameters such as crop size, type, harvesting schedule, irrigation pump, and nutrient soil are set based on programmed software. The major downside of hydroponics is the high initial cost due to the system setup. Another drawback is applying the system to the farm because plants grow in an artificial environment, such as light, nutrients, and monitored water [5, 10].

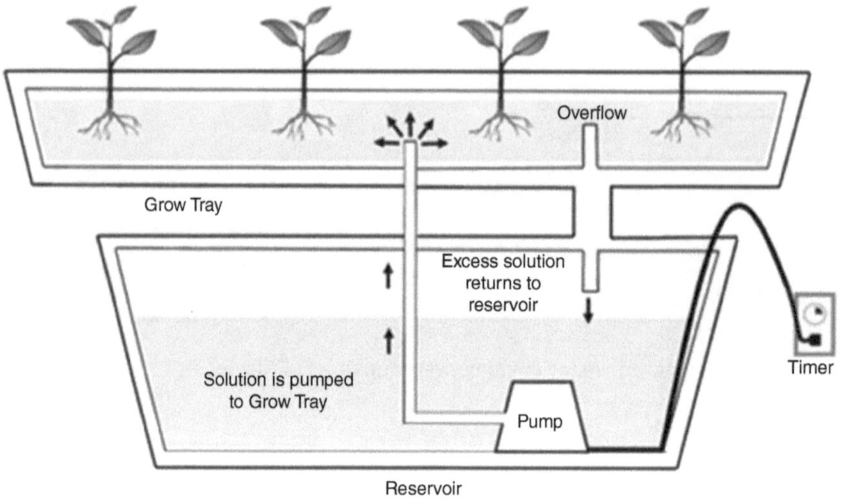

Figure 6.4 Schematic diagram of a hydroponic system.

(Source: Ref. [10])

6.2.2 Aeroponics

Plants hang in the air within a closed container without soil in the aeroponics farming method (Figure 6.5). Continuously, crop roots are sprayed with a fine mist through a spray nozzle. Plants obtain enough resources to grow up by being sprayed with a nutrient solution. The aeroponics farm method consumed 90% less water than hydroponic systems [10]. Moreover, fertilizer consumption decreased by 60% compared to the conventional farming style. In addition, it has been noticed that crops grow 45–75% faster than other farming methods [10]. According to literature studies, aeroponics crops are healthier and rich in vitamins and minerals. Consequently, they boost consumer health [5, 13].

6.2.3 Aquaponics

The aquaponics farming method integrates hydroponic growing with fish production (Figure 6.6). Fish is grown in a closed fish tank. Fish waste is used as a supplement to raise the plants in grow tray. The water rich in ammonia is pumped from the fish tank to the plant grow tray. The grow tray contains nitrifying bacteria that is used to transform ammonia to nitrite. Nitrates are transformed into manure which is used as crop fertilizer. The aquaponics method uses a closed-loop ecosystem, where the water used to manure the plants is recycled back to the fish tank. This method is not commonly used because it is costly and complicated. The aquaponics system

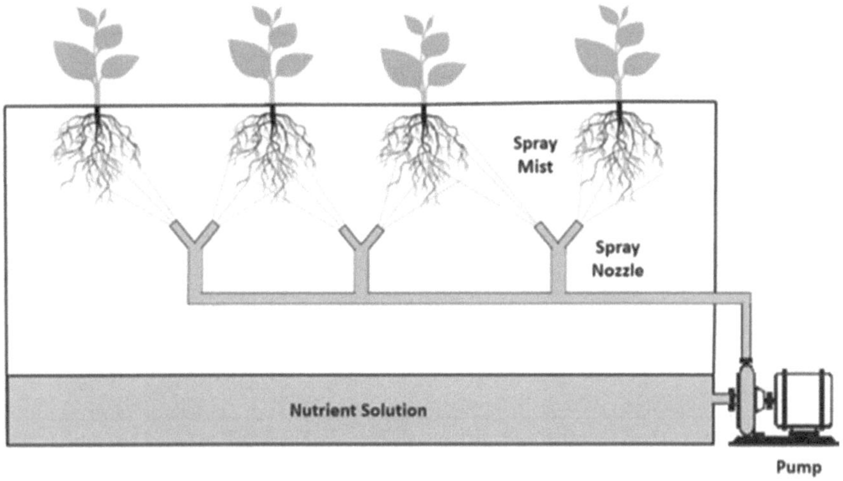

Figure 6.5 Schematic diagram of an aeroponic system.

(Source: Ref. [10])

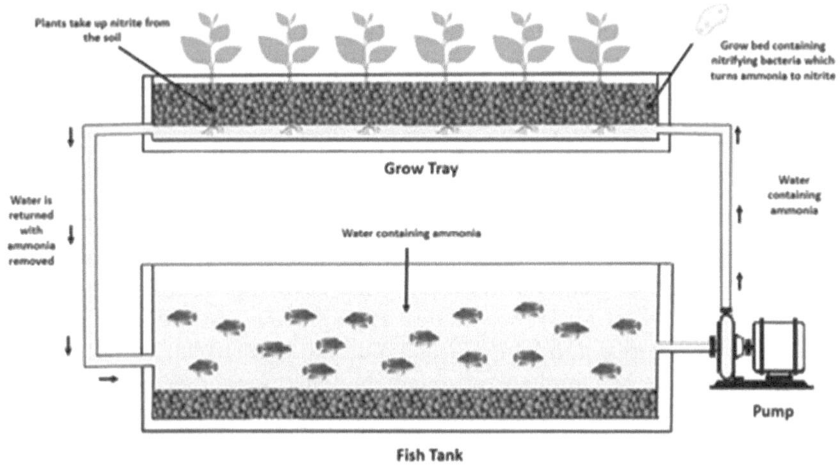

Figure 6.6 Schematic diagram of an aquaponic system.

(Source: Ref. [10])

becomes stable whenever established and monitored to assure both ammonia and pH levels [5, 10].

Table 6.1 shows the various types of vertical farming agriculture technologies farming example [14].

Table 6.1 Vertical farming agriculture technologies

Type/Technology	*Farming example*
Hydroponics	Oh, Chin Huat Hydroponic Farms: Singapore. FertiCorp TM:Vancouver, Canada. Nuvege Plant Factory: Kyoto. Japan.
Aeroponics	AeroFarms: USA. Planned Vertical Farm: Linkoping, Sweden. Ky Greens Farms: Singapore.
Aquaponics	Ecoponics: Iceland. The Plant Vertical Farm: Chicago, USA.
Greenhouses	Green Sense Farms: Indiana USA & Shenzhen, China.
Controlled Environment Agriculture	EuroFarms: Newark, New Jersey.

(Source: [14])

6.3 TYPES OF VERTICAL FARMING

There are three types of vertical systems:

- Despommier Skyscrapers, Mixed-use Skyscrapers, and Stackable Shipping Containers.

6.3.1 Despommier skyscraper

Researchers assumed that conventional farming methods could not feed the world population growth. The crops grow in trays that are stacked vertically. Crops grow in a closed and controlled environment. Thus, they will not be affected by external weather. A Despommier skyscraper farm (Figure 6.7) can be constructed at any place despite any agrarian restriction. This farm consumes less energy and causes less contamination to the environment. Moreover, it may be supplied by renewable energy sources such as wind turbines and photovoltaic cells. Another advantage of a skyscraper farm is the creation of job opportunities in the local community [10].

6.3.2 Mixed-use skyscraper

A mixed-use skyscraper farm integrates vertical farming concepts with conventional agriculture (Figure 6.8). Crops grow on the rooves of apartments and office buildings. Therefore, it uses sunlight as a source of light. This farm is considered a cheap investment compared with the Despommier skyscraper [10].

Figure 6.7 Despommier skyscraper.

(Source: Aerofarms)

Figure 6.8 Mixed-use skyscraper.

(Source: Skyscraper Farm)

6.3.3 Stackable shipping container

Shipping containers are used as small vertical farms to produce leafy crops (Figure 6.9). A shipping container operates as a standalone farm. Leafy green vegetables and tiny fruits like strawberries can be grown using this method. Plants grow in a closed and controlled environment. Shipping containers consist of various farming systems such as ventilation, climate control, irrigation, LED lighting, and so on [10].

Figure 6.9 Stackable shipping container.

(Source: Freight Farms)

6.4 SMART FARMING OPERATION

Smart farming has three pillars [14]:

1. Involves improving urban food security,
2. Enhances the domestic economy's economic capacities, and
3. Decreases carbon emissions due to food transportation.

In theory, smart vertical farming (Figure 6.10) is an agriculture mechanism concerning the massive scale of crop production in skyscraper buildings [9]. This technique can control the crops' environmental conditions and nutrient solutions. Thus, it enables rapid growth and controlled production. Organizations may utilize the unused space at the top of skyscraper buildings to create smart farming. Furthermore, smart rooftop farming can be sustainable by using the waste heat from the high-rise building to warm the greenhouse, while CO_2 emissions produced by residents may be used to assist crop growth [15]. In addition, a considerable ratio of the water demand by the smart farm may be supplied from stormwater that lands on the roof or through greywater treatment systems [15].

Smart farming produces a massive amount of data collected from intelligent devices. Collected data is used to analyze the influential crop growth factors. The collected data from each crop type is used for agriculture simulation [16]. Consequently, specific simulations adjust for each model based on crop type. Moreover, simulation can operate the farm with minimal human interference. Information technology (IT) and automation have an essential role in allowing data collection and assisting decision-making

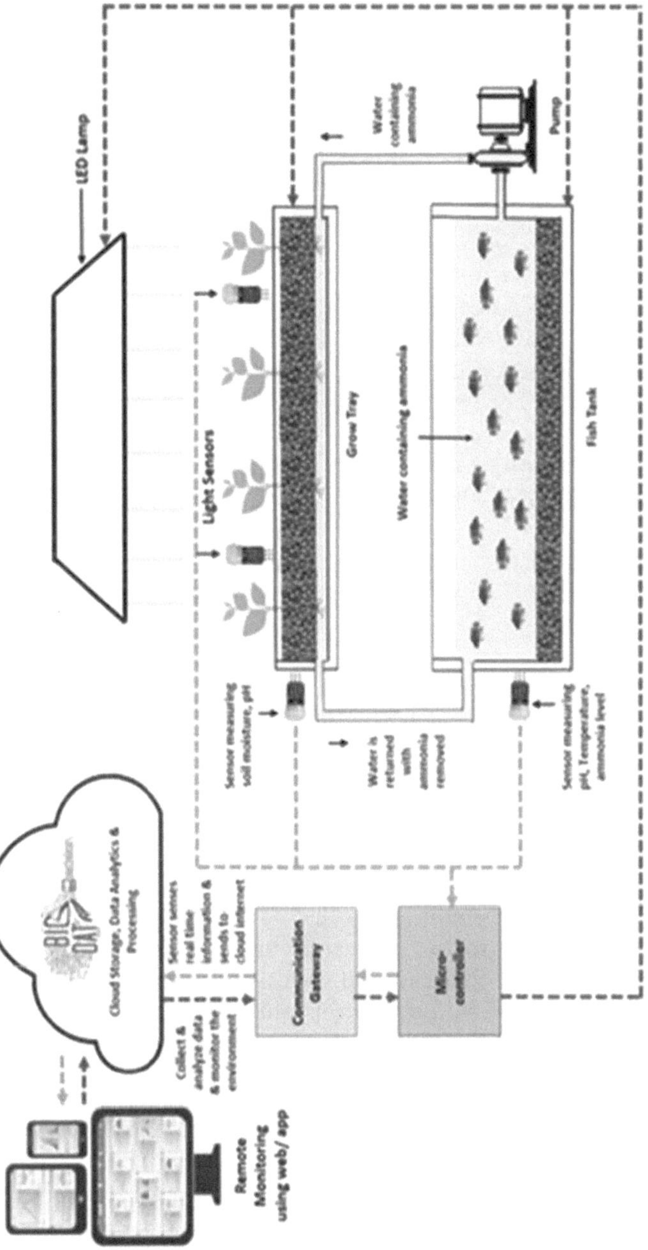

Figure 6.10 Schematic diagram of IoT in vertical smart farming.
(Source: Ref. [10])

processes [15]. IoT is now the smart farming backbone. It gathers data from various smart components to establish a crop simulation model. IoT and data analytics are used to monitor and control the growing environment. Therefore, it can be improved by predictive analysis. Smart sensors and actuators collect data from crop farming climates [10]. Smart farming sensors can evaluate the plant's needs. Therefore, the simulation crop model will change the farming environment to enhance growing factors. Farming environment factors may include artificial light, CO_2, temperature, soil moisture, nutrition level, moisture content, and pH level [17].

Collected data analysis and decision-making processes are taken on an actual-time basis [10]. Video sensors are used to monitor the plant's status to define the required farming environments. Smart farming uses different protocols viz. WiFi, ZigBee, Zwave, NFC, Bluetooth low energy, mobile wireless system (2G/3G/4G), and Wireless HART. The following section shows the data flow in vertical farming [10].

6.5 SMART FARMING LANDSCAPE AND ARCHITECTURE

Smart farming architecture includes four layers:

1. Physical layer,
2. Edge layer
3. Cloud layer and
4. Network Communication layer.

Figure 6.11 shows the Smart farming typical architecture layer.

6.5.1 Physical layer

The physical layer connects multiple Smart IoT components such as sensors, gateways, and actuators to the edge and cloud layers through the internet connection [18]. The physical layer collects data and information from IoT devices to form, process, and control collected data to produce diverse applications and user access [19]. Sensors collect data in real-time about various factors such as internal weather conditions, CO_2 level, crop status, and soil moisture level [20].

6.5.2 Edge layer

The edge computing layer contains various edge nodes. It is close to the end-users and ends with Smart IoT components. The edge layer plays an essential role in decision support systems, computations, data processing in real-time visualization, and monitoring. It decreases the localized cloud

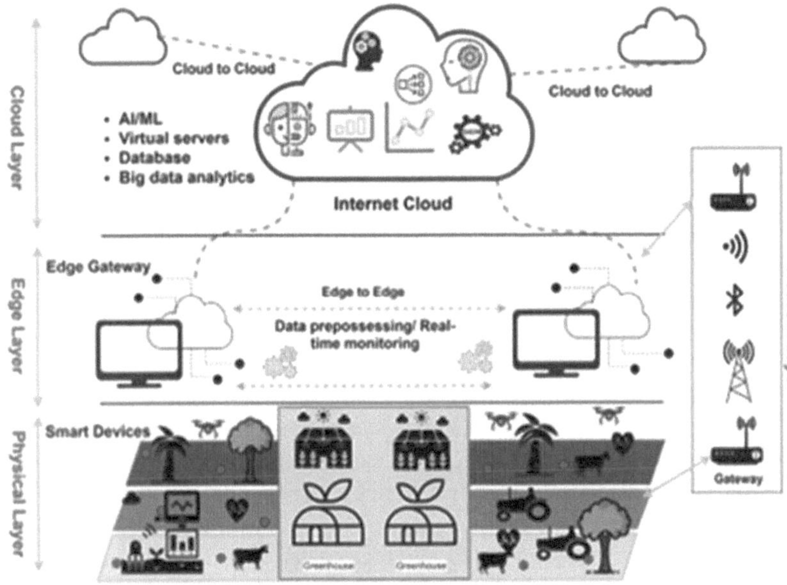

Figure 6.11 Multi-layer smart farming architecture.

(Source: Ref. [18])

layer's computation and network layers' communication load. Every node illustrates a gateway that provides different services such as collecting data, security monitoring, safety detection, and decision support in real-time. Data capturing services consist of data collection, data filtering, coding, and decoding of real-time data streams. Services depend on machine learning models at the edge layer. These are used to predict specific events such as crops production, fertilizer, and water level. Security and monitoring mechanisms can classify smart farming events as harmless or harsh [20].

6.5.3 Cloud layer

Cloud computing enhances Precision Agriculture (PA) connectivity. Cloud layers usually operate in data centers and connect with other layers through the internet connection. Users save smart farms data and run applications at the cloud layer. Collected data from smart farms are saved in the Distributed File System (DFS). Those data are used by farm software to enhance artificial intelligence knowledge [20].

6.5.4 Network communication layer

Connectivity among smart devices becomes a reality through the internet. IoT [21] enables connected appliances to be controlled, monitored, and allows the exchange of data among each other. Exchanged data may be analyzed and used by various applications. The network communication layer allows an interface between the physical, edge, and cloud layers. The network layer has two essential roles in smart farming [20]:

1. The network layer offers a secure and efficient connection. Various network connections such as wireless, wired, and mobile networks can consistently communicate across the layer. Accordingly, it improves access.
2. Maintain connectivity. A network layer is required for cyber-system communication.

Precision Agriculture aims to improve crop efficiency and enhance farmer decision-making. This enables the farmer to manage and control small parts or even grow trays on their farms individually [18]. IoT is considered as one of the leading technologies for smart farming because it can provide remote detection and operation without human interference [19]. On the other hand, IoT technology faces cyber security threats that can control all IoT field devices and related data and applications [22]. Cyber security attacks may have severe consequences on a smart farm.

6.6 SMART FARMING AGRICULTURE COMMUNICATION PROTOCOLS

A communication and networking block is responsible for the data exchange against farming environment uncertainties. Various networking technologies are suitable for smart farming applications (Table 6.2).

The following list defines the most common ones.

- **Wireless PAN/LAN**: is used for data transfer between field sensor devices and base station for short-range. Different communication technologies are used based on communication range [21, 23].
- **ZigBee**: is a wireless technology standard for short-range transmission protocols. It covers around 300 m for outdoor and 75 m for indoor use. This technology is used to control IoT devices and monitor smart sensors. Zigbee uses a low power source. Moreover, sleep mode is available most of the time for power saving. It is commonly used for IoT systems due to open-source standards. Thus, it is easy to implement because they do not have to pay any license fee. It uses mesh topology. Therefore, in case of any communication fault, a backup route may be used to transfer data. Finally, Zigbee wireless technology may be used for long ranges of between 10 and 100 meters by using mediate stations [21, 23].

Table 6.2 Smart farming networking technologies

Communication technology	*Data rate*	*Frequency band*	*Range*
IEEE 802.15.4	20–250 Kbps	2400/915/868 MHz	10 m
IEEE 802.15.4-ZigBee	20–250 Kbps	2400/915/868 MHz	10–100 m
Wi-Fi-IEEE 802.11	450 Mbps	2.4 GHz–5 GHz	100 m
GPRS-2G GSM	64 Kbps	900 MHz–1800 MHz	100 m
3G	14.4 Kbps-2 Mbps	1.6–2 GHz	100 m
4G-LTE	100 Mbps–1 Gps	2–8 GHz	100 m
LoRa	0.3–50 Kbps	433,868,780,915 MHz	2–5 km
Bluetooth LE	1 Mbps	2.4 GHz–2.485 GHz	>100 m
RFID	400 Kbps	125 KHz–915 MHz	3 m

(Source: [21])

- **Long-range-wide area networks (LoRaWAN)**: is a wireless radio network platform with long-range and low-power Wide Area Networks (LPWANs) (Figure 6.12). It covers 3 km in urban areas and up to 40 km in the countryside. It has become the wireless platform of the IoTs. Moreover, LoRaWAN manages communication between LPWAN gateways (base station) and nodes (sensors). It is inexpensive and efficient for energy use for IoT solutions, including smart agriculture. Battery life for sensors connected through LoRaWAN is more than ten years [21, 23].

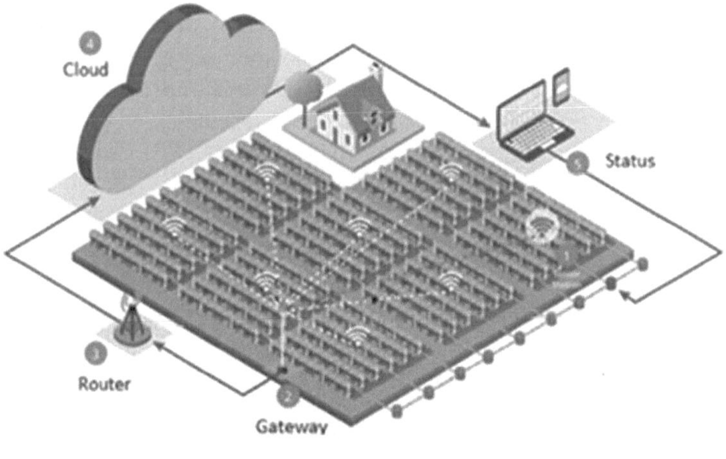

Figure 6.12 LoRaWAN based smart soil monitoring.

(Source: Ref. [23])

Figure 6.13 Radio frequency identification soil monitoring system.
(Source: Ref. [23])

- **RFID:** RFID (Radio Frequency Identification) technology automatically uses electromagnetic fields to identify and track tags stuck to items (Figure 6.13). It is used to monitor the objects in real-time. It includes tags, readers, and a host. Various tags can be tracked and read simultaneously, which helps collect data about whole farms, thus saving time. There are three types of RFID tags:
  1. Active tag
  2. Passive tag, and
  3. Semi-passive tag.

  Semi-passive and active tags are heavier than passive tags since they have power sources. The passive tag uses the power of the RFID reader to transmit data back to the reader since it does not have its own power supply. RFID is contact-free, has memory to store information, has a significant recognition rate, is secure, and can integrate with other systems and standards [21, 23].
- **Bluetooth Technology:** Bluetooth Standard (IEEE 802.15.1) is an open wireless technology standard. Bluetooth is a short-range wireless technology standard used for exchanging small data with low power consumption. It operates in microwave band 2.4GHz by creating personal area networks (PANs). It is suitable wireless technology over a distance range (8–10m). Data transfer speed could be achieved from 1–24Mbps. This technology is easy to use. Moreover, new technology has been issued such as Bluetooth low energy (BLE) technology [21, 23].

Additionally, the following communication technologies are used between the base station and sensor nodes.

- **WiFi** protocol according to IEEE 802.11 standard. It uses media access control (MAC) and physical layer through the local area network (WLAN). Wi-Fi is commonly used in various devices like a Smartphone and laptops. Therefore, it is used frequently in IoT smart farming systems. It operates at 2.4 GHz frequency of 20–100 m with an 11 Mbps data rate.
- **Cellular networks** are used by mobile devices such as tablets and mobile phones.
- **GPRS technology** GPRS (General Packet Radio Service) technology provides fast transmit and receive data over mobile networks relying on packet switching.
- **2nd, 3rd, and 4th generation** (LTE) **of GSM technology** provide higher communication speeds.

The comparison graph (Figure 6.14) shows different technologies, power consumption and coverage range.

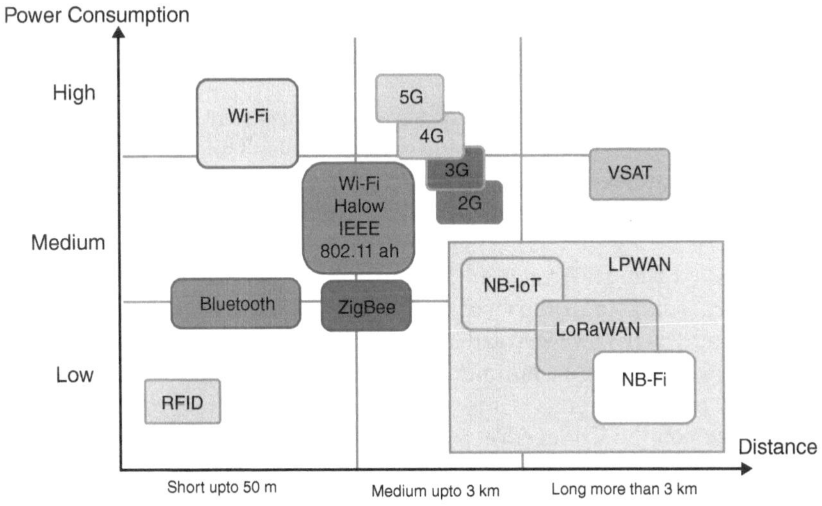

Figure 6.14 Comparison between different technologies.

(Source: Ref. [23])

6.7 CYBER-ATTACKS ON SMART FARMING

The primary security aspects of smart farming are listed as follows:

- **Privacy:** Various research has concentrated on the privacy of smart farming. Privacy is needed to control information access. Thus, keeping unauthorized users away from information access. Privacy attacks such as physical attacks, masquerade attacks, and replay attacks may cause a breach of privacy [24, 25].
- **Integrity:** Data assurance during the storage or transmission from any change or modification. This affect the smart farming operation, and it is a subject of great interest to various researchers [26].
- **Confidentiality:** Data breach to protect data from unauthorized access [25, 27].
- **Availability:** Provide services continuity to ensure smart farming permanent operation [28].
- **Non-Repudiation:** Record users' actions toward the smart farming system. Thus, keeping users from repudiating what has been done in the system [29].
- **Trust:** To ensure user identity and make it impossible to fraud another identity to control smart farming [18].

Cyber-attacks categorize according to the attacked components.

- **Attacks on Hardware:** Unprotected and unknown IoT and hardware devices vulnerabilities by using special tools. This attack may cause jamming attacks to the communication channel and related radiofrequency. Frequency jamming violate confidentiality and privacy [18, 30, 31].
    - **Side Channel Attack:** It aims to collect unauthorized information regarding the smart farming system, such as monitoring details and water and electrical systems [32].
    - **Radio Frequency Jamming Attack:** It occurs at wireless networking communication systems within the smart farming system [31].
- **Attacks on the Connection Network and Related Components:**

Network and attached devices category of attacks target the following:

- **Denial of Service (DoS):** It happens when the system prevents an authorized person from accessing system resources such as servers or communication equipment [33, 34]. This attack may occur due to radiofrequency jamming to the used communication network [35].
- **MITM (Man-In-The-Middle) Attacks:** It happens due to modifying, saving, and replicating the data transmitted among the system. Thus, breach system integrity [35, 36].

- **Botnets:** It is a set of internet components such as sensors used to control the farm. It is used for various purposes such as DoS attacks, information breaches, as a result, this violates system availability, trust, and integrity [18, 37].
- **Cloud Computing Attacks:** Infections could spread quickly into the system throughout the cloud resources. These attacks may target non-repudiation, trust, or integrity [20].
  - **Attacks on related data transmitted, processed, and stored in the smart system.**

  This category can include the following attacks (Figure 6.15) [38]:

- **Data breach:** This includes the transfer of smart Farms data by a source or device to an unauthorized foreign destination. This causes the confidentiality of the smart farm data [20, 36].
- **Ransomware:** Set Codes for smart farm storage data. This led to asking the owner to pay a ransom to access the data. This affects system privacy, integrity, and trust [35, 39, 40].

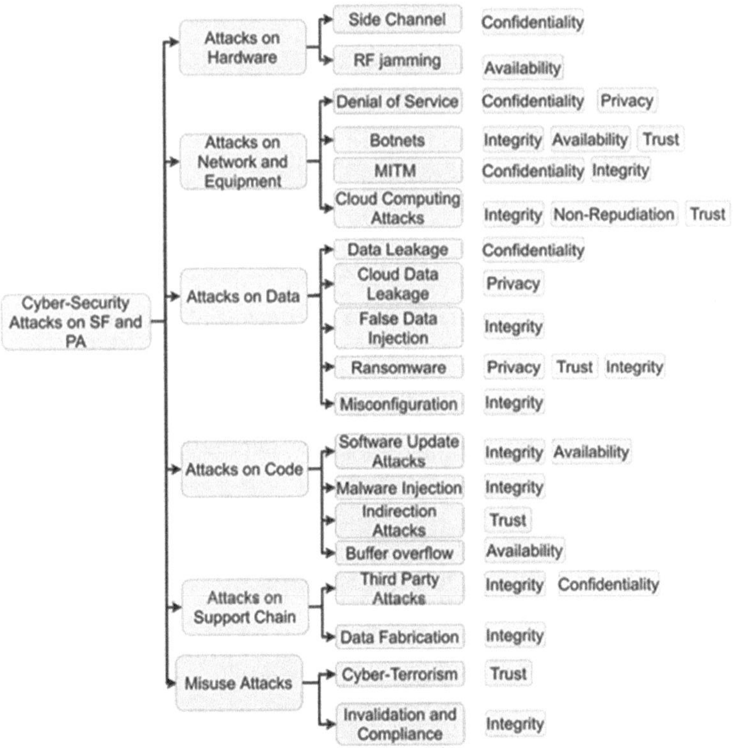

Figure 6.15 Classification of attacks on smart farming.

(Source: Ref. [18])

- **Cloud data leakage:** This includes smart farm data system breach. Thus, breach of system privacy [36, 41].
- **False data injection:** This includes feeding the smart farm system hurtful information or control commands. This attack affects system integrity [20, 42].
- **Misconfiguration:** This provides invalid information about the smart farms' system. This data may cause a wrong response, affecting the plants and increasing the operation cost. Misconfiguration attacks affect system integrity [20].
  - **Attacks on smart farms code** (Figure 6.15):
- **Software update attacks:** Affects system integrity via disturbing the smart farm system software or the operating system update [20].
- **Malware injection:** Infects the system with malicious codes. Thus, affects the system's integrity [20, 35, 43].
- **System buffer overflow:** This is a software coding fault that can cause unauthorized hackers to access the smart farm system data [35, 43].
- **Indirect attacks:** Injects fault codes to affect smart farms' database server. Thus, affects system trust and system operation [20, 44].
  - **Attacks on support chain include smart farms' support system chain components.**
- **Third-party attacks:** This happens when an attack occurs through a partner who has access to the system data. This attack affects system integrity [20, 43].
- **Data fabrication:** This creates fraudulent data or processes. Thus, affects system accessibility and leads to the violation of the system's integrity [45].
  - **Misuse attacks:** This includes mishandling smart farms' physical resources and attacking other institutions.
- **Cyber terrorism:** IoT systems and smart farms physical devices may be used to attack people or other system organizations. Thus, it can lead to abuse of smart system trust [20]
- **Invalidation and compliance:** This will affect system integrity via creating system certification based on false data [20, 46].

6.8 VERTICAL FARMING CHALLENGES

Vertical smart farming faces several challenges which need to be considered (Table 6.3). Thus, it may solve food global shortage.

Table 6.3 Vertical farming challenges

S/N	*Challenges*	*Propose solution*
1	The primary concern is economic efficiency. The start-up cost of setting up smart vertical farms requires enormous investments. Moreover, smart vertical farming operation costs are high due to skilled labor costs and the required energy costs. There is high energy demand due to artificial lighting, heating, and air conditioning [10].	Smart farming startups consider costly businesses. It involves different systems such as sensors, field devices, IoT components, batteries, and data information systems. It is necessary to ensure proper operation. Therefore, those systems shall be maintained regularly by providing the required maintenance by skilled operators and replacing the defective components. All of this adds to the cost of farming. The above shows that smart farming initially required significant capital, which may improve through new technology such as LED technology or enhanced crop type. However, operation costs could be reduced, mainly energy prices, by using solar energy to power the farm. Water usage may be reduced and recycled through a hydroponic system, as well as fertilizer, herbicides, and pesticides can be decreased since it is indoor farming. Another factor that may be considered is that farmers may reduce labor costs and increase crop production with less time [23].
2	Smart vertical farms are not suitable for all crop types. Green leafy vegetables like herbs, chives, and basil are commonly grown considering the cost of production and investment performance [10].	Crop variety in smart farming is selected according to diverse factors such as vertical farming agriculture technologies, soil, and marketing. Crop variety selection method (CVSM) is the algorithm used to select crops. This helps to achieve maximum yield. As a result, the rate yield of each variety will be known which helps to define the best crop to be farmed and gives the highest profit [23].
3	Enhancing employees' skills is another challenge that smart farms face, requiring human building capacity [10].	Smart farming requires staffing with various skill sets. A low level of skills may lead to substantial loss. Manpower should be trained to work with technology, IT, and agriculture. Workforces shall know about multiple systems to operate smart agriculture. Smart agriculture contains repeated activities and due to limited manpower resources thus automation is the solution. Thus, this saves time for critical tasks and actions. Moreover, this can lead to increased profit, efficiency, and productivity [23].

(Continued)

Table 6.3 (Continued)

S/N	*Challenges*	*Propose solution*
4	Smart vertical farms have a particular farming method that cannot be easily modified or changed. As a result, it makes it irrelevant for some crops due to specific environmental requirements. Despite that, any farming method change will have a considerable colossal cost impact [10].	conventional farming. Moreover, the additional operation costs include lighting and heat energy, water, and skilled laborers. Consequently, leafy vegetables and herbs are the most efficient crops at vertical farms. Slower-growing vegetables are not beneficial for vertical farms since they are slower-growing crops. However, innovations and technologies in smart farm infrastructure may reduce operating costs and infrastructure costs, thus, making smart farms more successful. Another term that may be considered is government support to start up the farms, thus making smart farms feasible at scale [23].
6	Vertical farming is a new technology that is not common. Moreover, farms consist of various complicated components such as controllers, sensors, etc., which require proficient maintenance procedures that are very costly and under development [5, 10].	The establishment cost of smart farms is high in comparison to conventional farming. Moreover, there are the additional operation costs such as lighting and heat energy, water, and skilled labors. Consequently, leafy vegetables and herbs are the most efficient crops at vertical farms. Slower-growing vegetables are not beneficial for vertical farms since they are slower-growing crops. "You can't feed the planet with lettuce alone." However, new innovations and technologies in smart farm infrastructure may reduce operation costs and infrastructure costs, thus, making smart farms more successful. Another area to consider is government support to start up a farm thus making smart farms feasible at scale. In addition, the loss of compatibility between IoT sensors and devices due to various venders is considered another reason for smart farms' high costs. The use of IoT has the potential to revolutionize the industry by increasing productivity, reducing costs, and improving sustainability. However, there are also challenges associated with the implementation of IoT, including the cost of devices, the need for reliable internet connectivity in rural areas, and the need for farmers to have the skills and knowledge required to use the technology effectively [23].

8	Risk of cyber security and information theft [47].	Security and privacy are among the main challenges of smart farming, which requires more work. Smart farming uses components from different suppliers with different protocols. Those devices use machine-to-machine (M2M) communication. Security and privacy solutions shall consider compatibility with various vendors and protocols. Security and privacy are developed mainly through TCP/IP networks. However, TCP/IP networks may reduce the communication speed of smart farming components and decrease system efficiency [23].
9	Failure of any IoT components may cause heavy losses due to crop death [48].	Deep learning of machine learning by using various layers from many data sets. Deep learning allows computer software to enhance smart farms' performance. Day by day, machine learning can improve through complex daily activities. Deep learning machines learn how to act like humans to do smart farms activities. Ultimately, machine learning acquires artificial intelligence and thus performs the activity just like a human. This will minimize IoT failure components which prevent smart farms' crops losses [23].
10	Due to lack of soil, the product does not get an organic label, even though it costs as much.	Organic wastes can produce organic manure in several ways in food waste, food leftovers, and plant leaves. Moreover, In this paper, there are three types of farming agriculture technologies [23]. Soil quality degraded because of misuse of fertilizers and using the same type of crops. Soil quality may be improved by recycling the building compost produced locally, chopped leaves, and residual vegetables and fruits. These materials may enhance the quality of the soil. However, recycling nutrients is next to impossible [23].

(*Continued*)

Table 6.3 (Continued)

S/N	*Challenges*	*Propose solution*
11	The challenge of sensor nodes to obtain efficient and continuous operations for an extended period. Moreover, the battery life of sensor nodes is not sufficient [21].	Sensor nodes are commonly supplied with batteries. As a result, those sensors and IoT nodes may turn off in the long run due to dead batteries. As a solution, energy is harvested using a solar panel that may be used to recharge the battery continuously or using a certain setup as standby mode (Low power mode) [23].
12	Communication among smart farm components, e.g. sensors and controllers are manufactured from different suppliers and various different protocols. Thus, it makes communication between them hard [21].	The absence of compatibility of IoT sensors and devices at smart farms due to various vendors. Decreasing IoT devices' complexity, unifying the communication protocol, and improving data transfer between smart farms components will increase smart farm efficiency [23].
13	The limited memory of the sensor nodes can deactivate a large amount of collected data. As a result, there can be delayed data transmission between smart farms components [21].	In smart farming, monitoring system routing protocols must operate with a minimum delay to provide adequate services in system sensor nodes and consider information security and privacy. The network system performance functions according to network size and communication protocol used. Therefore, new technologies of transmission protocols will enhance communication network performance [23].

6.9 CONCLUSION/SUMMARY

Global warming, weather changes, and the current pressure on natural resources make traditional farming system unsecured. IoT is a magnificent mechanism that make our lives easier. Smart agriculture is not applicable for all types of crops. In addition, it requires high investment costs. However, smart agriculture is one of the best sustainable farming methods and enhances global food security. Smart agriculture involves various systems that require integration. Thus, it helps to grow indoor crops in a simulated environment. Data analysis and IT technologies play an essential role in simulating master models for each type of plant. Smart farms use various technologies and computing which require cyber security protection. Thus, it can sustain cyber-attacks. Cyber-attacks can cause critical interruptions to the farming process thus leading to massive financial impact and a lack of supply chain. The IoT agricultural applications enable farmers to collect and analyze meaningful data.

NOMENCLATURE

Abbreviation	Description
SOS	System of system
PoE	Power over ethernet
CPS	Cyber-physical system
BAN	Body area network
LEDs	Light-emitting diodes
IT	Information technology
IoT	Internet of Things
LEED	Leadership in Energy and Environmental Design
NFC	Near-field communication
SOS	system of system
PoE	Power over ethernet
CPS	Cyber-Physical System
BAN	body area network
LEDs	light-emitting diodes

REFERENCES

[1] P. D. Mullins, "The ubiquitous-eco-city of Songdo: An urban systems perspective on South Korea's green city approach", *Urban Planning*, vol. 2, no. 2, pp. 4–12, 2017.

[2] S. Heitlinger, N. Bryan-Kinns, and R. Comber, "The right to the sustainable smart city", in *2019 CHI Conference on Human Factors in Computing Systems*, pp. 1–13, 2019.

[3] A. Shah, "Sustainable consumption and recycling practices", in *Sustainable Production and Consumption Systems: Springer*, pp. 191–204, 2021.
[4] C. Gnauer et al., "Towards a secure and self-adapting smart indoor farming framework", *e & i Elektrotechnik und Informationstechnik*, vol. 136, no. 7, pp. 341–344, 2019.
[5] D. Despommier, "Vertical farms, building a viable indoor farming model for cities", *Field Actions Science Reports* [Online], Special Issue 20, pp. 68–73, 2019.
[6] K. Benke and B. Tomkins, "Future food-production systems: Vertical farming and controlled-environment agriculture", *Sustainability: Science, Practice and Policy*, vol. 13, no. 1, pp. 13–26, 2017.
[7] S. Sivamani, N. Bae, and Y. Cho, "A smart service model based on ubiquitous sensor networks using vertical farm ontology", *International Journal of Distributed Sensor Networks*, vol. 9, no. 12, pp. 1–8, 2013.
[8] R. Dagar, S. Som, and S. K. Khatri, "Smart farming–IoT in agriculture", in *2018 International Conference on Inventive Research in Computing Applications (ICIRCA)*, pp. 1052–1056: IEEE, 2018.
[9] I. A. Lakhiar, G. Jianmin, T. N. Syed, F. A. Chandio, N. A. Buttar, and W. A. Qureshi, "Monitoring and control systems in agriculture using intelligent sensor techniques: A review of the aeroponic system", *Journal of Sensors*, vol. 2018, Article ID 8672769, 2018.
[10] M. K. Gupta and S. Ganapuram, "Vertical farming using information and communication technologies", White Paper, InfoSys, 2019.
[11] R. J. F. Burton, J. Forney, P. Stock, and L. Sutherland, *The Good Farmer: Culture and Identity in Food and Agriculture*, Routledge, 2021.
[12] I. Marcu, G. Suciu, C. Bălăceanu, A. Vulpe, and A. Drăgulinescu, "Arrowhead technology for digitalization and automation solution: Smart cities and smart agriculture", *Sensors*, vol. 20, no. 5, 2020.
[13] J. Vydra, "More food, less earth", https://m.youtube.com/watch?feature=youtu.be&v=nM1L_9djWSY, 2017.
[14] F. Kalantari, O. M. Tahir, R. A. Joni, and E. Fatemi, "Opportunities and challenges in sustainability of vertical farming: A review", *Sciendo*, vol. 11, no. 1, pp. 35–60, 2018.
[15] J. Wood, C. Wong, and S. Paturi, "Vertical farming: An assessment of Singapore City," *eTropic: Electronic Journal of Studies in the Tropics*, vol. 19, pp. 228–248, 2020.
[16] J. Birkby, "Vertical farming", *ATTRA Sustainable Agriculture* pp. 1–12, 2016.
[17] I. Haris, A. Fasching, L. Punzenberger, and R. Grosu, "CPS/IoT ecosystem: Indoor vertical farming system", in *2019 IEEE 23rd International Symposium on Consumer Technologies (ISCT)*, IEEE, pp. 47–52, 2019.
[18] A. Yazdinejad et al., "A review on security of smart farming and precision agriculture: Security aspects, attacks, threats and countermeasures", *Applied Sciences*, vol. 11, no. 16, 2021.
[19] C. Kamienski et al., "Smart water management platform: IoT-based precision irrigation for agriculture", *Sensors*, vol. 19, no. 2, p. 276, 2019.
[20] M. Gupta, M. Abdelsalam, S. Khorsandroo, and S. Mittal, "Security and privacy in smart farming: Challenges and opportunities", *IEEE Access*, vol. 8, pp. 34564–34584, 2020.

[21] A. Triantafyllou, P. Sarigiannidis, and S. Bibi, "Precision agriculture: A remote sensing monitoring system architecture", *Information*, vol. 10, no. 11, p. 348, 2019.
[22] K. Demestichas, N. Peppes, and T. Alexakis, "Survey on security threats in agricultural IoT and smart farming", *Sensors*, vol. 20, no. 22, p. 6458, 2020.
[23] P. Tomar and G. Kaur, *Artificial Intelligence and IoT-based Technologies for Sustainable Farming and Smart Agriculture*, IGI Global, 2021.
[24] G. Idoje, T. Dagiuklas, and M. Iqbal, "Survey for smart farming technologies: Challenges and issues", *Computers & Electrical Engineering*, vol. 92, p. 107104, 2021.
[25] A. F. Ametepe, S. A. R. M. Ahouandjinou, and E. C. Ezin, "Secure encryption by combining asymmetric and symmetric cryptographic method for data collection WSN in smart agriculture", in *2019 IEEE International Smart Cities Conference (ISC2)*, IEEE, pp. 93–99, 2019.
[26] R. Chamarajnagar and A. Ashok, "Integrity threat identification for distributed IoT in precision agriculture", in *2019 16th Annual IEEE International Conference on Sensing, Communication, and Networking (SECON)*, IEEE, 2019.
[27] N. Zhang, R. Wu, S. Yuan, C. Yuan, and D. Chen, "RAV: Relay aided vectorized secure transmission in physical layer security for Internet of Things under active attacks", *IEEE Internet of Things Journal*, vol. 6, no. 5, pp. 8496–8506, 2019.
[28] F. Bisogni, S. Cavallini, and S. D. Trocchio, "Cybersecurity at European level: The role of information availability", *Communications and Stategies*, no. 81, pp. 105–124, 2011.
[29] A. Nesarani, R. Ramar, S. Pandian, and Innovation, "An efficient approach for rice prediction from authenticated Block chain node using machine learning technique", *Environmental Technology & Innovation*, vol. 20, p. 101064, 2020.
[30] M. Xue, C. Gu, W. Liu, S. Yu, and M. O'Neill, "Ten years of hardware Trojans: A survey from the attacker's perspective", *IET Computers & Digital Techniques*, vol. 14, no. 6, pp. 231–246, 2020.
[31] A. N. Jahromi, H. Karimipour, A. Dehghantanha, and K.R. Choo, "Toward detection and attribution of cyber-attacks in IoT-enabled cyber-physical systems", *IEEE Internet of Things Journal*, vol. 8, no. 17, pp. 13712–13722, 2021.
[32] A. Ukil, J. Sen, and S. Koilakonda, "Embedded security for Internet of Things", in *2011 2nd National Conference on Emerging Trends and Applications in Computer Science*, IEEE, pp. 1–6, 2011.
[33] C. Kolias, G. Kambourakis, A. Stavrou, and J. Voas, "DDoS in the IoT: Mirai and other botnets", *Computer*, vol. 50, no. 7, pp. 80–84, 2017.
[34] M. Antonakakis et al., "Understanding the mirai botnet", in *26th {USENIX} security symposium*, pp. 1093–1110, 2017.
[35] D. Ye and T. Zhang, "Summation detector for false data-injection attack in cyber-physical systems", *IEEE Transactions on Cybernetics*, vol. 50, no. 6, pp. 2338–2345, 2020.
[36] B. Purohit and P. P. Singh, "Data leakage analysis on cloud computing", *International Journal of Engineering Research and Applications (IJERA)*, vol. 3, no. 3, pp. 1311–1316, 2013

[37] A. Al Shorman, H. Faris, and I. Aljarah, "Unsupervised intelligent system based on one class support vector machine and Grey Wolf optimization for IoT botnet detection", *Journal of Ambient Intelligence and Humanized Computing*, vol. 11, no. 7, pp. 2809–2825, 2020.
[38] S. M. H. Fard, H. Karimimpour, A. Dehghantanha, A. N. Jahromi, and G. Srivastava, "Ensemble sparse representation-based cyber threat hunting for security of smart cities", *Computers & Electrical Engineering*, vol. 88, p. 106825, 2020.
[39] E. M. Dovom, A. Azmoodeh, A. Dehghantanha, D. E. Newton, R. M. Parizi, and H. Karimipour, "Fuzzy pattern tree for edge malware detection and categorization in IoT", *Journal of Systems Architecture*, vol. 97, pp. 1–7, 2019.
[40] P. N. Bahrami, A. Dehghantanha, T. Dargahi, R. M. Parizi, K.-K. R. Choo, and H. H. Javadi, "Cyber kill chain-based taxonomy of advanced persistent threat actors: Analogy of tactics, techniques, and procedures", *Journal of Information Processing Systems*, vol. 15, no. 4, pp. 865–889, 2019.
[41] C. Priebe et al., "Cloudsafetynet: Detecting data leakage between cloud tenants", in *Proceedings of the 6th edition of the ACM Workshop on Cloud Computing Security*, pp. 117–128, 2014.
[42] A. S. Musleh, G. Chen, and Z. Y. Dong, "A survey on the detection algorithms for false data injection attacks in smart grids", *IEEE Transactions on Smart Grid*, vol. 11, no. 3, pp. 2218–2234, 2020.
[43] N. Gruschka and M. Jensen, "Attack surfaces: A taxonomy for attacks on cloud services", in *2010 IEEE 3rd International Conference on Cloud Computing*, pp. 276–279: IEEE, 2010.
[44] I. Stoica, D. Adkins, S. Zhuang, S. Shenker, and S. Surana, "Internet indirection infrastructure", *IEEE/ACM Transactions on Networking*, vol. 12, no. 2, pp. 205–218, 2004.
[45] S. Khan, M. A. Bagiwa, A. W. A. Wahab, A. Gani, and A. Abdelaziz, "Understanding link fabrication attack in software defined network using formal methods", in *2020 IEEE International Conference on Informatics, IoT, and Enabling Technologies (ICIoT)*, IEEE, pp. 555–562, 2020.
[46] B. E. Cuker, *Livestock and Poultry: Other Colonists Who Changed the Food System of the Chesapeake Bay, Diet for a Sustainable Ecosystem*, Springer Link, pp. 219–244, 2020.
[47] B. Ozdogan, A. Gacar, and H. Aktas, "Digital agriculture practices in the context of agriculture 4.0", *Journal of Economics, Finance and Accounting*, vol. 4, no. 2, pp. 184–191, 2017.
[48] A. Y. Ding and M. Janssen, "Opportunities for applications using 5G networks: Requirements, challenges, and outlook", *Proceedings of the Seventh International Conference on Telecommunications and Remote Sensing*, pp. 27–34, 2018.

Chapter 7

Deep learning-based autonomous vehicle to vehicle detection for smart traffic monitoring in smart cities

Mohammad Amir, Ahteshamul Haque, and Zaheeruddin
Jamia Millia Islamia, New Delhi, India

7.1 INTRODUCTION

Earlier, the detection of vehicular type was realized by the histogram of oriented gradient (HOG) techniques [1–3]. Later, the classification of detection features using intelligence-based techniques such as support vector machine (SVM) accordingly achieved significant vehicular identification. Recent study [4] demonstrated the deformable part model (DPM) for recognition, which has achieved better vehicle identification than the SVM model.

This chapter proposes a deep learning-based approach to solve the urban traffic issues and compensate for the existing transportation security issues, such as lack of vehicle identification and lower accuracy of vehicular recognition speed. Today's vehicle identification is an indispensable research trend that is broadly analyzed by various researchers all over the globe. In the current scenario, vehicle identification methods are divided into two categories, such as the traditional type (refer to machine-learning approach) and the advanced machine-learning approach (refer to deep learning-learning approach). The traditional technique referred to as a complex learning algorithm provides a composite approach for vehicle positioning accuracy and recognition of their type. Such techniques involve complex actions, such that those techniques need excessive human engagement, higher operational time, and more cost. Therefore, based on accuracy evaluation, traditional techniques are not much more suitable than the deep learning-learning approach for real-time vehicle detection.

7.1.1 Key challenges and recent developments

Nowadays deep learning-based [5] object detection has become a more popular research field for traffic monitoring and control. Vehicle detection using various sensors [6–8] typically demonstrates a better performance than the traditional approach. To achieve better vehicle identification features, existing research [9–11] demonstrates a CNN-based significant vehicle identification approach. These approaches did not need any human involvement for feature extraction, although many tagged vehicular pictures are required for training the network with a supervised approach automatically.

DOI: 10.1201/9781032669809-7

Malik, Haque, and Amir [12] proposed a vehicular classification using the softmax tool and based on an unsupervised learning (with spare coding) network for training purposes. The first object detection model [13] was the Region-Based Convolutional Neural Networks (R-CNN). That algorithm utilized appropriate research to generate a regional-based interest that created an object identification approach using deep learning, as employed in SPPN [14], RFCN [15], and faster R-CNN [16]. Konoplich, Putin, and Filchenkov [17] employed a hybrid neuro-fuzzy approach that divides the two-region proposal-based sub-networks for vehicle identification. The simulation results demonstrated that the network accuracy was much enhanced, and the computational approach was superior greatly to R-CNN. Moreover, MS-CNN is employed for the detection of frames at a per second rate. Azam, Rafique, and Jeon and Tang et al. [18, 19] applied the faster R-CNN-based method to vehicle detection and achieved good detection. In He et al. [20] a comparative analysis demonstrated using FRCNN and ResNet-152 for vehicular identification that attained better vehicular identification accuracy. However, the detection speed was much slower and could not meet the constraints for vehicular identification in real-time scenarios. Normally, the vehicle detection speed using the deep learning technique is slower and cannot meet the appropriate detection accuracy with better generation capabilities. Therefore, to enhance the detection accuracy and speed, transform direct object identification to an end-to-end object identification method using the YOLO technique [20]. Later in 2017, Redmon et al. [21] demonstrated the YOLOv2 based object identification tool, which significantly enhanced the object detection speed as well as accuracy.

7.1.2 Key contributions

This research focuses on the designing of intelligent vehicular detection framework platforms that are an integral part of the Intelligent Transportation System (ITS) [22]. The key components of ITS have brought lots of detection challenges and opportunities with advanced machine-learning compatibility. Major components of STS are classified as:

- Automatic self-driving system.
- Real-time vehicle to vehicle detection for traffic and accident control.
- Advanced Vehicle Control and Security System (AVCSS).
- Platooning of vehicles, where each vehicle in the platoon communicates detailed diagnostic messages, such as nearby vehicles over speed. These messages are beneficial for V2V telematics information.

To enhance the YOLO-based detection accuracy, Sang et al. [23] demonstrated an upgraded version of YOLOv2. A novel framework Darknet-19 shows by eliminating the maximum interconnection layers, YOLOv2 is

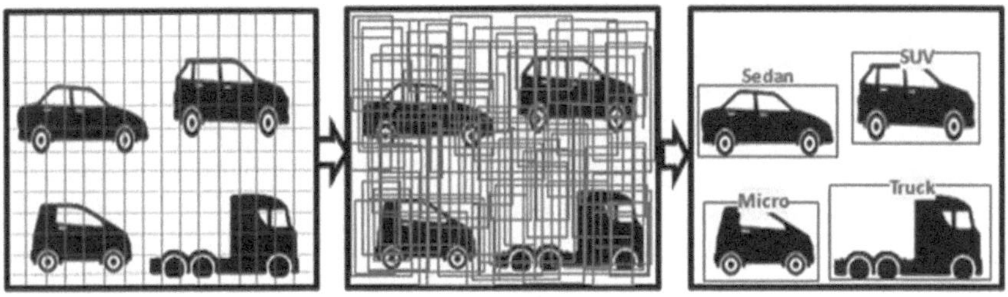

Figure 7.1 Flow diagram of V2V detection.

around 1000 times faster than R-CNN. The batch normalization is employed in every layer [24]. The estimated boxes are retrained with direct estimation.

Based on the comparison of YOLO, the proposed YOLOv2 has significantly enhances the speed of object identification. It can detect various classes using the YOLOv2 technique. This chapter demonstrates the YOLOv2-based vehicle identification technique for better vehicle detection and has acquired better performance on the speed response. The proposed YOLOv2 approach fits the different vehicle dataset, where many varieties of vehicular class are to be identified such as SUV, sedan, truck, micro cars, and so on. But normally the main differences for vehicle identification are in local configurations, such as headlights, tires, and so on, as shown in Figure 7.1. This chapter suggests an enhanced YOLOv2-based vehicle identification technique for better vehicle detection and acquired better performance on the speed response.

7.2 FEATURED V2V DETECTION FOR INTELLIGENT TRANSPORTATION

7.2.1 Car model classification and transfer learning

In most practical cases, vehicular computation systems experience a loss of their generalization capability over time or when dealing with large datasets. This V2V recognition can be implemented using deep learning-based computation for better learning accuracy. To estimate and predict the accurate V2V identification Residual Neutral Network (RNN) can be used. The RNN has better function approximations because they can easily identify the function in a consecutive manner (see Equation 7.1). So the output of each function becomes the input of the next function.

$$f(x) = x \tag{7.1}$$

$$g(x) = f(x) + x \tag{7.2}$$

When the RNN has too much data then the problem of vanishing gradient arises. This problem can be rectified by using the chain rule.

The Featured Visual Geometry Group (VGG-19) is an advanced convolutional-based neural network, that has 19 deep layers. The VGG-19 is an enhanced version of a trained network on more than a million pictures dataset. Figure 7.2 demonstrates the example of the "VGG-19" [24] framework, having 19.6 billion FLOPs. On the other hand, "34-layer residue" and

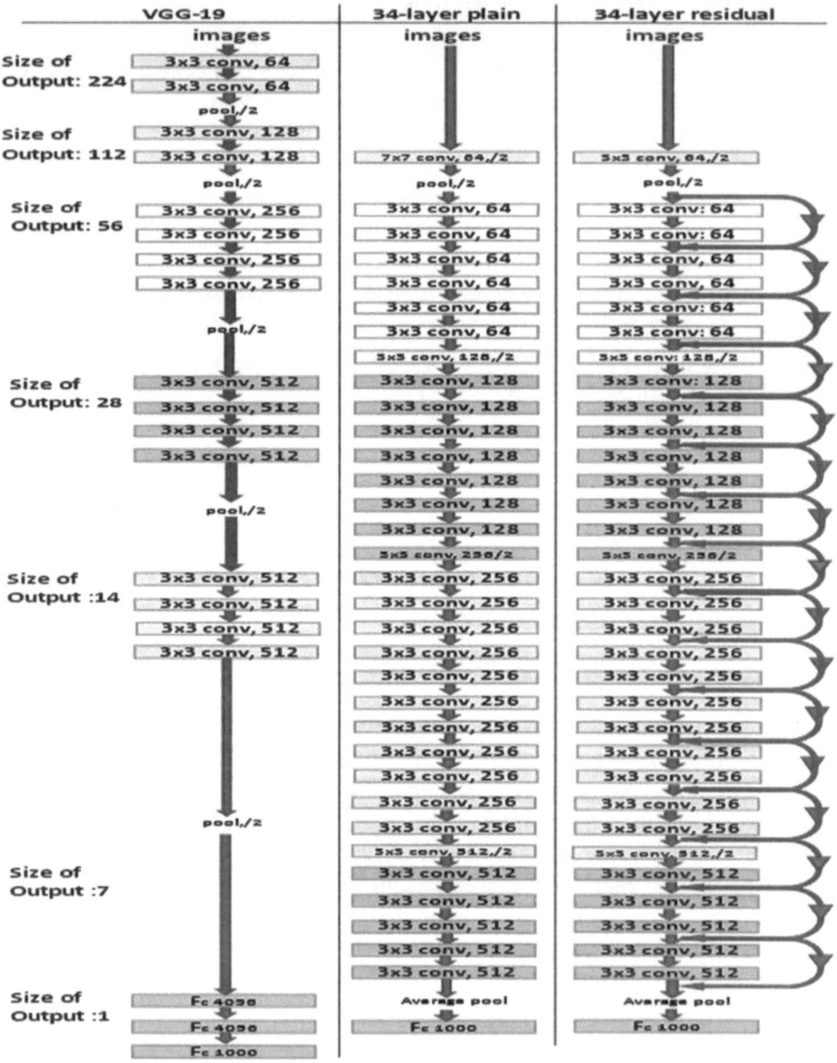

Figure 7.2 Framework of YOLOv2 for feature extractor using ResNet layer architecture.

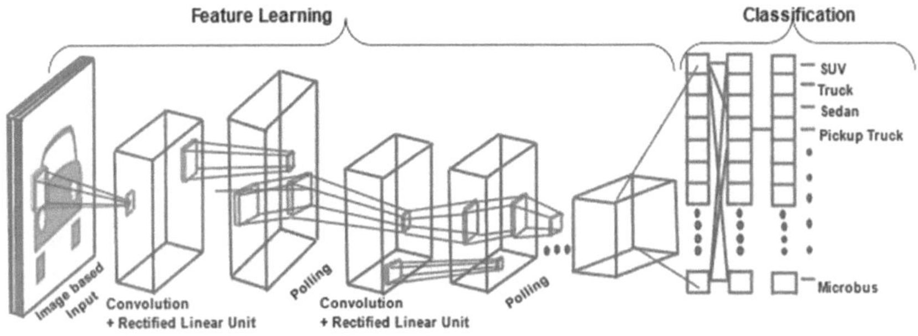

Figure 7.3 Converting the vehicle image data set to real-time visualization for vehicle classification using deep learning.

"34-layer plain" have 3.6 billion FLOPs. The ResNets tool has variable sizes that depend on the actual size of each layer of the proposed model [25]. In this chapter, we are designing the ResNet model that comprises the convolution as well as the pooling step (followed by four layers of similar behavior). Every layer supports a similar pattern. They perform convolution (3×3) with a fixed framework of 64, 128, 256, and 512 to prevents the input variables in adjacent next convolution layers. During the entire layer, dimensions of width and height remain constant. The red arrow line shows the change in the reduction of convolution. At the 1st convolution of every layer, the reduction in between the layers is accomplished by an increment of stride from 1 to 2 in its place by using a pooling operation. Further, that can be implemented to get the down samplers.

Figure 7.3 illustrates the conversion of a vehicle image dataset into a feature classification dataset from the image dataset. Here, we have taken the Kaggle data set [12]. The data set contains 21,495 images of 240 classes of cars. This data set is split into 12,350 training images and 9,145 testing images. The ResNet-based V2V recognition is carried out in a parallel manner and that represents the information that was received during the training period. Adding one more layer in RNN will form the convolutional layer. Additionally, we got 240 classes from 4-wheeled cars as the total outputs. Thus, changing the desired output layer up to 240.

7.3 MATHEMATICAL MODELING OF PROPOSED VEHICULAR DETECTION

7.3.1 Anchor box selection

In 2016, Redmon et al. suggested the end-to-end YOLO-based object detection method [26]. In Equation 7.3, Prob(vehicle) signifies the estimation of the vehicle type that depends on the related framework cell as intersection over union (IOU). It also demonstrates the estimated bounding boxes; subsequently, these bounding boxes with minimal object confidence in the given threshold limit must be eliminated [27].

$$\text{Confidance} = \text{Prob}(\text{vehicle}) \times \text{IOU}_{\text{Estim}}^{\text{Real}} \tag{7.3}$$

After data collection, we conducted *k-means++*-based cluster analysis for the actual size of vehicular bounding boxes. For vehicular detection selection of anchor, boxes must be performed. In this chapter, YOLOv2-based distance function over Euclidean-based distance function for effective k-means++, this distance function is referred to in Equation 7.4.

$$d(\text{box}, \text{centroid}) = 1 - [\text{IOU}.(\text{box}, \text{centroid})] \tag{7.4}$$

The IOU is used as an estimation metric to determine the error in selecting the actual sizes of related anchor boxes. Figure 7.4 analyzes the cluster-based results, where the value of k is set to 8 (that means eight different size anchor boxes can be employed for identification of position). Figure 7.4 shows the clustering anchor boxes (certain clustering anchor boxes were rectangular or maybe square) with different shapes. Each shape matched the existing shapes of the individual eight vehicle types, although this information incorporated the distance from the camera. So, based on clustering analysis using the training dataset, the proposed sizes of the anchor boxes are certainly capable of vehicle detection and further enhance the positioning accuracy.

7.3.2 Loss function improvement

For vehicular identification, vehicle images can be taken from different road surveillance cameras, which means that during identification the camera approached the vehicle. If the vehicle is far from the road surveillance

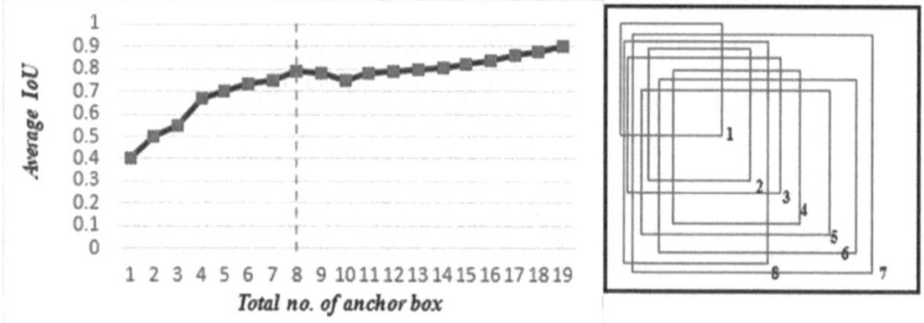

Figure 7.4 Analyze of clustering-based results for the anchor boxes information.

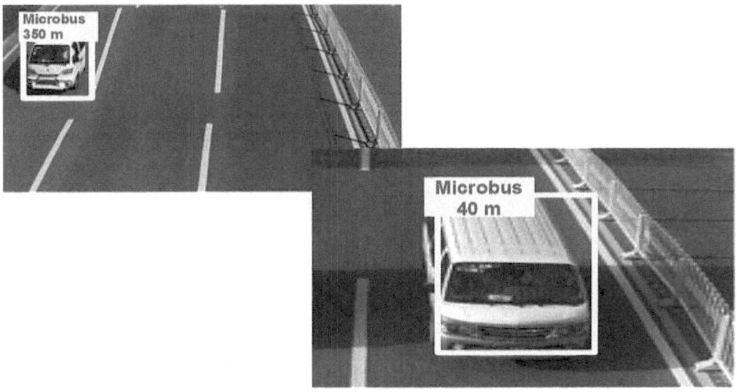

Figure 7.5 Comparison of the same vehicle with different distance.

camera, it seems like the smaller picture as demonstrated in Figure 7.5. When a vehicle nears the road surveillance camera, then it gets a larger picture area. Thus, even when the car type is the same, the size of the image may be different.

During training in the YOLOv2-based model, various vehicle sizes can have distinct effects on the entire model, which result in more errors for a high-sized vehicle than for a small-sized vehicle. To reduce loss functions, we can improve the dimensions of bounding boxes with the help of normalization. The improvement of the loss function is demonstrated in Equation 7.10.

Calculation of coordinate losses

$$= \lambda_{\text{coordinate}} \sum_{i=0}^{G^2} \sum_{j=0}^{V} \prod_{ij}^{\text{vehicle}} \left[(x_i - \hat{x}_i)^2 + (y_i - \hat{y}_i)^2 \right] \tag{7.5}$$

Calculation of box size losses

$$= \lambda_{\text{coordinate}} \sum_{i=0}^{G^2} \sum_{j=0}^{V} \prod_{ij}^{\text{vehicle}} \left[\left(\frac{w_i - \hat{w}_i}{w_i} \right)^2 + \left(\frac{h_i - \hat{h}_i}{h_i} \right)^2 \right] \tag{7.6}$$

Equation 7.6 compared with YoLOv2, further used $\left(\frac{w_i - \hat{w}_i}{w_i} \right)$ and $\left(\frac{h_i - \hat{h}_i}{h_i} \right)$ in place of $(w_i - \hat{w}_i)$ and $\left(h_i - \hat{h}_i\right)$ and because it reduces the effect of various sizes of same vehicle type in a picture. Calculation of bounding box-based confidence losses with vehicles is as follows:

$$= \lambda_{\text{coordinate}} \sum_{i=0}^{G^2} \sum_{j=0}^{V} \prod_{ij}^{\text{vehicle}} \left(\text{Con}_i - \widehat{\text{Con}}_i \right)^2 \tag{7.7}$$

Whereas calculation of bounding box-based confidence losses without vehicles is shown as:

$$\sum_{i=0}^{G^2}\sum_{j=0}^{V}\prod_{ij}^{\text{vehicle}}\left(\text{Con}-\widehat{\text{Con}_i}\right)^2 \tag{7.8}$$

$$\text{Calculation of class losses} = \sum_{i=0}^{G^2}\prod_{ij}^{\text{novehicle}}\sum_{c\in\text{Classes}}\left(C(P_i)-C(P_i)\right)^2 \tag{7.9}$$

$$\begin{aligned}&\lambda_{\text{coordinate}}\sum_{i=0}^{G^2}\sum_{j=0}^{V}\prod_{ij}^{\text{vehicle}}\left[\left(x_i-\hat{x}_i\right)^2+\left(y_i-\hat{y}_i\right)^2\right]\\&+\lambda_{\text{coordinate}}\sum_{i=0}^{G^2}\sum_{j=0}^{V}\prod_{ij}^{\text{vehicle}}\left[\left(\frac{w_i-\hat{w}_i}{w_i}\right)^2+\left(\frac{h_i-\hat{h}_i}{h_i}\right)^2\right]+\sum_{i=0}^{G^2}\sum_{j=0}^{V}\prod_{ij}^{\text{vehicle}}\left(\text{Con}_i-\widehat{\text{Con}_i}\right)^2\\&+\lambda_{\text{novehicle}}\sum_{i=0}^{G^2}\sum_{j=0}^{V}\prod_{ij}^{\text{novehicle}}\left(\text{Con}-\widehat{\text{Con}_i}\right)^2+\sum_{i=0}^{G^2}\prod_{ij}^{\text{novehicle}}\sum_{c\in\text{Classes}}\left(C(P_i)-C(P_i)\right)^2\end{aligned} \tag{7.10}$$

For ith grid cell, the dimensions are w_i (width) and h_i (hight). The center coordinates are represented by x_i and y_i. The class probability is $C(P_i)$ and confidence is Con for ith grid cell. The estimated values are $\widehat{x_i}, \widehat{y_i}, \widehat{w_i}, \widehat{h_i}, \widehat{C(P_i)}$, and $\widehat{Con_i}$ corresponding to the x_i, y_i, w, h_i, $C(P_i)$, and Con_i. The weight of bounding boxes without vehicle loss ($\lambda_{\text{novehicle}}$) and weight of coordinate loss ($\lambda_{\text{coordinate}}$) cells. The G^2 represented the GXG grid. V represents the vehicles, while $\prod_{i}^{\text{vehicle}}$ vehicle is located in ith cell and $\prod_{ij}^{\text{vehicle}}$ vehicle is located in jth estimator box in ith cell.

7.3.3 Multi-layer fusion and elimination of repeated higher ordered convolution layers

For vehicle identification, there are certain differences among the same vehicles as such differences may be in vehicle color, headlight, shape of tires, and so on. To get helpful vehicle information, we implemented a multi-layer fusion operation. In this chapter, the CNN model was employed. In part (i) we conduct the operation through 3×3 and 1×1 convolution layers and further employ down sampling based on Reorge/4. While in part (ii) we conducted the same operation but down sampling performed based on Reorg/2. The advantage of Reorg is to improve the network understanding of local information among minor differences in vehicle types.

The proposed YOLOv2 model is designed as a typical object identification network. Therefore, the total number of classes is identified by this network and that may be high or the large difference among the classes of 2-wheeled, 4-wheeled, human, and obstacle, and so on. In the proposed YOLOv2 based network, three continuous and repeated (3×3×1024) higher convolution layers are implemented. These higher layers do not enhance the model's performance and are only responsible for making the network extra complex. Thus, we must eliminate the repeated higher convolutional layers. In Figure 7.6, the total continuous 3×3×1024 convolution layers are reduced to a single layer and the final layer is marked as the yellow box. For 4-wheeler identification, the total number of 4-wheeler types identified as six, and the feature variations among all the 4-wheeler types are very small.

After employing multi-layer fusion and removal of the repeated convolution layers for the higher order, the proposed YOLOv2 vehicle identification network was finally developed. Thus, to validate the efficiency of eliminated repeated higher convolutional layers, an additional reference Model_Comp network was employed for comparative analysis. Table 7.1 shows the comparison of the proposed YOLOv2 model with Model_Comp to check the effectiveness of removed one convolution (3×3×1024) layer.

The initial stage is data preparation for vehicle detection using YOLOv2 in MATLAB-2020b simulation tool. The next stage is modeling using supervised deep learning, test, and ultimately deployment in real-time vehicle identification.

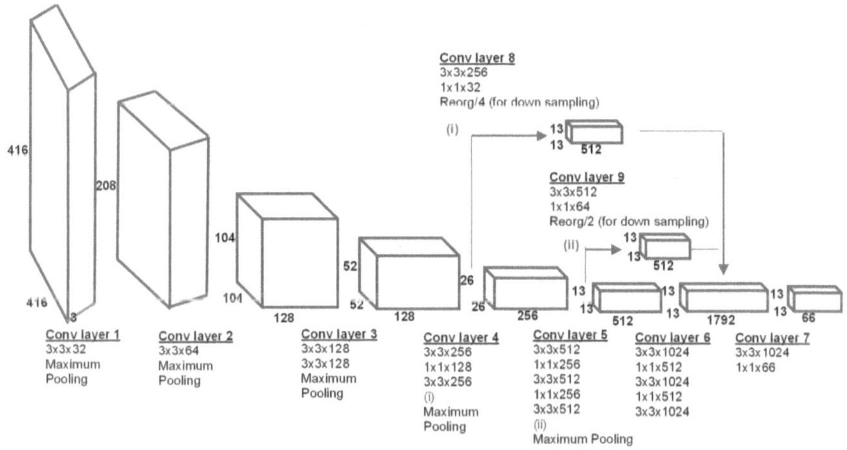

Figure 7.6 Proposed multi-layer fusion YOLOv2 framework model for vehicle identification.

Table 7.1 Comparative network analysis of proposed YOLOv2 vs. Model_Comp (reference model [28])

No. of layer	*Comp_model*	*YOLOv2*
0	Convolution3 => 32	Convolution3 => 32
1	MaximumPooling/2	MaximumPooling/2
2	Convolution3 => 64	Convolution3 => 64
3	MaximumPooling/2	MaximumPooling/2
4	Convolution3 => 128	Convolution3 => 128
5	Convolution1 => 64	Convolution1 => 64
6	Convolution3 => 128	Convolution3 => 128
7	MaximumPooling/2	MaximumPooling/2
8	Convolution3 => 256	Convolution3 => 256
9	Convolution1 => 128	Convolution1 => 128
10	Convolution3 => 256	Convolution3 => 256
11	MaximumPooling/2	MaximumPooling/2
12	Convolution3 => 512	Convolution3 => 512
13	Convolution1 => 256	Convolution1 => 256
14	Convolution3 => 512	Convolution3 => 512
15	Convolution1 => 256	Convolution1 => 256
16	Convolution3 => 512	Convolution3 => 512
17	MaximumPooling/2	MaximumPooling/2
18	Convolution3 => 1024	Convolution3 => 1024
19	Convolution1 => 512	Convolution1 => 512
20	Convolution3 => 1024	Convolution3 => 1024
21	Convolution1 => 512	Convolution1 => 512
22	Convolution3 => 1024	Convolution3 => 1024
23	Convolution3 => 1024	Route 10
24	Route 16	Convolution3 => 256
25	Convolution3 => 512	Convolution3 => 32
26	Convolution1 => 64	Reorg/4
27	Reorg/2	Route 16
28	Route 27 23	Convolution3 => 512
29	Convolution3 => 1024	Convolution1 => 64
30	Convolution1 => 66	Reorg/2
31	Identification	Route 30 26 22
32	-	Convolution3 => 1024
33	-	Convolution1 => 66

7.4 DATASET PREPARATION AND SIMULATION FOR VEHICULAR CHARACTERIZE LABELS

7.4.1 Data collection and normalization

To compare the performance analysis of the proposed YOLOv2 model with ResNet feature extraction model, first it is necessary to train the sub-dataset for further YOLOv2 model training. The datasets were gathered from Kaggle and CompCars [29]. Images in both datasets are collected from the BIT, China [30] that includes 27550 vehicular images.

These image datasets are for the 4-wheeled type training data set and the total number of images are 21,450. The 4-wheeler data set contains sports utility vehicle, hatchback, sedan, coupe, CUV, MPV, wagon, hybrid cars, pickup truck, mini car, electric cars, and cranes, and so on images. The heavy vehicle contains various types of images of truck, loaded truck, and so on. Further, this CompCars dataset is divided into two sub-datasets. The first sub-dataset includes pictures of a commercial vehicle with 8887 4-wheeler types. The other sub-dataset includes the collection of vehicle pictures using road surveillance-based recorded cameras. Only this dataset includes two 4-wheeled types (SUV and sedan) with a collection of more than 6,250 images. These datasets involve both night scenes and day scenes (sunny days, foggy, rainy day, snow, shades, etc.). The ratio of 7:3 of vehicular images dataset has been taken for both 4-wheeled (training (15,015) and validation (6,435). To validate, the generalization ability and testing of proposed model, data sets of 750 vehicular images randomly selected using CompCars [31] manual dataset. As per classified features from Figure 7.7, some dataset images of the CompCars are shown in Figure 7.8.

In Figure 7.8, first, create a label called "Four-wheeler" and add several images associated with the dataset to the labeler app tool in the MATLAB Image Processing Toolbox. We can employ another proposed dataset where

Figure 7.7 Various images dataset [28, 29] and feature classification based on different scenario.

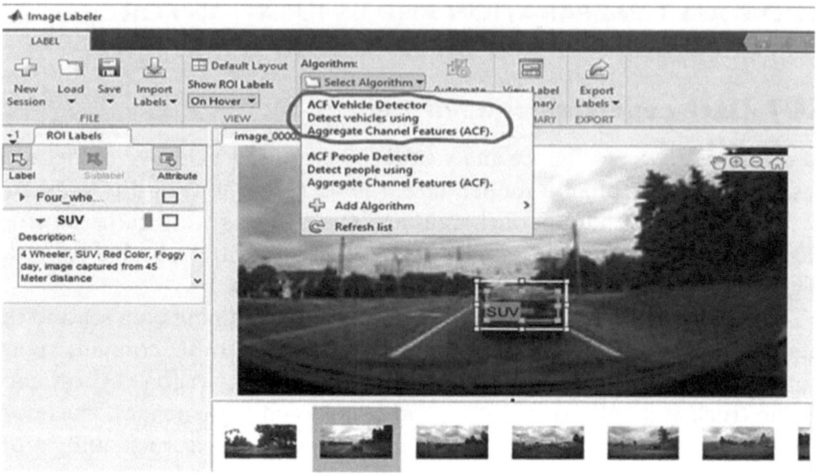

Figure 7.8 Simulation of a vehicular image data set export.

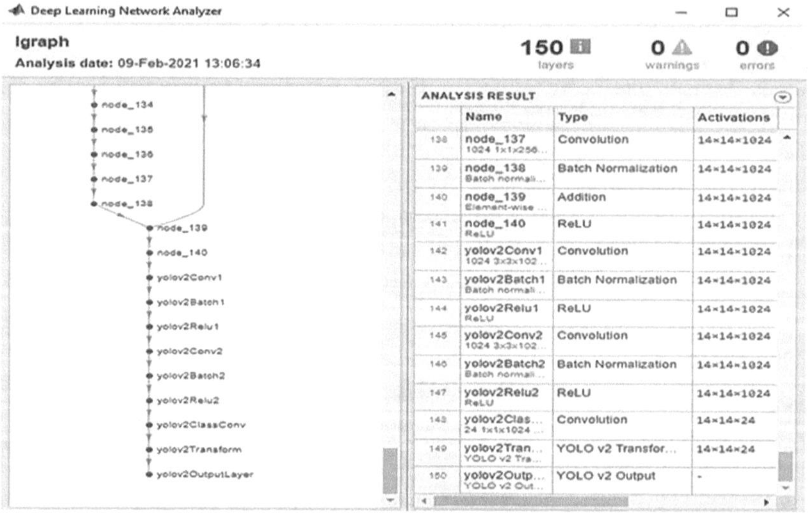

Figure 7.9 Trained proposed YOLOv2 using ResNet featured model.

the “vehicle face” was different and then operating the “ACF Vehicle Detector” in the algorithm section to automate the processing of the 4-wheeler images. After successfully exporting labels, and once an adequate number of pictures are labeled, can be export the desired data. Thus, it can be utilized for training purposes referred to Figure 7.9.

7.5 PERFORMANCE EVALUATION OF PROPOSED VEHICULAR DETECTION MODEL

7.5.1 Evaluation of training stage

Figure 7.10 demonstrates a comparative analysis of the proposed YOLOv2 model and Comp_model for the mean loss curves during training. The y-axis represents the average loss values, while the x-axis represents the number of iterations being employed for training. From Figure 7.10, we can conclude that the average loss had a downhill trend, and in the end, tended in a stable manner. For these two models, the average loss decreases based on the YOLOv2 model over the Comp_model. So, more vehicle information could be achieved by faster convergence during training. And the YOLOv2 model fits the vehicle identification task in an effective manner.

In Figure 7.11, the IOU level of Comp_model and YOLOv2 model is considered to check the training accuracy for vehicle-detected bounding boxes. From Figure 7.11, the avg IOU based on a comparative model shows gradually upward characteristics. They tended to stable IOU in between 0.8 and 0.95 values, which means the proposed YoLOv2 model had better execution time during detection. Even though in the early-stage IOU the characteristics of the Comp_model and YOLOv2 models were close together, these characteristics rise in a much faster manner and instantly reach between 0.8 and 0.9 in the earlier iteration. The YOLOv2 model required less training time, which shows that the YOLOv2 model could speed up the convergence for vehicle identification.

From Figures 7.10 and 7.11, the comparative analysis of the training stage shows that the characteristics of IOU of the proposed YOLOv2 model had better capabilities than the reference Comp_ model.

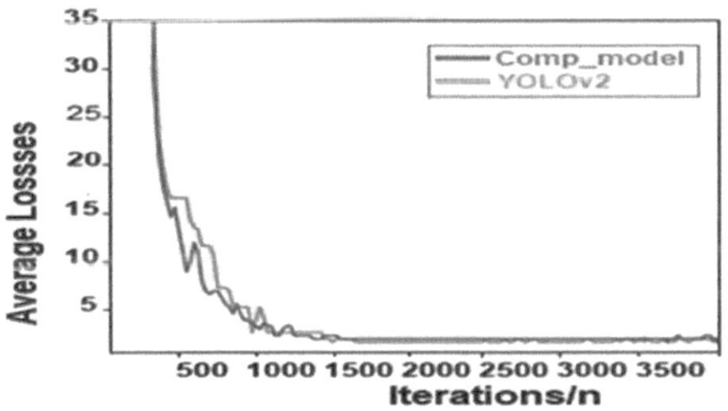

Figure 7.10 Comparison of the average loss values with proposed model.

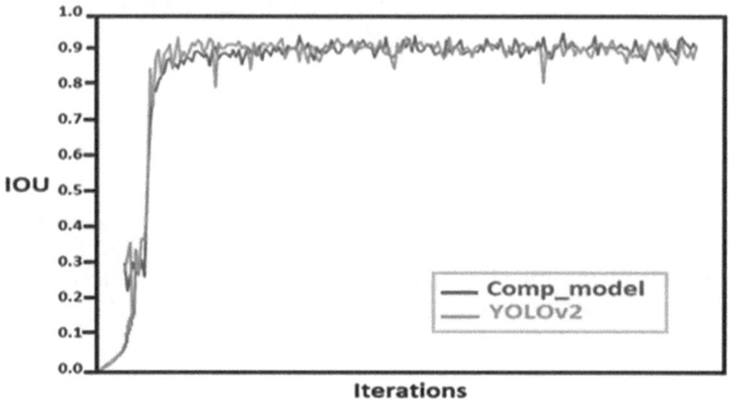

Figure 7.11 Analysis of average IOU for the comparison of YOLOv2 and Comp_model.

7.5.2 Evaluation of testing stage and validation accuracy

The performance evaluation of Model_Comp and YOLOv2 are successfully evaluated based on Kaggle data set with an 0.5-threshold by recall accuracy for avg IOU as the assessment of test dataset. From Table 7.2, the proposed YOLOv2 model had better detection accuracy. The recall, precision, and average IOU of proposed models were superior to the Comp_ model (reference model [28]).

For the evaluation of proposed model performance, "*mAP*" parameter is used for model performance measurement. The *mAP* of the proposed YOLOv2 based model is the highest noticed at 97.45% and the average value is 96.41%. The average speed of vehicle identification detection is 0.074 seconds, which means the proposed YOLOv2-based model deals with around 39 pictures per second in a real-time scenario. This feature can be crucial in intelligent transportation systems for real-time vehicle identification systems [31]. Based on the *mAP*, *AP*, and detection speed results, the YOLOv2 model is better in comparison with the Comp_model. The comparative results of YOLOv2 are much better than the Comp_model. In

Table 7.2 The recall, precision, and average IOU based on the Kaggle vehicular data set [12]

Model	*Recall (in %)*	*Average IOU (in %)*	*Precision (in %)*
Comp_model	98.60	84.97	95.70
YOLOv2	99.29	85.40	96.41

addition, since the proposed YOLOv2 model was much faster than R-CNN [32], the average vehicle identification speed was 0.9 seconds, though the classes of *AP* for YOLOv2 and Comp_model were too close. The proposed YOLOv2 network detached the repeated convolution layers of higher order. Therefore, it has been concluded that in the initial stage of the design network these layers were not effective but more effective for high convergence rate during vehicle identification.

Figure 7.12 displays the vehicle identification layers, whether it was a sunny day, cloudy day, or night scenario using the proposed YOLOv2 model. The YOLOv2 model had better performance for either single or multiple vehicle types and positioning identification. From Figure 7.12, the three vehicle identification images demonstrate that weather conditions did not significantly impact vehicular recognition accuracy, thanks to YOLOv2's ability to track vehicle positions and identify their categories. In Figure 7.12, the YOLOv2 model has accurately identified the vehicle type even in the night scenes.

Table 7.3 depicts the comparative results validation for vehicle detection using a Kaggle data set. The proposed YOLOv2 model implemented with the ResNet feature fusion approach. To validate the accuracy of the proposed strategies for vehicle identification, use the CompCars dataset to analyze two models: YOLOv2 and ResNet. A total of 450 vehicular pictures were arbitrarily chosen from the CompCars sub-dataset and named Random_Comp. The comparative results of two models are based on the random dataset, which is suitable for accurate identification, because the maximum value of mAP is around 97.45%.

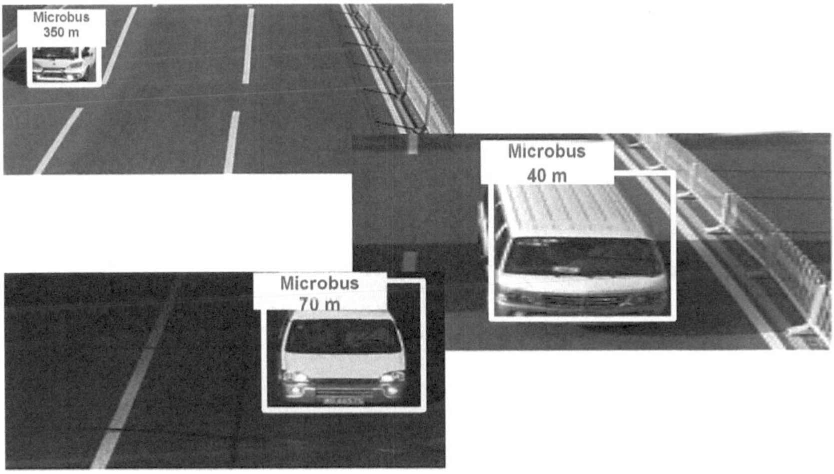

Figure 7.12 Vehicle detection results of YOLOv2.

Table 7.3 Comparative results analysis with other model during validation for vehicle detection using Kaggle data set

Model comparison	*Bus (in %)*	*Sedan (in %)*	*Truck (in %)*	*SUV (in %)*	*Microbus (in %)*	*mAP (in %)*
R-CNN+ResNet [32]	90.62	90.63	90.67	91.25	94.42	91.28
YOLOv2	97.68	98.89	92.09	95.97	93.50	95.80

7.5.3 Deployment of proposed YOLOV2 model for V2V detection

The testing of V2X detection was performed on MATLAB-20 toolbox using the GPU server (with Nvidia tesla K80x4) with compute capability at least 3.0. In addition, the CPU of Intel® (with 6.00 GHz) and memory size 64GB must be required. Initially the experimentation of learning rate was set as 0.001, then divided by 10 (if the epochs moved to 60 and 90). The maximum epoch of the batch size and the momentum are taken as to 160, 8, and 0.9 respectively. During training, after ten epochs a new vehicle image is randomly selected so that a new input image size is randomly selected. This chapter considers the down sampling factor of 32, thus all randomly input sizes of vehicle images are the multiplication of 32 (min size taken as 352×352, and max size is taken as 608×608). This training analysis supports the development of a proposed model to improve vehicle detection based on different sizes, although the similar proposed model can be utilized for vehicle identification based on distinct resolutions, which can improve the V2X identification tendency.

From Table 7.4 we have taken certain parameters and keep running the epochs many times. The simulation gives 89.66% validation accuracy at the

Table 7.4 Validation accuracy of proposed model

Epochs number	*Training losses*	*Validation losses*	*Validation accuracy*
1	4.1865359238312389	4.6938262742497688	0.9876003744584022
2	3.4677120937103516	3.8446787446446899	0.7231940823481753
3	2.4569362706623892	2.9874273889293723	0.6973197047661011
4	1.9248666914562393	2.8652839070018217	0.5136983104578355
5	1.5455935737065538	1.8852567904278391	0.4769130139131615
6	0.8884399853895480	1.5698032137187101	0.4183710264667792
7	0.6607466839475439	1.7612053573219831	0.3899531253918313
8	0.9599548835673898	0.0126161381754678	0.3345529539264930
9	0.4533357846599754	0.9812351831066427	0.3137141641049116
10	0.1991235477408463	0.9562351903582056	0.2537154648130136

end of 20 epochs. The transfer learning allows the use of knowledge gained while detecting individual vehicular image and applying to V2V detection for intelligent transportation applications.

7.6 CONCLUSIONS AND FUTURE RESEARCH DIRECTIONS

This chapter has successfully implemented the YOLOv2-based simulation model of vehicle identification for smart traffic monitoring in smart cities. The testing results of the proposed simulation showed that the vehicle remained within the anchor boxes, following the BIT-China training dataset. Further performed clustering analysis to get improved anchor box selections and eight anchor boxes with distinct sizes were chosen. Afterward, the loss function increased to higher values with stabilization to mitigate the uneven scaling impact during vehicle detection. Subsequently, to achieve improved feature classification and better extraction capability, the YOLOv2 model was implemented with the high layer feature fusion approach and further elimination for the convolution layer that is again repeated. Further, the total numbers of proposed network parameters for YOLOv2 (3.98 million) and reference Comp_model (4.52 million) were examined. That means the Comp_model is more complex than the YOLOv2. On the other hand, YOLOv2 has the greater precision to realize the identification of vehicles with better generalization capability, and it is further appropriate for vehicle identification in different weather conditions. Based on the enhanced generalized results, the mAP of YOLOv2 reached around 94–96% using Kaggle and CompCars. Images in both datasets are collected from the BIT, China that included different vehicular images dataset. Hence, the proposed model is much more suitable for vehicle identification than other real-time detection applications. This chapter successfully achieved better accuracy in the MATLAB simulation and recommends further developments in real-time V2V identification. In this chapter, the vehicular types and the quantity of image dataset are comparatively low in number as compared to real-time transportation scenarios. Future research will be based on large vehicular fleet datasets with the help of advanced optimization algorithms to enhance the vehicle speed as well as increase the accuracy of vehicular detection.

NOMENCLATURE

CPU	Central Processing Unit
CAN	Controller Area Network
EV	Electric Vehicle
ECUs	Electronic Control Unit
FRCNN	Faster R-CNN

CAN-FD	Flexible Data Rate Controller Area Network
IOU	Intersection Over Union
LSTM	Long Short-Term Memory
LIN	Local Interconnect Network
MOST	Media Oriented Systems Transport
ML	Machine Learning
mAP	mean Average Precision
OBUs	On Board Units
RNN	Recurrent Neural Network
R-CNN	Region-Based Convolutional Neural Networks
SVM	Support Vector Machine
ITS	Intelligent Transportation System
V2V	Vehicle to Vehicle
VeLoc	Vehicle Localization
VGG-19	Visual Geometry Group
V2TN	Vehicle-to-Transportation Networks

REFERENCES

[1] Cao, X.; Wu, C.; Yan, P.; Li, X. Linear SVM classification using boosting HOG features for vehicle detection in low-altitude airborne videos. In *Proceedings of the 2011 IEEE International Conference Image Processing (ICIP)*, Brussels, Belgium, 11–14 September 2011; pp. 2421–2424.

[2] Guo, E.; Bai, L.; Zhang, Y.; Han, J. Vehicle detection based on Superpixel and improved HOG in aerial images. In *Proceedings of the International Conference on Image and Graphics*, Shanghai, China, 13–15 September 2017; pp. 362–373.

[3] Laopracha, N.; Sunat, K. Comparative study of computational time that HOG-based features used for vehicle detection. In *Proceedings of the International Conference on Computing and Information Technology*, Helsinki, Finland, 21–23 August 2017; pp. 275–284.

[4] Pan, C.; Sun, M.; Yan, Z. The study on vehicle detection based on DPM in Traffic scenes. In *Proceedings of the International Conference on Frontier Computing*, Tokyo, Japan, 13–15 July 2016; pp. 19–27.

[5] LeCun, Y.; Bengio, Y.; Hinton, G. Deep learning. *Nature* 2015, 521, 436–444.

[6] He, K.; Zhang, X.; Ren, S.; Sun, J. Deep residual learning for image recognition. In *Proceedings of the IEEE Conference on Computer Vision and Pattern Recognition*, Las Vegas, NV, 27–30 June 2016; pp. 770–778.

[7] Huang, G.; Liu, Z.; Van Der Maaten, L.; Weinberger, K.Q. Densely connected convolutional networks. In *Proceedings of the IEEE Conference on Computer Vision and Pattern Recognition*, Honolulu, HI, 22–25 July 2017; pp. 2261–2269.

[8] Pyo, J.; Bang, J.; Jeong, Y. Front collision warning based on vehicle detection using CNN. In *Proceedings of the International SoC Design Conference (ISOCC)*, Jeju, Korea, 23–26 October 2016; pp. 163–164.

[9] Tang, Y.; Zhang, C.; Gu, R. Vehicle detection and recognition for intelligent traffic surveillance system. *Multimed. Tools Appl.* 2017, 76, pp. 5817–5832.

[10] Gao, Y.; Guo, S.; Huang, K.; Chen, J.; Gong, Q.; Zou, Y.; Bai, T.; Overett, G. Scale optimization for fullimage-CNN vehicle detection. In *Proceedings of the IEEE Intelligent Vehicles Symposium (IV)*, Los Angeles, CA, 11–14 June 2017; pp. 785–791.

[11] Huttunen, H.; Yancheshmeh, F.S.; Chen, K. Car type recognition with deep neural networks. In *Proceedings of the IEEE Intelligent Vehicles Symposium (IV)*, Gothenburg, Sweden, 19–22 June 2016; pp. 1115–1120.

[12] Malik, J. A.; Haque, A.; Amir, M. Investigation of intelligent deep convolution neural network for DC-DC converters faults detection in electric vehicles applications. In *2023 International Conference on Recent Advances in Electrical, Electronics & Digital Healthcare Technologies (REEDCON)*, New Delhi, India, 2023; pp. 139–144.

[13] Girshick, R.; Donahue, J.; Darrell, T.; Malik, J. Rich feature hierarchies for accurate object detection and semantic segmentation. In *Proceedings of the 2014 IEEE Conference on Computer Vision and Pattern Recognition*, Columbus, OH, 24–27 June 2014; pp. 580–587.

[14] He, K.; Zhang, X.; Ren, S.; Sun, J. Spatial pyramid pooling in deep convolutional networks for visual recognition. In *Proceedings of the 2014 IEEE International Conference of European Conference on Computer Vision*, Zurich, Switzerland, 6–12 September 2014; pp. 346–361.

[15] Dai, J.; Li, Y.; He, K.; Sun, J. R-FCN: Object detection via region-based fully convolutional networks. In *Proceedings of the 2016 IEEE International Conference of Advances in Neural Information Processing Systems*, Barcelona, Spain, 5–8 December 2016; pp. 379–387.

[16] Ren, S.; He, K.; Girshick, R.; Sun, J. Faster R-CNN: towards real-time object detection with region proposal networks. arXiv 2016, arXiv:1506.01497v3, 2016.

[17] Konoplich, G.V.; Putin, E.O.; Filchenkov, A.A. Application of deep learning to the problem of vehicle detection in UAV images. In *Proceedings of the 2016 XIX IEEE International Conference on Soft Computing and Measurements (SCM)*, St. Petersburg, Russia, 25–27 May 2016; pp. 4–6.

[18] Azam, S.; Rafique, A.; Jeon, M. Vehicle pose detection using region based convolutional neural network. In *Proceedings of the International Conference on Control, Automation and Information Sciences (ICCAIS)*, Ansan, Korea, 27–29 October 2016; pp. 194–198.

[19] Tang, T.; Zhou, S.; Deng, Z.; Zou, H.; Lei, L. Vehicle detection in aerial images based on region convolutional neural networks and hard negative example mining. *Sensors* 2017, 17, 336.

[20] He, K.; Zhang, X.; Ren, S.; Sun, J. Deep residual learning for image recognition, in *CVPR*, 2016.

[21] Redmon, J.; Farhadi, A. YOLO9000: Better, faster, stronger. In *Proceedings of the IEEE Conference on Computer Vision and Pattern Recognition (CVPR)*, Honolulu, HI, 21–26 July 2017; pp. 6517–6525.

[22] Arthur, D.; Vassilvitskii, S. k-means++: The advantages of careful seeding. In *Proceedings of the eighteenth annual ACM-SIAM symposium on Discrete algorithms*, New Orleans, LA, 7–9 January 2007; pp. 1027–1035.

[23] Sang, J.; Wu, Z.; Guo, P.; Hu, H.; Xiang, H.; Zhang, Q.; Cai, B. An improved YOLOv2 for vehicle detection. *Sensors (Basel, Switzerland)* 2018, 18(12), 4272.

[24] Pradeep, D.J.; Kumar, Y.V.P.; Siddharth, B.R.; Reddy, C.P.; Amir, M., Khalid, H.M. Critical performance analysis of four-wheel drive hybrid electric vehicles subjected to dynamic operating conditions. *World Electric Vehicle Journal* 2023, 14(6), 138.

[25] Sang, J.; Guo, P.; Xiang, Z.; Luo, H.; Chen, X. Vehicle detection based on faster-RCNN. *J. Chongqing Univ. (Nat. Sci. Ed.)* 2017, 40, 32–36.

[26] Neubeck, A.; Van Gool, L. Efficient non-maximum suppression. In *Proceedings of the International Conference on Pattern Recognition (ICPR)*, Hong Kong, China, 20–24 August 2006; pp. 850–855.

[27] Mateen, M.; Wen, J.; Song, S.; Huang, Z. Fundus image classification using VGG-19 architecture with PCA and SVD. *Symmetry* 2019, 11(1), 1.

[28] Vehicle dataset: available on https://www.kaggle.com/jutrera/stanford-car-dataset-by-classes-folder. (last access on July 2021).

[29] Dong, Z.; Wu, Y.; Pei, M.; Jia, Y. Vehicle type classification using a semisupervised convolutional neural network. *IEEE Trans. Intel. Transp. Syst.* 2015, 16, 2247–2256.

[30] Amir, M.; Zaheeruddin; Haque, A., Kurukuru, V.S.B., Bakhsh, F.I., Ahmad, A. Agent based online learning approach for power flow control of electric vehicle fast charging station integrated with smart microgrid. *IET Renew. Power Gener.* 2022, 00, 1–13.

[31] Redmon, J.; Divvala, S.; Girshick, R.; Farhadi, A. You only look once: Unified, real-time object detection. In *Proceedings of the IEEE Conference on Computer Vision and Pattern Recognition (CVPR)*, Las Vegas, NV, 27–30 June 2016; pp. 779–788.

[32] Sang, J.; Guo, P.; Xiang, Z.; Luo, H.; Chen, X. Vehicle detection based on faster-RCNN. *J. Chongqing Univ. (Nat. Sci. Ed.)* 2017, 40, 32–36.

Chapter 8

Integration of power electronics in renewable energy for smart cities

Himanshu Sharma
Noida Institute of Engineering & Technology, Greater Noida, India

Ahteshamul Haque
Electrical Engineering Department, Jamia Millia Islamia, New Delhi, India

8.1 INTRODUCTION

Power electronics have evolved fast over the last many years. There is an increase in the number of applications, due to technological advancements such as semiconductor technology. Furthermore, the power electronics components are becoming cheaper all the time [1]. A smart grid is an electricity network that employs digital technologies to monitor and control the transportation of electricity from all generation sources to satisfy end-users shifting electricity demands. The role of the smart grid in integrating power electronics & renewable energy is shown in Figure 8.1.

Figure 8.2 shows the power generation, transmission, and distribution for the residential and commercial sectors. There are various types of power generation sources such as Nuclear Power plants, Fossil Fuels (oil, coal, gas),

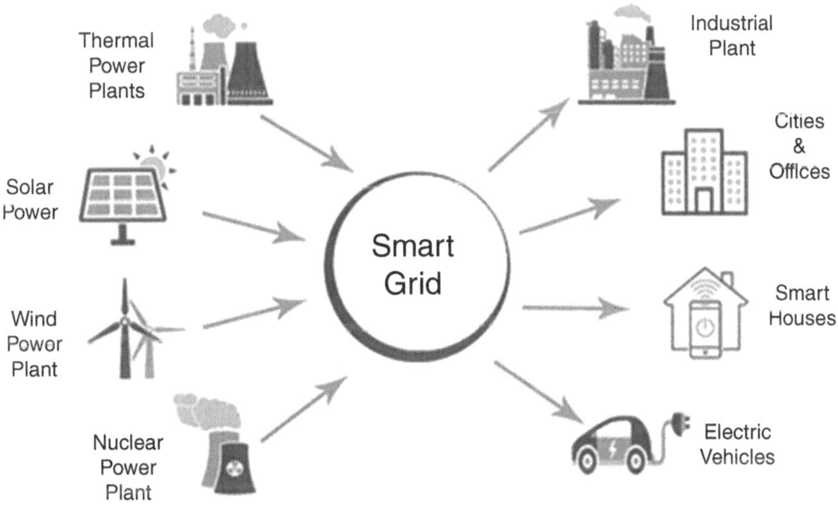

Figure 8.1 Role of smart grid to integrate power electronics & renewable energy.

DOI: 10.1201/9781032669809-8

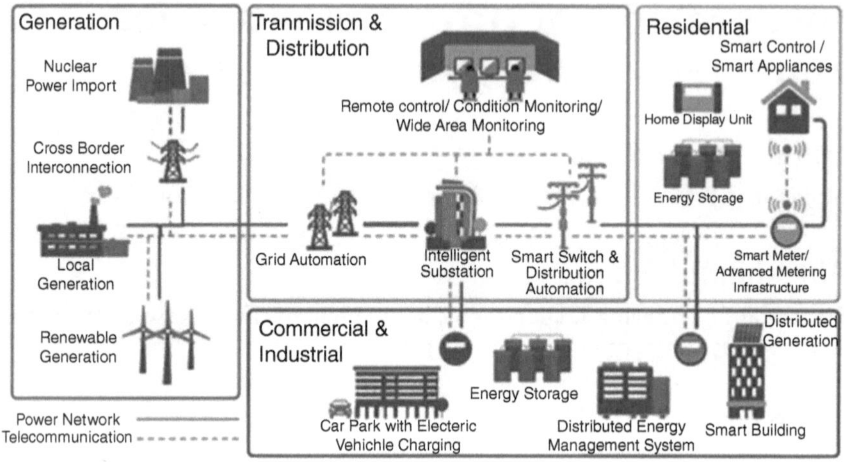

Figure 8.2 Power generation, transmission, distribution for residential & commercial sector.

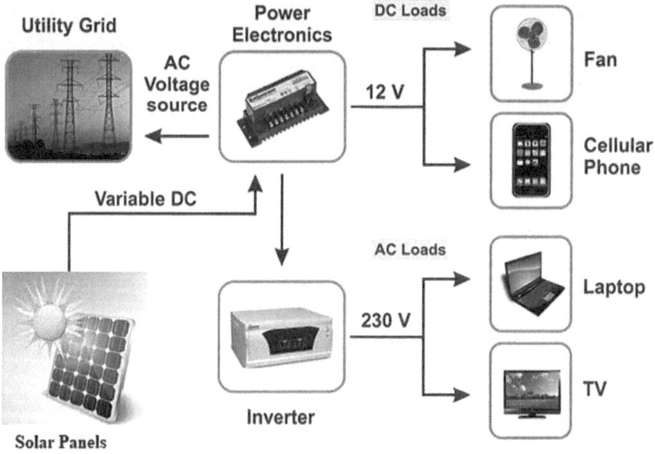

Figure 8.3 Integration of power electronics and renewable energy systems.

and renewable energy sources. The next task is the transmission and distribution of power to the residential and commercial sectors.

In Figure 8.3, the Integration of Power Electronics and Renewable Energy Systems is shown. Most renewable energy methods only give a particular voltage and current density to the power converter depending on the load. The voltage available from the fuel cell must then be adjusted by the power converter to a voltage high enough to run the load. A DC-DC Boost converter is necessary to enhance the voltage level for the inverter, as shown in Figure 8.4. This Boost converter adjusts the inverter input voltage and separates the low and high voltage circuits in addition to increasing the fuel cell voltage.

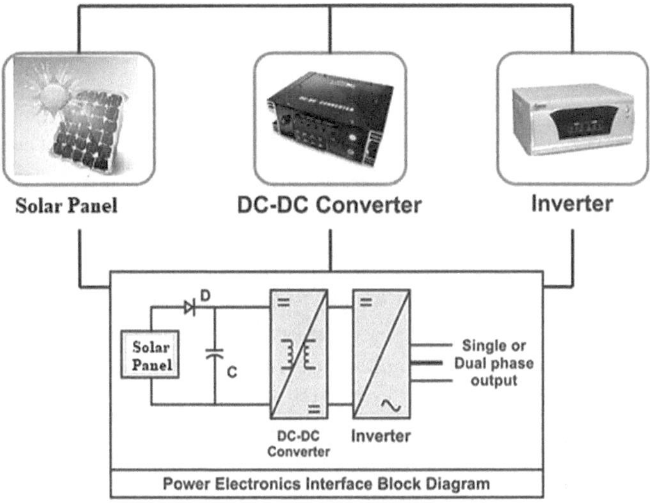

Figure 8.4 Power electronics interface block diagram.

An hybrid power system consists of the fuel cell, lithium-ion battery, constant voltage regulation system, and smart battery charger.

8.2 POWER SYSTEMS CONVERTERS

Power regulation and inverters are the two primary power electronics aspects that must be handled in renewable energy applications. Fuel cells, solar cells, and wind turbines all provide varying amounts of electrical power. Voltage regulators, DC/DC converters, and other circuits often keep the fuel cell voltage constant, which can be greater or lower than the fuel cell working voltage. Multilevel converters are of relevance in the distributed energy resources field because they can link many batteries, fuel cells, solar cells, and wind turbines to serve a load or grid without causing voltage balance concerns. The multilevel inverter's main purpose is to generate the desired AC voltage by combining different levels of voltage. As a result, multilayer inverters are suitable for connecting an AC grid to renewable energy sources like photovoltaics or fuel cells, as well as energy storage devices like capacitors or batteries, in series or parallel. In addition, multilevel converters have lower switching frequencies than typical converters, resulting in fewer switching losses and higher efficiency [6].

Fuel cell technology [7] advancements necessitate power converter technology advancements as well. A tiny, affordable converter may be designed to accompany a moderately large solar panel, wind turbine, or fuel cell for high system power and energy density by addressing power conversion design parameters early in the overall system design.

8.3 DC-TO-DC CONVERTERS

Because the output of a renewable energy system fluctuates with load current, a DC-to-DC converter is employed to adjust the voltage. Because many fuel cell and solar cell systems are intended for lesser voltage, a DC-DC Boost converter is frequently used to raise the voltage to greater levels. Because the voltage fluctuates with the amount of electricity required, these renewable energy systems require a converter. At full load, the voltage of a typical fuel cell drops from 1.23 V DC (no-load) to less than 0.5 V DC. As a result, a converter must operate with a wide variety of input voltages.

DC-to-DC converters [8] are essential in portable electronic devices that require batteries, such as mobile phones and laptop computers. These sorts of electrical devices frequently comprise numerous subcircuits, each with its own voltage level need that differs from that supplied by the battery or an external source. As the battery's stored power is depleted, a DC-to-DC converter provides a technique to Boost voltage from a partially-lowered battery voltage, saving space instead of utilizing numerous batteries to perform the same purpose. Figure 8.4 shows the use of a DC-to-DC converter device.

8.4 INVERTERS

Both houses and businesses may utilize renewable energy as their primary power source. The AC grid will be required for these energy technologies to function. In certain grid-independent systems, the output of renewable energy systems will have to be converted to AC. To do this, an inverter can be employed. The generated AC may be used with the proper transformers and control circuits since it is at the required voltage and frequency. Inverters are utilized in a wide range of applications, from computer switching power supplies to high voltage direct current bulk power applications. Inverters [9] are often used to convert DC electricity from fuel cells, solar panels, and batteries to AC power. An inverter is seen in Figure 8.4.

8.5 POWER ELECTRONICS CONVERTERS INTEGRATION WITH RENEWABLE ENERGY

The design and operation of power electronics converters for both wind turbine and PV systems strongly rely on the grid requirements and the energy demand. It can be seen from the evolution of wind turbine power converters, which has changed from non-power-electronics-based topologies to full-scale power converters with increasing power ratings of individual wind turbine. As the demand for higher power ratings and efficiency

Table 8.1 Design challenges in renewable energy harvesting technologies

S. No.	*Energy harvesting technologies*	*Design challenges*
1.	Solar Panel	Solar Cell Efficiency, Shading Effects, PV Fault Diagnosis
2.	DC-DC Converter	Types of DC-DC converters (Buck Converter, Boost Converter, Buck-Boost Converter, SEPIC Converter etc.)
3.	DC-DC Converter Control Techniques	PWM or MPPT, Types of MPPT: • Perturbation and Observation (P&O) technique, • Incremental Conductance (INC) technique and • Fraction Open Circuit Voltage (OCV)
4.	Storage Technologies	Choice of Battery or Super capacitor, Types of Battery (Li-ion, Nicd etc.)
5.	Energy prediction Algorithms	Choice of Solar energy prediction algorithm, Machine Learning based algorithms

increases for PV systems, the PV power converters also experienced a clear change, and they are mostly transformerless nowadays. Besides developing more advanced control strategies, the power electronic converters can also be customized as an active damper for stabilizing the integration of renewable power plants (Table 8.1).

8.6 DESIGN CHALLENGES IN INTEGRATING POWER ELECTRONICS CONVERTERS IN TO SOLAR RENEWABLE ENERGY

- To expand the battery charging-discharging life cycle.
- Designing of simple and inventive solar charger.
- To shrink the overall power consumption.
- To enhance the stability of the overall Solar Energy Harvesting Smart Grids systems.
- Energy harvester circuits should be compatible with existing grid industry communication standards like IEEE 802.15.4 (ZigBee) and IEEE 1451.5 standards.
- To achieve the highest power from the sun.
- To ensure small power consumption for DC-DC Boost converter operation.
- DC-DC converter output voltage, inductor ripple current at the output.
- To convey maximum power to the Solar energy harvesting (SEH) smart grids using the harvested energy.
- To start up (or bootstrap) the SEH-smart grids.

8.7 SOME DESIGN ISSUES IN INTEGRATING POWER ELECTRONICS CONVERTERS INTO SOLAR RENEWABLE ENERGY

- To determine the effect of irradiance & temperature on the solar panel output voltage, current & power? How to simulate it? Which simulation software should be used for low power solar panels?
- To compare the PWM and MPPT? Which MPPT is suitable for an energy harvesting smart grid? Which simulation software is best suited for PWM and MPPT comparison?
- Analysis of output voltage, inductor ripple current, efficiency of a DC-DC Buck converter considering various power losses.
- Design and analysis of a DC-DC Buck converter power stage for less output voltage ripples & to observe the effect of inductor (L), capacitor (C), and duty cycle (D) values on output Ripples voltage in DC-DC Buck converter.
- To know the effect of variation of irradiance level (W/m^2) on the network Throughput (bits/sec.) in an SEH smart grid scenario.
- To calculate the efficiency of a real-life commercial DC-DC converter and apply this knowledge to SEH systems.

8.8 SOLAR PV PANEL

A solar PV cell is a semiconductor device, which converts light energy into electrical energy. When a photon of light energy ($h\nu > E_g$) is incident over a solar cell the electron-hole pair (EHP) is generated. This newly generated EHP contributes to the electric current called light generated current denoted by (I_L). The ideal theoretical current–voltage (I–V) equation of solar cell is given as [13]:

$$\text{Solar cell current } (I) = I_L - I_o\left[\exp\left(\frac{qV}{kT}\right) - 1\right] \quad (8.1)$$

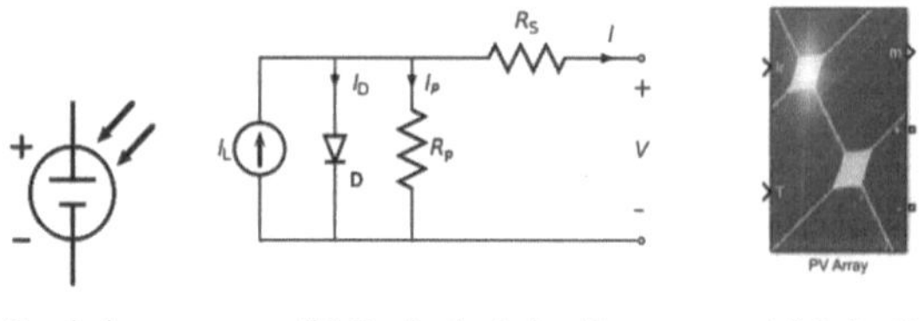

(a) Symbol (b) Equivalent circuit (c) Solar Panel

Figure 8.5 Modeling of solar cell.

Where I = total output current of solar cell, I_L = Light generated current by the solar cell, I_o = reverse saturation current due to recombination, q = charge of electron (1.6×10^{-19} C), V = open circuit voltage of solar cell, k = Boltzmann's constant (1.38×10^{-23} J/K), T = Temperature of Solar cell (300 K). The symbol of the solar cell is shown in Figure 8.5(a). The solar cell equivalent circuit model can be represented as shown in Figure 8.5(b). It consists of a light generated current source (I_L), a diode (D) modeled by the Shockley equation, and two series and parallel resistances. By applying Kirchhoff's current law (KCL), we can get the output current for this equivalent circuit:

$$\text{Output Current of Solar Cell Equivalent Model } (I) = I_L - I_D - I_p \quad (8.2)$$

Where I_p = current in parallel resistance, I_L = Light generated current, and I_D = diode current.

$$\text{Diode Current } (I_D) = I_0\left[\exp\left(\frac{V + IR_s}{nV_T}\right) - 1\right] \quad (8.3)$$

Where I_o = reverse saturation current due to recombination, V = open circuit voltage of solar cell, I = solar cell output current, R_s = series resistance, n = diode ideality factor (1 for ideal, 2 for practical diode), V_T = Thermal voltage (kT/q), k = Boltzmann's constant (1.38×10^{-23} J/K), T = Temperature of Solar cell (300 K), and q = charge of electron (1.6×10^{-19} C). The current in parallel resistance is given as:

$$\text{Current in parallel resistance } (I_p) = \frac{V + IR_s}{R_p} \quad (8.4)$$

Now, put the value of I_D and I_p in current eq. 8.2 we get the complete IV equation of equivalent circuit of single solar cell, which related all parameters with output current and voltage given as:

$$\text{Solar cell current } (I) = I_L - I_o\left[\exp\left(\frac{q(V + IR_s)}{nkT}\right)\right] - \left(\frac{V + IR_s}{R_p}\right) \quad (8.5)$$

Where R_p = parallel resistance and remaining parameters I_L, I_o, q, V, I, R_s, n, k, T have been already defined in eq. (8.3). The efficiency (η) of the solar cell is given as:

$$\text{Solar Cell Efficiency } (\eta) = \frac{V_{oc}.I_{sc}.\text{FF}}{P_{in}} \quad (8.6)$$

Where V_{oc} is called open circuit voltage, I_{sc} is short circuit current, FF is Fill Factor, and P_{in} = incident optical power. The Fill Factor (FF) of a solar cell is given as

$$\text{Fill Factor}(\text{FF}) = \frac{P_{\max}}{P_{\text{dc}}} = \frac{I_m.V_m}{I_{\text{sc}}.V_{\text{oc}}} \tag{8.7}$$

Where I_m is called maximum current and V_m is the maximum voltage of the solar cell. Practically, there are many types of solar cells like monocrystalline silicon solar cell (c-Si), amorphous silicon solar cell (a-Si), polycrystalline solar cell (multi-Si), Thin-film solar cell (TFSC), and so on. However, the efficiency of a-Si solar cells is more than all others at up to 18% efficiency.

8.9 DC-DC CONVERTER USED IN PHOTOVOLTAIC SYSTEMS

There are generally three types of DC-DC converters used in the design of a photovoltaic system such as Buck converter, Boost converter, and Buck-Boost converter. Here, we have used a DC-DC Buck converter because its efficiency is high as compared to Boost and Buck-Boost converters. A DC-DC Buck Converter is a power electronics converter in which the output voltage is always less than the input voltage. The Buck converter consists of a DC voltage source (V_{dc}), an inductor (L), a switch (MOSFET), a diode (D), and a capacitor (C) as shown in Figure 8.6. When the MOSFET switch (S) is closed at time t_1, the input voltage V_s appears across the load resistor. If the MOSFET switch remains OFF for the time t_2, then the voltage across the load resistor is zero. The amplitude of output voltage (V_o) is less than

The input voltage V_o. The duty cycle (D) can be varied from 0 to 1 by varying time period (t_1). The duty cycle of the Buck converter is $D = V_o/V_{in}$. The average output voltage of buck converter is given as:

$$V_o = \frac{1}{T}\int_0^{t_1} v_o dt = \frac{t_1}{T} V_{\text{in}} = f.t_1.V_{\text{in}} = V_{\text{in}}.D \tag{8.8}$$

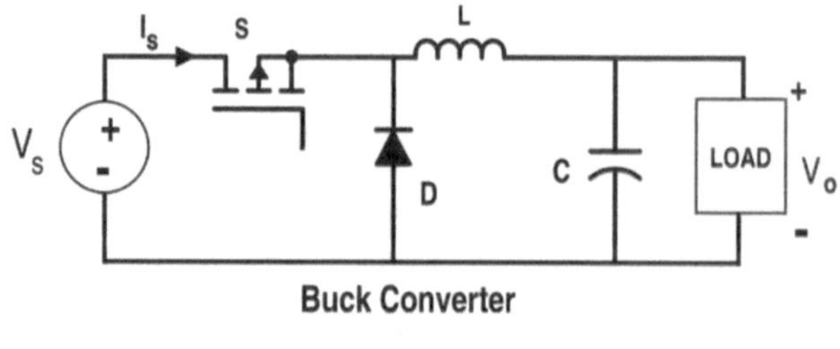

Figure 8.6 Circuit model of a DC-DC Buck Converter.

Where V_o is the output voltage, V_{in} is the input voltage, t_1 = MOSFET switch ON time duration, T = Total Time period, f is the frequency of operation, and D is the duty cycle. The average load current at the output is given as:

$$I_o = I_L = V_o / R = D.V_{in} / R \tag{8.9}$$

Where T = chopping period, $D = t_1 / T$ is duty cycle, f = chopping frequency.

8.10 POWER LOSSES IN DC-DC CONVERTERS

The three main causes of power dissipation in a DC-DC converter are as follows [14]:

- Inductor conduction losses
- MOSFET conduction losses
- MOSFET switching losses

In all types of DC-DC converters, the inductor consumes the highest amount of power. The value of MOSFET switching loss and diode conduction losses are very small as compared to inductor losses and can be neglected practically.

- **Inductor conduction loss:** The inductor charging and discharging waveform is shown in Figure 8.7. Here, the inductor conduction loss is given as:

$$P_L = I_{L(\text{rms})}^2 \times R_{L(\text{dc})} \tag{8.10}$$

Where P_L = Power loss in inductor (mW), $I_{L(\text{rms})}$ = inductor RMS current, $R_{L(\text{dc})}$ = DC resistance of inductor.
The r.m.s. the inductor current is given as:

$$I_{L(\text{rms})}^2 = I_o^2 + \frac{\Delta I^2}{12} \tag{8.11}$$

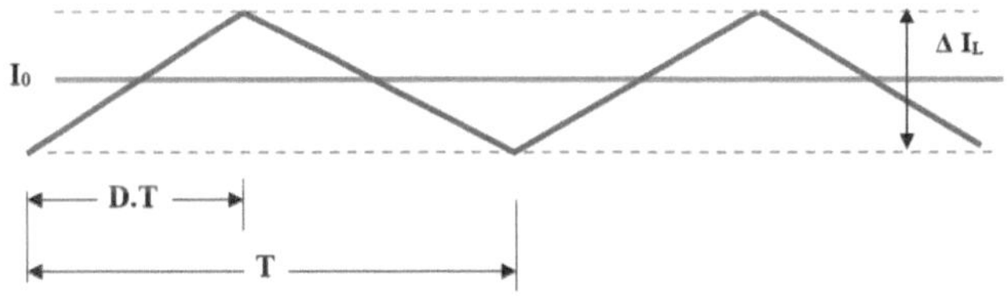

Figure 8.7 Inductor current during charging & discharging waveform.

Where ΔI = Inductor ripples current and given as:

$$\Delta I = \frac{(V_{in} - V_o) \times V_o}{L \times V_{in} \times f} \tag{8.12}$$

Where V_{in} is the input voltage, V_o is the output voltage, L is inductance, f is frequency.

Typically, ΔI is around 30% of the output current. Therefore, it can be calculated as:

$$I^2_{L(rms)} = I_o \times 1.00375 \tag{8.13}$$

As the ripple current is very small 0.375% of the total RMS inductor current, therefore we can ignore it. The power dissipated in the inductor now can be calculated as:

$$P_L = I_o^2 \times R_{L(dc)} \tag{8.14}$$

- **MOSFET conduction losses:**

The MOSFET waveforms during ON time and OFF time are shown in Figure 8.8. The power dissipated during conduction time in the MOSFET is given by [15]:

$$P_{Q(cond.)} = I^2_{Q(rms)} \times R_{DS(ON)} \tag{8.15}$$

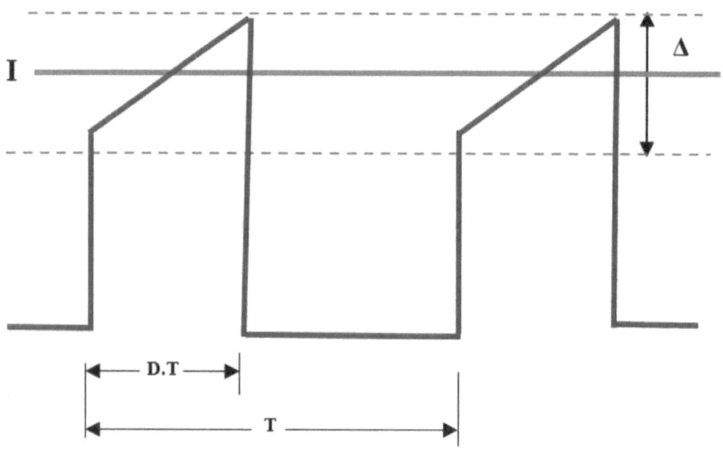

Figure 8.8 MOSFET conduction switching (ON-OFF) current waveforms.

Where $R_{DS(ON)}$ is the Drain to Source resistance of MOSFET during ON time. The MOSFET RMS current is given as:

$$I^2_{Q(rms)} = \frac{V_o}{V_{in}} \times \left(I_o^2 + \frac{\Delta I^2}{12} \right) \tag{8.16}$$

Put the value of MOSFET RMS current from eq. (8.16) into eq. (8.15) we get,

$$P_{Q(cond.)} = \frac{V_o}{V_{in}} \times \left(I_o^2 + \frac{\Delta I^2}{12} \right) \times R_{DS(ON)} \tag{8.17}$$

For typical buck power supply designs, the inductor's ripple current, ΔI, is less than 30% of the total output current, so the contribution of $\Delta I^2/12$ to this is negligible and can be dropped. From eq. (8.17), the MOSFET conduction losses at any output voltage can be calculated. The other losses such as switching losses and inductor conduction losses are independent of output voltage and remain constant with changes in output voltage. Hence, power dissipation P_D now can be computed as:

$$P_D = P_L + P_{Q(cond.)} + P_{Q(sw)} \tag{8.18}$$

For a DC-DC converter with known output power and power losses, the efficiency can be calculated as output power divided by output power plus loss power as:

$$\eta = \frac{P_0}{P_0 + P_D} \tag{8.19}$$

This relation is the same as the basic efficiency formula, that is, output power divided by input power, where input power must be equal to output power plus losses in the DC-DC converter.

For example, let us calculate the efficiency of a DC-DC Buck converter, having input voltage = 5 v, input current = 0.8 A, input power = 4 watts, output voltage = 3.6 v, output current = 1 A, output power = 3.6 w, converter power loss or power dissipation P_D = 4 – 3.6 = 0.4 watts. The efficiency can be calculated by using eq. (8.19) as:

$$\text{Efficiency}\,(\eta) = \frac{P_0}{P_0 + P_D} = \frac{3.6}{3.6 + 0.4} = 0.9 \text{ or } 90\% \tag{8.20}$$

8.11 MAXIMUM POWER POINT TRACKING (MPPT) TECHNIQUE

The MPPT techniques are widely used in the design of photovoltaic (PV) solar systems to maximize the power extraction from the sun under varying solar irradiance conditions. It is an algorithm which continuously measures the voltage (V_{pv}) and current (I_{pv}) from the solar panel and calculates the amount of duty cycle (D) to be fed to the MOSFET switch of the DC-DC buck converter [16]. The following algorithms are generally used in photovoltaic applications:

- Perturbation and Observation (P&O) technique,
- Incremental Conductance (INC) technique, and
- Fraction Open Circuit Voltage (OCV).

The P&O technique is mostly used in all types of solar energy harvester systems. A flow chart for the P&O algorithm is shown in Figure 8.9. The output of this algorithm is a varying duty cycle (ΔD) which depends on input solar irradiance (W/m^2). When solar irradiance changes then a change in duty cycle occurs and the solar panel voltage and current change. The MPPT algorithm senses these changes and adjusts the impedance of the solar panel to the maximum power point. Thus, maximum power (P) can still be

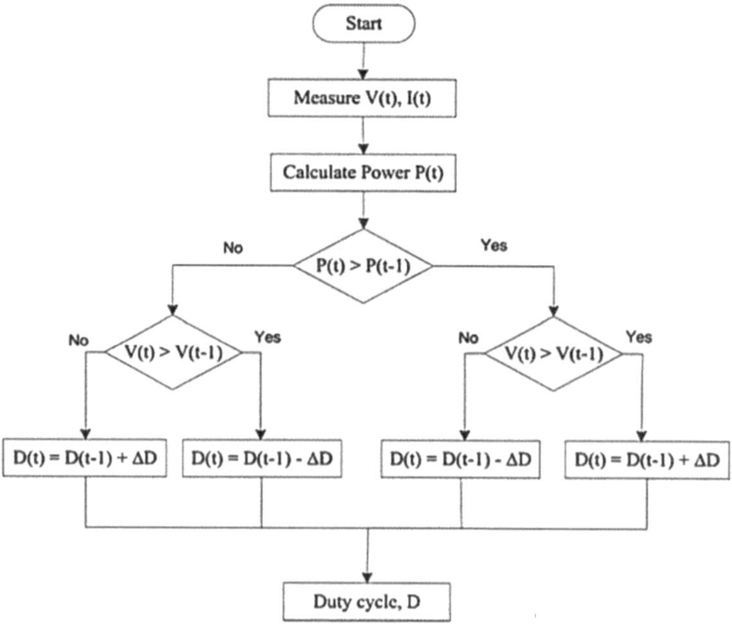

Figure 8.9 Flowchart of perturb and observation (P&O) algorithm for MPPT.

extracted from the solar panel even if the irradiance changes. It generates a PWM waveform whose initial duty cycle (D) is 0.7, provided arbitrarily (in the range of 0 to 1) as a seed value during the simulation.

The P&O algorithm works on the principle of impedance matching between load and solar panel. For maximum power transfer, impedance matching is necessary. This impedance matching is achieved by using a DC-DC converter. By using a DC-DC converter, the impedance is matched by changing the duty cycle (ΔD) of the MOSFET switch. The relation between the input voltage(V_{in}), the output voltage (V_o), and duty cycle (D) is given as

$$V_o = V_{\text{in}}.D \tag{8.21}$$

and,

$$R_{\text{in}} = R_L \,/\, D^2 \tag{8.22}$$

Therefore, if the duty cycle changes (ΔD), then the solar energy harvester output voltage (V_o) changes. If the duty cycle (D) is increased, the output voltage (V_o) also increases and vice-versa. By changing the duty cycle (D), the impedance of the load resistance (R_L) can be matched with input solar panel impedance for maximum power transfer to the load for optimum performance.

- **Algorithm: P&O MPPT**

```
Function D= PandO(V_pv, I_pv)
Persistant D_prev   P_prev   V_prev
If is empty (D_prev)
            D_prev  = 0.7;
            P_prev  = 190;
            V_prev  = 2000;
End
deltaD = 0.0025;
P_pv = V_pv * I_pv;
If (P_pv - P_prev) ~ = 0
              If (P_pv - P_prev) > 0
                          If (V_pv - V_prev) > 0
                          D = D_prev - deltaD;
                          Else
                          D = D_prev + deltaD;
                          End
             End
Else
             D = D_prev ;
End
```

```
Dprev = D;
Vprev = Vpv ;
Pprev = Ppv ;
```

The P&O algorithm is shown by a flowchart and MATLAB codes are shown above. Here, the Boost converter is used to charge the 3.6 volts battery. The battery parameters state of charge (SoC), battery current, and voltages are observed under various charging conditions in MATLAB Simulink 2017. Figure 8.10 shows an MPPT controlled DC-DC Boost converter battery charger.

8.12 DESIGN AND ANALYSIS OF A DC-DC BUCK CONVERTER POWER STAGE FOR LESS OUTPUT VOLTAGE RIPPLES

The flowchart of design and analysis of a DC-DC Buck converter power stage for less output voltage ripples is shown in Figure 8.11 [12]. It has the following steps:

- To mathematically model all parameters of DC-DC Buck converter in MATLAB.
- To plot output ripple voltage (*V*) w.r.t. frequency (*f*) with variable capacitor (*C*) while inductor (*L*) and duty cycle (*D*) are constants.
- To plot output ripple voltage (*V*) w.r.t. frequency (*f*) with variable inductor (*L*) while capacitor (*C*) and duty cycle (*D*) are constants.
- To plot output ripple voltage (*V*) w.r.t. frequency (*f*) with variable duty cycle (*D*) while capacitor (*C*) and inductor (*L*) are constants.
- To observe the result graphs and find the values of *L*, *C*, and *D* for minimum output voltage ripples.

The DC-DC converter is a power electronic device, which is used to convert the voltage amplitude from one level to another level. The input-output voltage levels may range from very low (mV) to high (kV) values. The main types of DC-to-DC converters topology are given as Buck converter, Boost converter, Buck-Boost converter, Cuk converter, and Single End Primary Inductor Current (SEPIC). The buck converter is used to decrease the amplitude of harvested energy if the excess amount of energy is received by the solar panel. The Boost converter increases the amplitude of harvested energy by using capacitors and inductors and the MOSFET switch.

Generally, in outdoor photovoltaic applications, the buck converter is mostly used because of its simple construction, low power dissipation, and high efficiency. The remaining converter's topologies can be implemented using buck and Boost converter configurations. A Solar Powered DC-DC Buck converter for Energy Harvesting IoT nodes is shown in Figure 8.12.

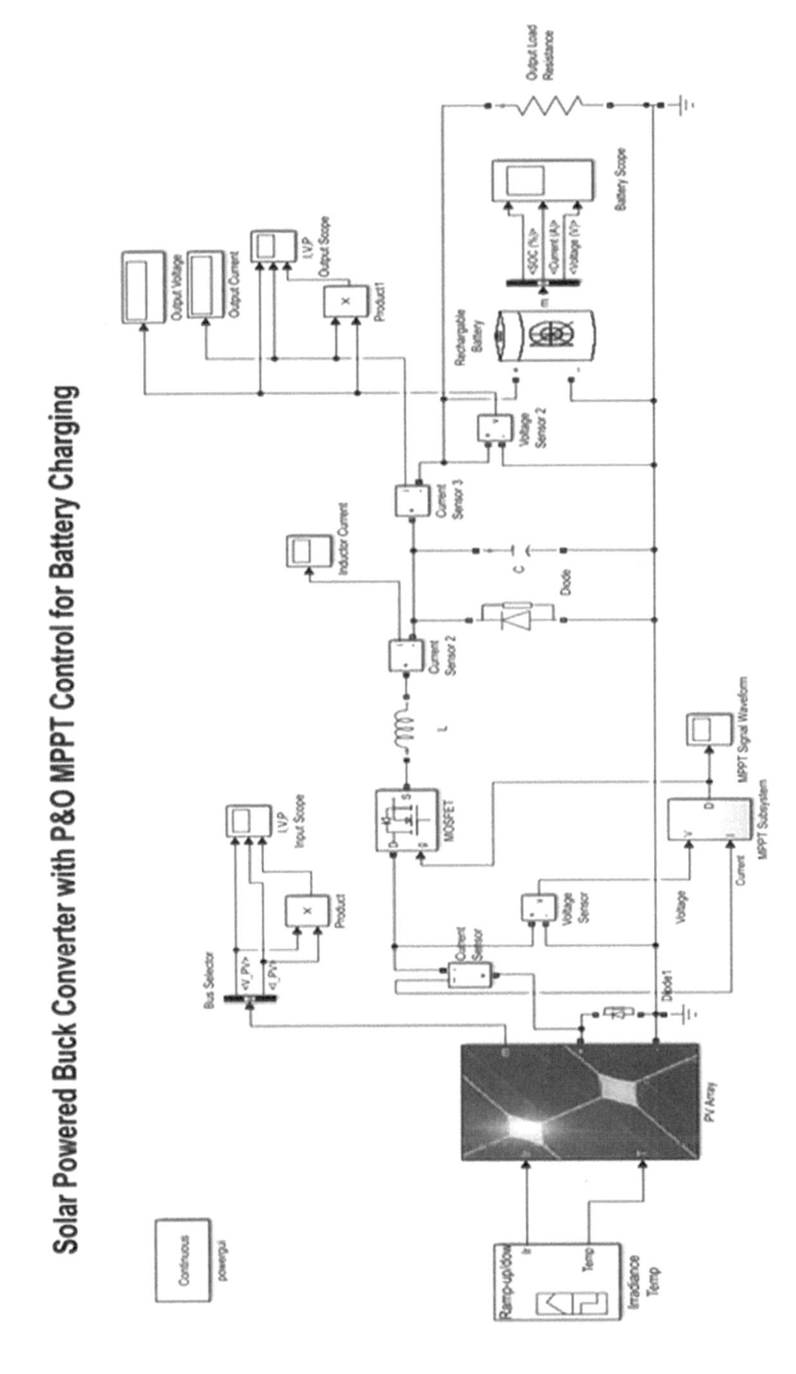

Figure 8.10 Solar powered DC-DC buck converter with MPPT control.

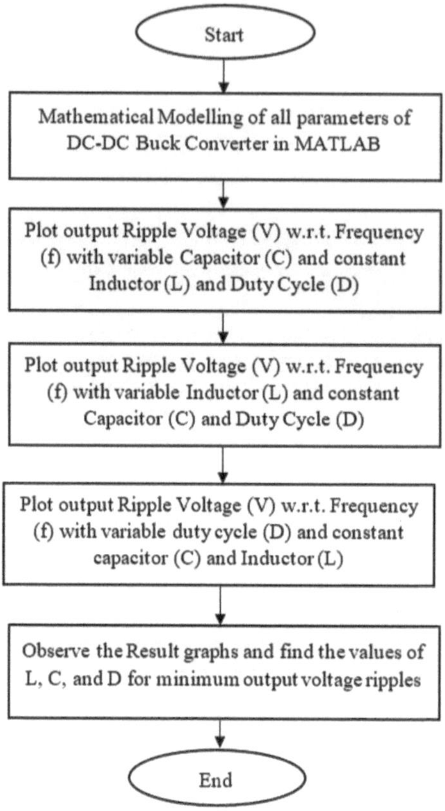

Figure 8.11 Flowchart for design and analysis of a DC-DC buck converter power stage for less output voltage ripples.

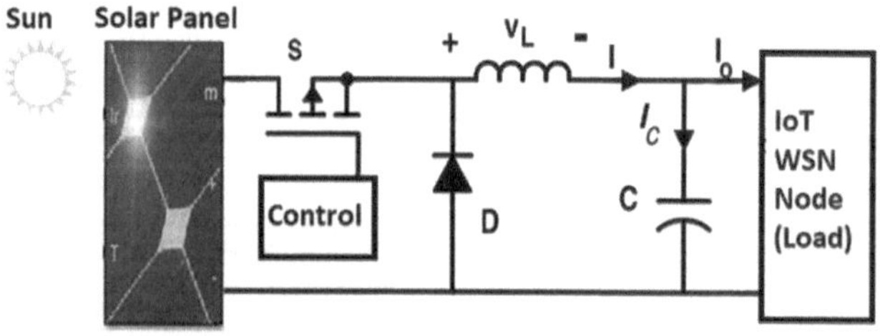

Figure 8.12 Solar powered DC-DC buck converter.

Here, it takes energy from the ambient light energy using the solar panel. The solar panel converts light energy directly into DC. But, this DC has noise and ripples. To remove this a buck converter is used. The buck converter removes ripples from the DC and supplies them to the IoT WSN node battery. Generally, all WSN nodes are battery-operated. Therefore, the battery lifetime in WSN is of critical importance. In Buck converter the MOSFET gate is supplied by a pulse width modulated (PWM) control signal. The output capacitor acts as a filter and stores the energy to supply the IoT WSN node. When the MOSFET switch is closed (ON condition), the input voltage V_{in} appears across the battery (or WSN load). If the MOSFET switch is open (OFF condition), then the voltage across the load is zero. The amplitude of output voltage V_o is less than the input voltage V_{in}.

8.13 ANALYSIS OF OUTPUT RIPPLE VOLTAGE IN BUCK CONVERTER

When the switch is closed (ON state)

$$V_L = V_i - V_o \tag{8.23}$$

When the switch is opened (OFF state)

$$V_L = -V_o \tag{8.24}$$

Energy stored in Inductor (L) is given as

$$E = \frac{1}{2} \times L \times I_L{}^2 \tag{8.25}$$

$$\text{Inductor voltage}, V_L = L\frac{dI_L}{dt} \tag{8.26}$$

The Increase in Inductor current (I_L) in ON state is given as

$$\Delta I_{L\,\text{on}} = \int_0^{t_{\text{on}}} \frac{V_L}{L}\, dt = \left(\frac{V_i - V_o}{L}\right) t_{\text{on}} \tag{8.27}$$

The decrease in inductor current in OFF state is given as

$$\Delta I_{L\,\text{off}} = \int_{t_{\text{on}}}^{t_{\text{off}}} \frac{V_L}{L}\, dt = \left(\frac{-V_o}{L}\right) t_{\text{off}} \tag{8.28}$$

For Steady State conditions, the energy stored at end of commutation cycle T is equal to the beginning of the cycle. Equating eqs (8.27) and (8.28)

$$\left(\frac{V_i - V_o}{L}\right)t_{\text{on}} - \left(\frac{V_o}{L}\right)t_{\text{off}} = 0 \tag{8.29}$$

In terms of duty cycle (D),

$$t_{\text{off}} = (1 - D)T \tag{8.30}$$

$$(V_i - V_o)D \times T - V_o(1 - D)T = 0 \tag{8.31}$$

$$V_o - DV_i = 0 \tag{8.32}$$

$$D = \frac{V_o}{V_i} \tag{8.33}$$

From eq. (8.27) we have,

$$\Delta I_{L\,\text{on}} = \left(\frac{V_i - V_{\text{HS}} - V_o}{L}\right) \times \frac{D}{F_s} \tag{8.34}$$

$$(\because t_{\text{on}} = DT, \text{and}\, T = 1/f)$$

where V_{HS} = High Side Voltage

The ripple current can also be expressed in terms of the inductor current ripple ratio (LIR) as:

$$\Delta I_L = LIR.I_{\text{out}} \tag{8.35}$$

$$V_{L(\text{off})} = V_{\text{out}} + V_{LS} \tag{8.36}$$

$$D = \frac{V_o}{V_i} = \frac{V_{\text{out}} + V_{LS}}{V_{\text{in}} - V_{\text{HS}} + V_{LS}} \tag{8.37}$$

Putting in eq. (8.34) and using the value from eq. (8.35)

$$L_{\text{min}} = \left(\frac{V_{\text{in}} - V_{\text{HS}} - V_o}{\text{LIR}}\right) \times \frac{D}{F_s} \tag{8.38}$$

Now introducing a new Peak to Peak Output Voltage which is dependent on Capacitor Voltage Ratio (CVR) as:

$$\Delta V_{PP} = CVR . V_{out} \tag{8.39}$$

The Equivalent Series Resistance (ESR) of the device will be maximum at ΔV_{PP}

$$\text{ESR}_{\text{max}} = \frac{\Delta V_{\text{PP}}}{I_{\text{out,max}}} = \frac{\text{CVR} \times V_{\text{out}}}{I_{\text{out,max}}} \tag{8.40}$$

The current through the output capacitor is given as

$$I_C = C \frac{\Delta V_C}{\Delta t} \tag{8.41}$$

$$\Delta t \times I_C = C \times \Delta V_C = \Delta Q_C \tag{8.42}$$

$$\Delta t = \left(\frac{1}{2}\right) t_{\text{on}} + \left(\frac{1}{2}\right) t_{\text{off}} \tag{8.43}$$

$$\Delta t = \left(\frac{1}{2}\right) \times \left(\frac{D}{F_S}\right) + \left(\frac{1}{2}\right) \times \left(\frac{1-D}{F_S}\right) = \frac{1}{2F_S} \tag{8.44}$$

and,

$$\Delta I_C = \frac{\Delta I_L}{2} \tag{8.45}$$

The minimum output Capacitance (C_{min}) due to ripple voltage is obtained by combining eq. (8.35) and (8.41) as:

$$C_{\text{min}} = \frac{\text{LIR}\, I_{\text{out,max}}}{8F_S \times \text{CVR} \times V_{\text{out}}} \tag{8.46}$$

8.14 RECHARGEABLE BATTERY

In general consumer applications, the rechargeable AA-size alkaline batteries have ratings of 1.5 V, 250 mAh. Our WSN nodes use two batteries to provide a 3 volts supply (2 × 1.5 = 3.0 Volts). As shown in Table 8.2, the

Table 8.2 Rechargeable battery specifications

S. No.	*Type of battery*	*Voltage rating*	*Current rating*
1	Alkaline Battery	1.5 volts	250–1000 mAh
2	NiCd and NiMH Battery	1.2 volts	500–2850 mAh
3	Li-Ion Battery	3.6 volts	600–850 mAh

Table 8.3 Li-ion battery life cycle

Depth of discharge (DoD)	*No. of charge/discharge cycles*
100%	~500
80%	~1000
60%	~1500
40%	~3000
20%	~9,000
10%	~15,000

Nickle Cadmium (Nicd) and Nickle Metal hydride (NiMH) batteries are rated at 1.20 V/cell or 1.25 V/cell with 500–2850 mAh ratings. The series combination of three Nicd/NiMH batteries can provide a 3.6 volts supply ($3 \times 1.2 = 3.6$ Volts) for WSN nodes. The nominal voltage of a single lithium-ion (Li-ion) battery is 3.60 V/cell or 3.7 V/cell with 600–850 mAh ratings. Thus, a single Li-ion battery is sufficient for supplying power to WSN nodes.

One strategy to increase the SEH-WSN lifetime can be considered to increase the battery lifetime. The battery charge/discharge cycles can be increased if the battery is not allowed to drain completely. For a Li-ion battery, the number of charge/discharge cycles as a function of Depth of Discharge (DoD) are shown in Table 8.3.

8.15 CONCLUSION

In this chapter we have discussed, smart grids, solar cells, DC-DC converters, MPPT algorithm, and rechargeable batteries. All these technologies are used for the integration of power electronics in renewable energy for smart cities. We also discussed design challenges for energy harvesting in to DC-DC converters.

REFERENCES

[1] PV Module DPS-10-1000, Manufacturer specifications datasheets, Dow Chemical Company, USA.

[2] https://www.smart-energy.com/features-analysis/getting-ready-to-operate-the-smarter-grid/ (Accessed 18/02/2022).

[3] Steve Roberts, *DC/DC Book of Knowledge and Practical Tips for the User*, RECOM, Austria, 2016.

[4] Domenico Balsamo, Davide Brunelli, "Sleep Power Minimization Using Adaptive Duty-Cycling of DC-DC Converters in State-Retentive Systems", *IET Circuits, Devices & Systems*, Voluem: 8, Issue: 6, 2014, pp. 478–486.

[5] C. S. Solanki, *Solar Photovoltaics: Fundamentals, Technologies, and Applications, Prentice-Hall of India (PHI)*, 3rd Edition, Delhi, India, 2015.

[6] M. H. Rashid, Narendra Kumar, Ashish Rajeshwar Kulkarni, *Power Electronics: Circuits, Devices & Applications*, 4th Edition, Pearson Education, Delhi, India, 2013.

[7] Ned Mohan, T. M. Undeland, W. P. Robbins, *Power Electronics: Converters, Applications, and Design*, 3rd Edition, John Wiley & Sons Inc., California, 2007.

[8] Texas Instruments, USA, data sheets "TPS54620 4.5-V to 17-V Input, 6-A, Synchronous, Step-Down Converter" specification datasheet, May 2017. Available online: http://www.ti.com/product/TPS54620

[9] Texas Instruments, USA, "Application Report on Calculating Efficiency of DC-DC Controllers SLVA 390" February 2010. Available online: http://www.ti.com/lit/an/slva390/slva390.pdf

[10] Texas Instruments Application Report, "Basic Calculation of a Buck Converter's Power Stage for Low Power DC/DC Applications", August 2015. (Available on www.ti.com).

[11] Ian Mathews, Paul J. King, Frank Stafford, Ronan Frizzell, "Performance of III–V Solar Cells as Indoor Light Energy Harvesters", *IEEE Journal of Photovoltaics*, Volume: 6, Issue: 1, January 2016, pp. 230–235.

[12] Denis Dondi, Alessandro Bertacchini, Davide Brunelli, Luca Larcher, Luca Benini "Modelling and Optimization of a Solar Energy Harvester System for Self-Powered Wireless Sensor Networks", *IEEE Transactions on Industrial Electronics*, Volume: 55, Issue: 7, 2008, pp. 2759–2766.

[13] Sheng Liu, Yi Zhao, Menglian Zhao, "A Burst-Mode Based Boost Converter Harvesting Photovoltaic Energy for Low Power Applications", *IEEE 57th International Midwest Symposium on Circuits and Systems (MWSCAS)*, 2014, pp. 49–52.

[14] Guolei Yu, KinWai Roy Chew, Zhuo Chao Sun, "A 400 nW Single-Inductor Dual-Input–Tri-Output DC-DC Buck-Boost Converter with Maximum Power Point Tracking (MPPT) for Indoor Photovoltaic Energy Harvesting", *IEEE Journal of Solid-State Circuits*, Volume: 50, Issue: 11, November 2015, pp. 2758–2772.

[15] Amzar Omairi, Zool H. Ismail, Kumeresan A. Danapalasingam, Mohd Ibrahim, "Power Harvesting in Wireless Sensor Networks and Its Adaptation with Maximum Power Point Tracking (MPPT): Current Technology and Future Directions", *IEEE Internet of Things Journal*, Volume: 4, Issue: 6, 2017, pp. 2104–2115.

[16] Farhan I. Simjee, Pai H. Chou, "Efficient Charging of Supercapacitors for Extended Lifetime of Wireless Sensor Nodes", *IEEE Transactions on Power Electronics*, Volume: 23, Issue: 3, 2008, pp. 1526–1536.

[17] Hai Chen, Bingqing Wei, Dongsheng Ma, "Energy Storage and Management System with Carbon Nanotube Supercapacitor and Multidirectional Power Delivery Capability for Autonomous Wireless Sensor Nodes" *IEEE Transactions on Power Electronics*, Volume: 25, Issue: 12, 2010, pp. 2897–2909.

[18] Carlo Bergonzi, Davide Brunelli, Luca Benini, "Algorithms for Harvested Energy Prediction in Batteryless Wireless Sensor Networks", *3rd IEEE International Workshop on Advances in Sensors and Interfaces*, Italy, 2009.

[19] Selahattin Kosunalp, "New Energy Prediction Algorithm for Energy-Harvesting Wireless Sensor Networks with Q-Learning", *IEEE Access*, Volume: 4, 2016, pp. 5755–5763.

Chapter 9

Integration of IoT in renewable energy for smart cities

Shrankhla Saxena
KIET Group of Institutions, Ghaziabad, India

Himanshu Sharma
Noida Institute of Engineering & Technology, Greater Noida, India

Abhinav Juneja and Khushboo Pandey
KIET Group of Institutions, Ghaziabad, India

9.1 INTRODUCTION

The increasing population of cities demands faster growth and fast growth of cities is difficult because of many reasons like overcrowding, limited resources, and many more [1]. Integration of technology can solve these issues and speed up the growth of cities. The use of smart services, sensors, and actuators can provide infrastructure for smart cities. Applications of the smart city like smart healthcare, smart homes, intelligent transportation, and energy control need central data storage and huge data communication [2]. Therefore, IoT and cyber systems are the most fundamental factors for the development of smart cities. Conservation of natural energy sources like heat, wind, and light is a prime factor in smart cities [3]. Smart grids are used for managing and distributing power and energy in smart cities. And these smart grids are controlled by IoT. Therefore, for the conservation and management of renewable energy in smart cities IoT integration with smart grids is needed [4]. To control the energy consumption in applications like smart homes and smart vehicles, smart grids are used. Energy management policies for cities are provided by the information management system in smart grids. To operate energy management systems smoothly in buildings, automatic detection of energy consumption is required. Many researchers have proposed diagnosis methodologies for this purpose. Due to a lack of knowledge about the usage of energy in buildings, smart cities are unable to take the maximum advantage of energy management systems. To be more secure, connected, intelligent, adaptable, and unceasing, smart cities need the latest technological advancements like information and communication technology (ICT), Artificial Intelligence & Machine Learning (AI & ML), blockchain, and IoT. Figure 9.1 shows the overview of IoT-based services in a smart city [5].

DOI: 10.1201/9781032669809-9

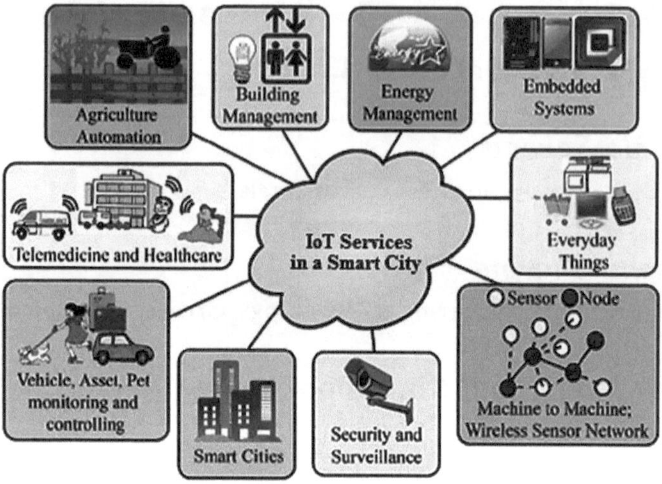

Figure 9.1 Overview of IoT-based services in a smart city.

9.2 USE OF BLOCKCHAIN IN SMART CITIES

Smart cities provide many services like smart transport, smart homes, smart healthcare systems, smart banking, and many more. These services demand high security as private data of citizens is used and handled. To provide increased security in smart cities, the use of blockchain technology is a better option. Blockchain and ICT can be used for the welfare of citizens, the environment, administrative services, and the economy in smart cities [6]. The very first purpose of the smart city is to provide reformed fundamental services such as residence, education, medical care, transportation, water, safe environment, and so on, and the infrastructure of a smart city should be suitable for dealing with population management and urbanization. To provide better connectivity and enhanced security in smart cities many AI technologies can be used like blockchain, big data, IoT, ML, and so on. Blockchain can record various transactions in smart cities. Smart contracts can make tedious legal procedures easy and simple, and data can be shared automatically. Decentralized applications in blockchain help in the automatic execution of transactions in the smart city. Blockchain provides better authentication, high protection, and easy maintenance for a smart city. Blockchain is a decentralized network in which distributed records are stored in a sequence of blocks. Blockchain uses the Merkle tree arrangement to store the records. Records are stored permanently in the Merkle tree using the cryptographic hashing technique. The root node of the Merkle tree is stored in the header block [7]. A header block keeps the record of the hash value of the previous block, which creates a time-stamped chain. To provide better authentication for the operations, blockchain uses digital signatures.

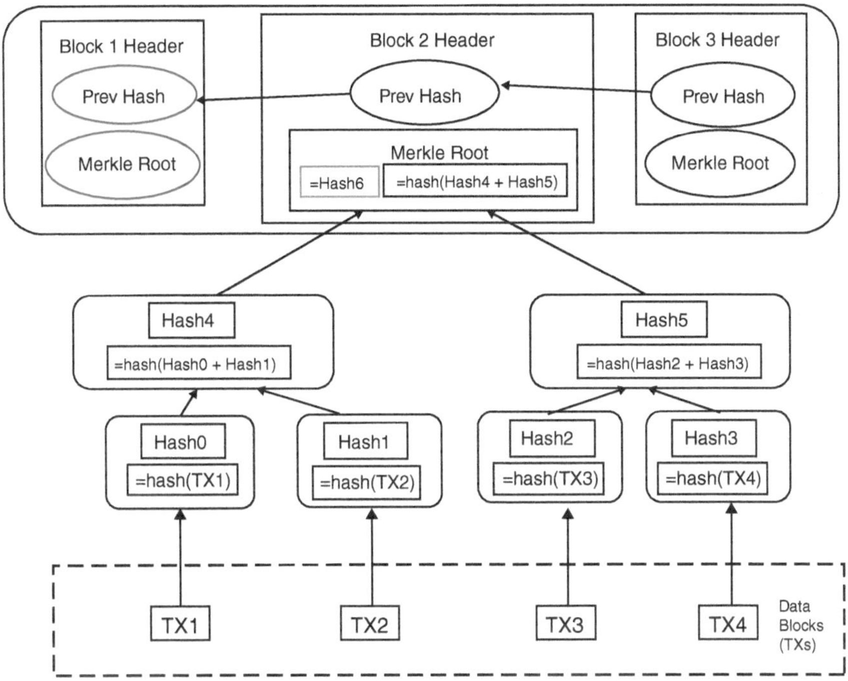

Figure 9.2 Blockchain architecture with Merkle tree and chain of blocks.

Hash is used to encode the input string of random size into a fixed size. The hash function takes the input string and outputs the corresponding hash value. In the blockchain, the hash function is used for secured storage, digital signature, and block extraction during transactions.

Merkle tree provides protected and fast storage for data. Large data can be verified securely in the Merkle tree. All leaf nodes of the Merkle tree are the hash functions with a combination of two child nodes. Merkle root is the root node of the Merkle tree. Updating any data block in the Merkle tree will update the root node also. There is no need to download the whole block for authentication. Partial nodes in blockchain enable verification by downloading only a branch of the block [8] (Figure 9.2).

9.3 APPLICATIONS OF IOT & BLOCKCHAIN IN SMART CITIES

- **Smart e-commerce:** In e-commerce, sellers and buyers sell and buy products through online platforms. Many times, these e-commerce platforms perform their transactions through trusted third parties. Blockchain removes the need for intermediate retailers for those transactions of e-commerce. To trace the intermediary carriers, logs of orders can be used in the blockchain. Some researchers proposed a blockchain-enabled framework for proof of delivery for all physical goods [9].

- **Smart e-voting:** Smart cities make use of ICT for automating the process of governance through an e-voting system. E-voting enables citizens to poll or vote digitally. The biometric method is used to check the authenticity of voters. Although the e-voting system suffers from cyber security threats, blockchain provides multi-point protection due to which a single point cannot cause failure to the e-voting system. With blockchain, all users get a private key to authenticate their poll through a digital signature which is then appended to a digital ledger. In a blockchain-enabled e-voting system, all voters get a wallet that consists of a private key. This wallet gets the credit of one coin which is used for voting. Blockchain technique can validate every voter by keeping his identity hidden at the time of final counting of votes. Figure 9.3 shows the conceptual framework of smart e-voting in a smart city [10].
- **Smart healthcare:** Smart healthcare is the main asset of the smart city. The quality and security of healthcare can be improved drastically with the use of blockchain. Electronic records of all patients can be stored with enhanced protection using blockchain. Smart contracts in blockchain can provide better protection for access and retrieval of patients' data. Sharing of private and health records of patients among

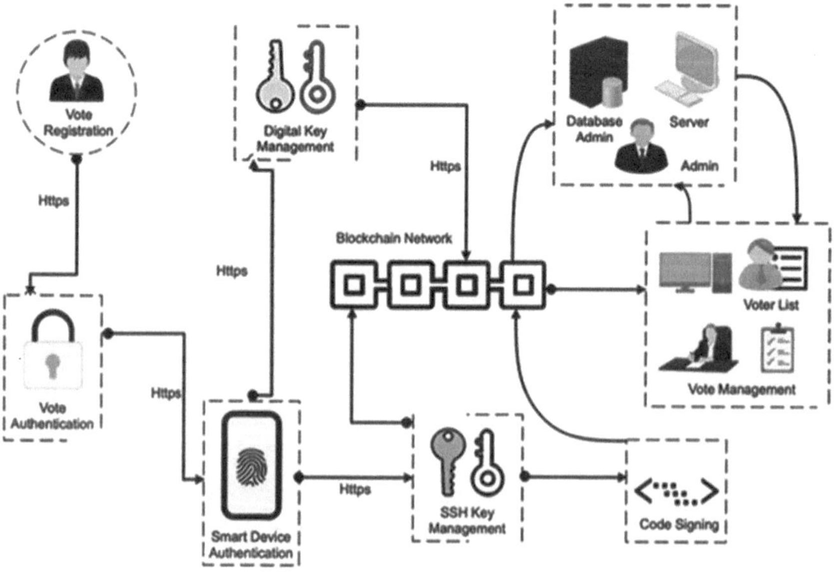

Figure 9.3 Smart e-voting in a smart city.

multiple parties is more secure due to the distributed architecture of blockchain [11]. Figure 9.4 represents the general architecture of a smart healthcare system [11].

- **Smart transport:** Smart transportation plays a critical role in a smart city. Smart transportation involves sensors, wireless networks, speed detection camera systems, automatic number plate recognition, and methods to manage traffic and make it safe and fast. Blockchain offers a solution to many transportation problems like communication among automobile devices, and the connection among roadside devices. Blockchain removes the need for a central mediator from the financial transaction with the help of feature-double-spending resistance. Blockchain offers a secure and efficient eco-system for ride-sharing transportation. Figure 9.5 represents the architecture of smart transportation using blockchain [12].
- **Smart grid:** Smart grid is a power grid capable of providing more efficiency and reliability. Smart grid is consisting of sensors, power generators, resources of renewable energy, smart appliances, and communication channels. Smart grids can produce and distribute power automatically in a controlled manner. Smart grids can sense, monitor, communicate, analyze, visualize, compute, control, automate, diagnose, and maintain the dump power delivery system [7]. Smart grids can fulfill the demand of power supply, and reduce loss of power with increased reliability and sustainability. Hence, smart grids are a necessary tool for a smart city. Blockchain eliminates the need for a central authority in the peer-to-peer networks for power supply infrastructure. Figure 9.6 represents the blockchain-based smart grid infrastructure [13].

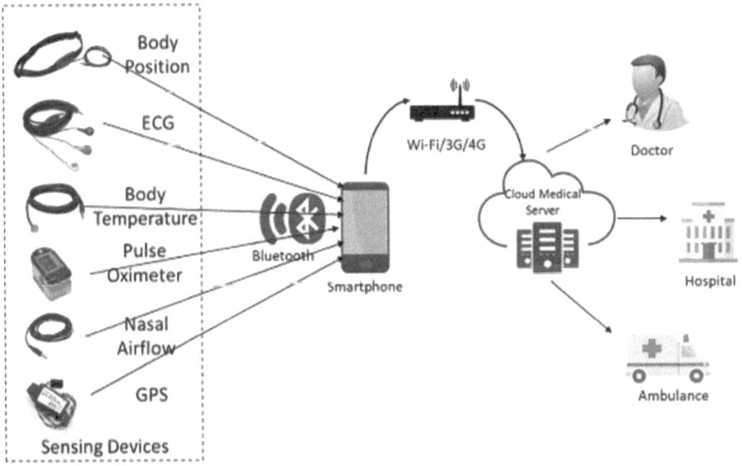

Figure 9.4 Smart healthcare system in a smart city.

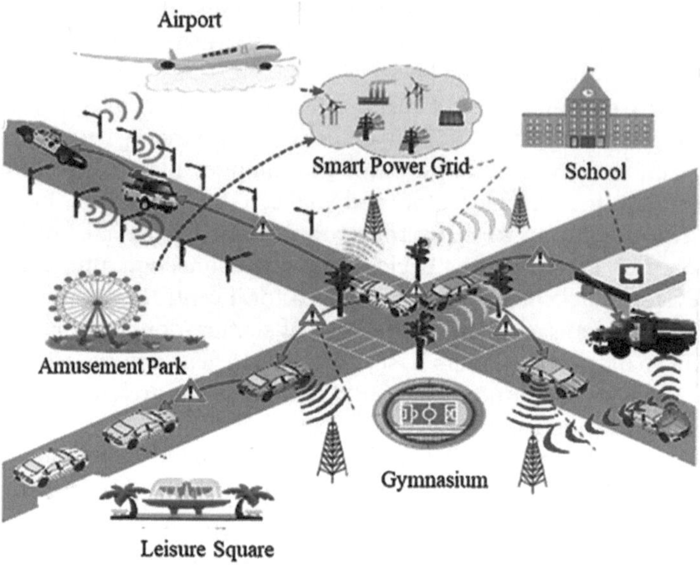

Figure 9.5 IoT and blockchain-based smart transportation.

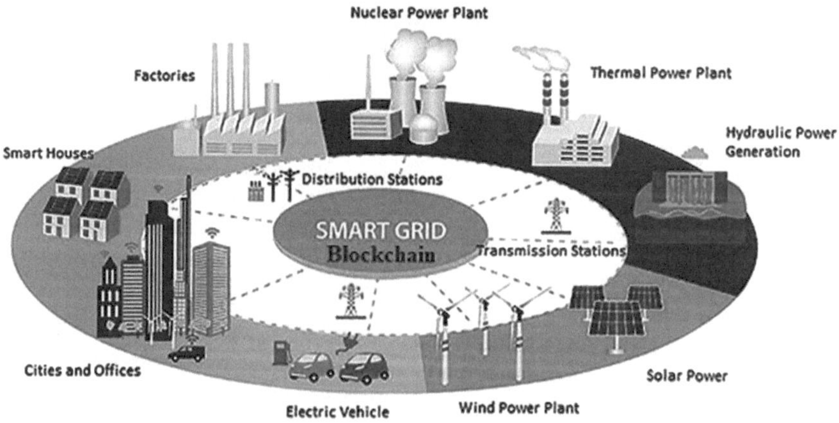

Figure 9.6 IoT and blockchain-based smart grid infrastructure.

9.4 ROLE OF IOT IN SMART CITIES

With the capability of automatic communication of smart devices with humans and other devices, citizens' life has become smarter. Internet of Things (IoT) made this possible so that devices and appliances can communicate and share information. Information can be shared globally over the internet [14]. IoT is a stream that involves systems, technologies, and principles of design for getting the devices interconnected. Nowadays, many companies utilize IoT services for their process optimization, for improvement in data collection, for cutting down the operational costs, and so on. "IoT is a smart earth", proposed by an international business machines corporation [15]. So, we can say that IoT can create a smart city. Today we are surrounded by connected and intelligent devices ranging from smart wearables to smart homes in smart cities. IoT enables smart devices to create their social network with other devices and humans.

To deal with the problems of energy management systems, IoT makes use of various AI techniques like edge computing. There is a variety of AI approaches existing to solve the issues of energy optimization. To save more energy, IoT provides various ICT and AI-based solutions. For the reliable power supply and for cutting down the cost of fuel, economic load dispatch (ELD) can be an effective method in IoT. To solve the issues of ELD, genetic algorithm and PSO (particle swarm optimization) methods can be used effectively [16]. Local control centers send the required data to the AI model to perform optimized calculations. Artificial neural networks (ANN) and support vector machines (SVM) can be efficiently used to forecast demand and consumption of energy. Intelligent AI models are used to develop energy management systems to save energy and ultimately to protect the environment and the quality of citizens. With the rapid growth of information technologies, energy sectors now have a digital energy system which generates a good amount of energy. The smart grid enables data transmission and collection parallelly. Integration of big data in smart grids is used to accurately predict electricity demand, find the patterns of energy consumption, and improve the production of real-time power. Smart grids integrated with big data analytics help in restoring the failures, responding more quickly against electricity demand, and supplying reliable energy. All operators of the smart grid make use of an efficient decision support system provided by big data analytics. Big data has brought enormous change in the generation and consumption of energy. Big data integrated IoT systems have been proved to be the most effective and cost-saving energy production and management systems.

The interoperability feature and ability to communicate among heterogeneous devices make IoT more significant and important for the development of smart cities. Smart cities utilize natural energy to establish green communication among devices and users. This way IoT helps to achieve a pollution-free environment in smart cities. IoT infrastructure controls the smart grids for managing and distributing power in the smart city environment. Thus, smart cities can improve their power management by integrating smart grids with IoT and ICT [17].

A smart grid is an electrical grid in which IT systems, automation, and communication capabilities are integrated so that flow of power can be tracked from production point to consumption points. Smart grids can regulate and limit the flow of energy. Smart grids can also control the flow of energy and can find out the loss. IoT-enabled energy management brings the benefits of distributed computing, storage, retrieval, and energy distribution to control the power grid. The energy management system needs some information like the requirement of power, deficiency, and rate of allocation. A smart grid is a network of electricity that allows bi-directional digital and electric transmission of data which helps in tracking the flow of data. Smart grids have the capability of healing themselves and involving consumers of electricity. Green communication helps in conserving power for a long time with the use of devices that are energy efficient. The basic building block for a smart city is a platform for sharing information through the combination of three networks. Three integrated networks are required to offer more speed of information services to all intelligent units [18]. For the construction of smart buildings three components perform a critical role: 1) data exchange, 2) automation of buildings, and 3) high security. The construction of a smart city needs to cover all aspects like the needs of daily life, business, entertainment, safety, traffic management, power supply management, environment protection, and many more. Figure 9.7 represents the components of the smart city.

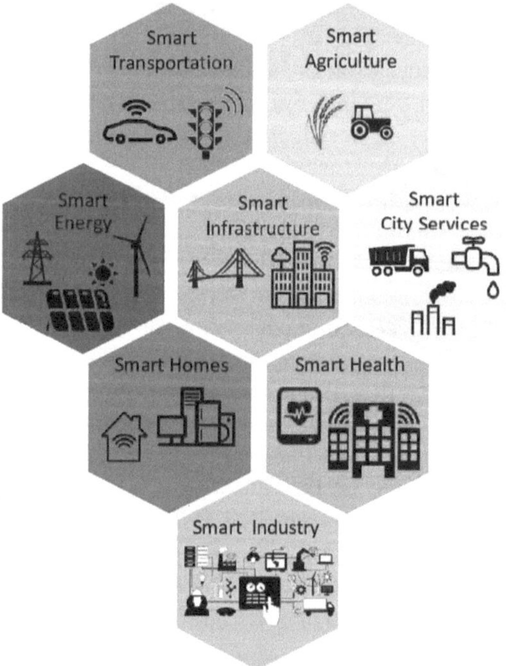

Figure 9.7 Important parts of a smart city.

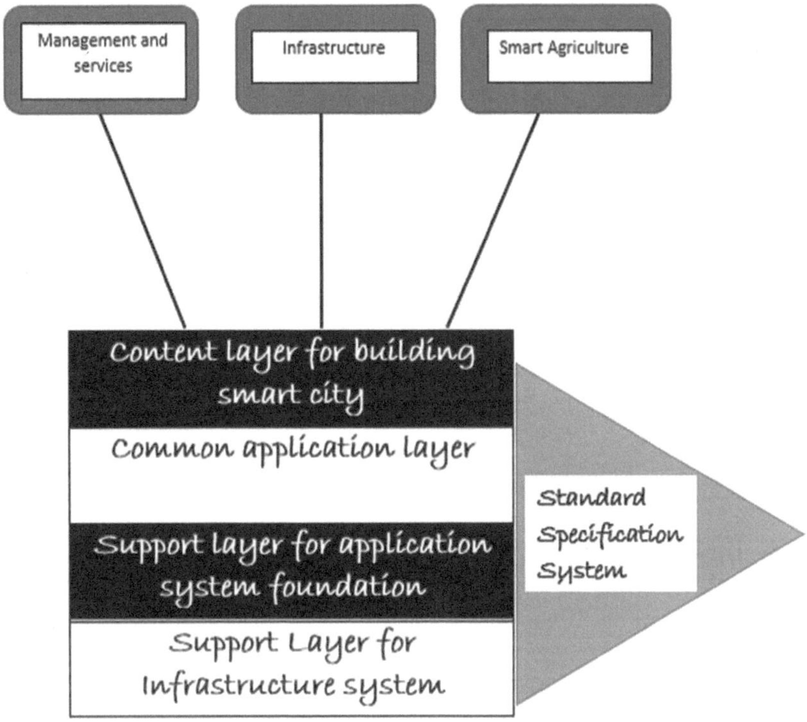

Figure 9.8 IoT architecture for building a smart city.

To provide better services to citizens, information and communication technology is the basic building block for the construction of the smart city. Automatic data collection and sharing is a crucial part of the smart city. Collecting data and creating an intelligent perception of all this intelligent IoT can be very helpful. The architecture of IoT can be divided into different layers for this purpose. Figure 9.8 represents the architecture of IoT needed in the construction of the smart city [19].

9.5 IOT-ENABLED SMART GREEN ENERGY

Sensors linked to generating, transmission, and distribution equipment are used in IoT applications in renewable energy production. These gadgets allow businesses to remotely monitor and manage the operation of equipment in real-time. This minimizes our reliance on already restricted fossil fuels while also lowering operational expenses. Renewable energy sources already have several advantages over traditional energy sources [20]. We will be able to use these renewable energy sources to a greater extent using IoT as shown in Figure 9.9.

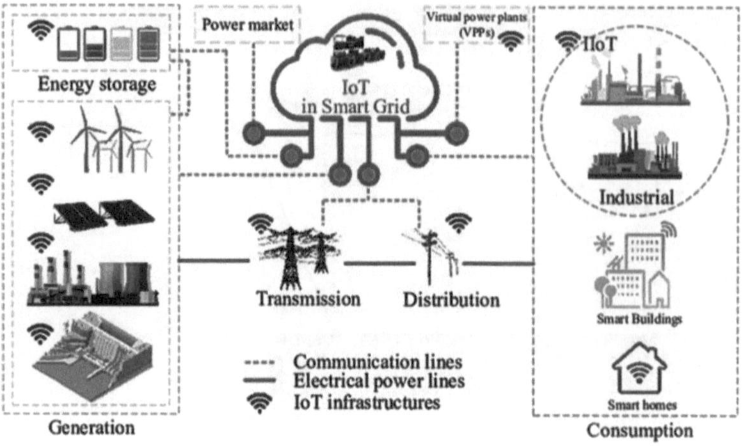

Figure 9.9 IoT-enabled smart green energy management.

9.6 ADVANTAGES OF IOT IN RENEWABLE ENERGY

- Automation
- Cost-efficiency
- Smart grid management
- Smart power distribution
- Residential solution

9.6.1 Automation

The usage of IoT devices for monitoring and efficient operation of wind turbines is a great example of automation utilizing renewable energy. By evaluating the data collected by IoT sensors, the orientation of the wind turbine may be changed to achieve optimal efficiency. Solar power usage is a comparable example. Solar panels can be angled such that they are perpendicular to the source of energy. This ensures that solar energy is fully used. To optimize power output, these modifications might be made manually or automatically. The power plant can function more effectively with automated controls. This increases power production while simultaneously lowering running expenses and raising safety requirements [21].

9.6.2 Cost-efficiency

The internet of things provides great power usage monitoring tools. IoT technology provides utility businesses and electrical suppliers greater control over their resources. Companies may then use this information to make data-driven business choices. Power distribution firms can utilize

IoT-generated data to evaluate and analyze consumers' power usage trends. Utilities can balance supply depending on customer demand [22].

9.6.3 Grid management

The GMS is a system of systems (SoS) that offers a complete grid management solution in an increasingly complicated distribution environment. It takes the place of the existing Outage Management System (OMS) and the old Distribution Management System (DMS). Integration of IoT technology not only allows for the addition of new equipment to the grid but also enhances grid management overall. Companies can acquire real-time power usage data by installing sensors at substations and along distribution lines. This data can help energy businesses make better judgments about voltage regulation, load switching, and network layout. Grid sensors can also assist operators in receiving real-time notifications regarding disruptions. Workers can swiftly switch off the electricity to damaged wires since real-time data is available. This minimizes the risk of electrocution, wildfires, and other dangers [23].

9.6.4 Distributed system

The Smart Grid Distribution Management System enables smart grids to be fitted with improved distribution strategies in order to reduce electricity expenditures. Furthermore, the suggested system uses MAS as a communication channel, which improves data transfer reliability and efficiency. Because of the increase in business and residential use, the smart energy grid is developing. For energy businesses, the increasingly dispersed power grid marks a significant shift. They must now manage an increasing number of tiny generators scattered over the grid in addition to their major units. IoT enables easy monitoring of these widely dispersed smart grids. Sensors positioned at numerous locations in the manufacturing and transmission phases can assist in the monitoring of a large number of points [24].

9.6.5 Residential solutions

The advantage of using IoT in renewable energy is that we can obtain efficient energy management for smart homes by properly scheduling household appliances, such as lowering peak demand and lowering electricity costs. Citizens may create "green energy" in their home gardens with the help of IoT devices to suit their home requirements. IoT infrastructure solutions aid in the optimal utilization of renewable energy. Users may integrate their solar panels, rainwater harvesters, smart roofs, and windows into a single system using IoT. Users can use a desktop or mobile application to control the operation of their electrical gadgets [25].

9.7 DIFFICULTIES IN IMPLEMENTING RENEWABLE ENERGY WITH IOT

- Cost of the original investment
- Susceptible to cyber-attack
- DC-DC converter design
- Choice of MPPT technique
- Data management

9.8 DESIGN CHALLENGES IN INTEGRATING RENEWABLE ENERGY WITH IOT

- **Sensing level:** At sensing level power management, the main design problems are duty cycling or adaptive sensing. Duty cycling means the transmission of measured data after a certain interval to reduce power consumption during transmission and reception. Thus the power consumption of IoT node is reduced by activating the sensors only at specific times at which the samples need to be taken. On the other hand, Adaptive sensing refers to the sensing the measuring quantity (temp., light, humidity, pressure etc.) only when desired. Figure 9.10 shows various design challenges in integrating renewable energy with IoT. Otherwise we keep the sensors in OFF or in the sleep condition. The

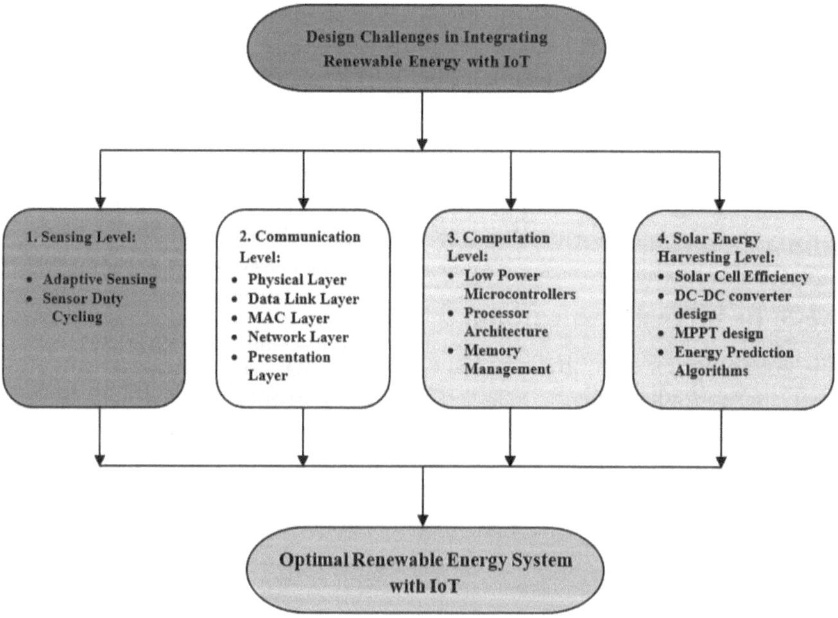

Figure 9.10 Design challenges in integrating renewable energy with IoT.

adaptive sensing and sensor duty cycling techniques save the energy of the IoT node at the cost of latency of data packets [26].

- **Computation level:** The main microcontrollers used in IOTs are Microchip PIC series, ATMEL AVR series, and Texas Instruments MSP430 series. The microcontroller provides power management features such as battery state of charge (SoC) information, MPPT algorithm, and duty cycling of sensors and a transceiver unit [27].
- **Communication level:** At the communication network level power management, the major design challenge is the suitable designing of OSI network layer parameters. At the physical layer, the network designers have the choice of energy harvesting compatible modulation techniques and channel conditions. The suitable design of open systems interconnect (OSI) model layers compatible for EHIOTs are also available [28].
- **Energy harvesting level:** The biggest challenge in the solar EHIOT is designing of its energy harvesting stage. The energy harvesting stage design dominates over other design levels. Here, solar cell efficiency (η), MPPT, DC-DC boost converter, solar energy prediction algorithms, and energy storage technologies are designed. The *I–V* characteristics equation of solar cell is given as [29–30]:

$$I = I_L - I_O\left[\exp\left(\frac{V}{V_T}\right) - 1\right] \tag{9.1}$$

Where

I = terminal current of solar cell,
I_L = current generated due to light radiation from the sun,
I_o = Reverse saturation current,
V = Diode voltage, V_T = Thermal voltage (kT/q), T is temperature (K), k is Boltzmann's constant (J/K), and q is the charge of an electron (Coulombs). The thermal voltage (V_T) is equal to 26 mV at room temperature (300 K).

At solar energy harvesting level, the following design parameters needed to be handled carefully as (as shown in Table 9.1):

- Variations in solar radiation level
- Solar cell efficiency (η)
- DC to DC converter
- Energy prediction algorithms
- Elimination of off-chip inductor in IC design
- Maximum Power Point Tracker (MPPT) design

Xiangdong et al. [31] introduced IoT-enabled smart green energy which can manage the requirements of energy and allocation of smart power systems.

Table 9.1 Design challenges in integrating renewable energy with IoT

S. No.	*Technology level*	*Design challenges/problems*	*Effects*	*Our proposed solution to the problem*	*Optimized results after solution*
1.	Sensing Level	Duty cycling Adaptive sensing	Data packets Latency (delay)	$t_{on} \geq t_{wakeup} + t_{acquire}$ where, t_{on} = sensor node on time t_{wakeup} = wake up time $t_{acquire}$ = time to acquire the measured information	No delay
2.	Computation Level	Low Power Consumption Microcontroller (CPU) architecture (MSP430 series)	Slow clock speed, No pipeline, no load sharing architecture	Texas instruments 16-bit RISC processor (MSP430) series microcontrollers	Lowest power consumption
3.	Communication Network level	Routing layer protocols design	Packet latency, less reliability	sLEACH protocol (an extension of LEACH) harvesting-aware clustering technique for routing data	Increase the overall lifetime by 10%–45%
		MAC layer protocols design: (Error control codes, packet framing, encapsulation)	Packet loss, channel noise, Less Bit Error Rate (BER)	CSMA-based and TDMA-based protocols	No channel loss, high BER
		Data collection: (information dissemination i.e. flow of data)	More energy consumption during information flow	Event-driven dissemination approach	Less energy consumption
		Physical layer (modulation, channel conditions, node cluster & deployment)	More energy consumption	BPSK, OQPSK	Less energy consumption

4.	Energy Harvesting Level	Solar cell device level efficiency	Less photon to electron conversion efficiency (η) of solar cell	Less electrical power output (P_o)	Thin film technology solar cells should be used to improve efficiency	Efficiency is improved
			Temperature effects on efficiency (η) Solar cCell	An increase in solar cell temperature of 1 °C results in a decrease in efficiency (η) of 0.45%.	A transparent silica crystal layer can be applied to a solar panel acting as heat absorber plate	Efficiency is improved by cooling the solar cell
		DC-DC converter design		Poor DC-DC converter design affects Output Voltage regulation	Single Shared inductor-based MOSFET DC-DC converters	Better output voltage regulation across load
		MPPT design		Less efficient MPPT design results in less solar power extraction	In MPPT, the Hill-Climbing or Perturb & Observe (P&O)	Max. solar power extraction

Energy requirements of users can be fulfilled by smart power systems. Smart power systems can conserve more energy by prohibiting wastage. At present, the biggest challenge is to decrease the use of energy. An IoT-integrated smart city makes use of sensors and other devices to collect the required information from various features of city infrastructure which involves highways, water, trains, communication, bridges, electricity, buildings, and so on and the analysis of gathered data is done through data science and a predictive analysis approach which helps citizens to be more considerate for the usage of energy. To develop the environment of a smart city, renewable energy resources or power grids are connected to an IoT platform that is autonomous and omnipresent. IoT platforms use their computational intelligence for the allocation and distribution of energy. This helps to reduce the wastage and loss of power hence enhancing energy conservation. ML algorithms are used which helps in balancing the requirements and allocation of power grids. IoT helps save natural energy resources, reducing the harmful effects on the environment, and minimizing the cost. Hence, green IoT emphasizes green manufacturing, green design, green exploitation of resources, and green ejection.

Green IoT solutions involve the reduction of CO_2 emissions and minimizing the usage of IoT energy. Green IoT focuses on creating and using green features. The design of green IoT comprises computing devices, communication protocols, the architecture of the network, and efficiency measures of energy. Some researchers proposed that green ICT can increase energy efficiency and decrease CO_2 emissions. Collecting the information from the environment of the smart city provides the important features of smart cities that must be used to create the intelligent model. Green ICT helps in decreasing cost, energy consumption, and pollution hence improving the life quality of citizens. Green IoT is a set of the latest technologies that make the IoT environment friendly and allow users to store, retrieve, and manage data. ICT plays an important role in greening the IoT by saving energy needed in designing and manufacturing.

9.9 IOT COMPONENTS FOR SMART ECO-SYSTEM

The IoT components of a smart eco-system is shown in Figure 9.11.

- **Sensors/actuators:** Sensors and actuators are required to establish interaction with other devices and humans in the physical world. These components take required data from the environment and provide the collected information to the respective processing unit. Temperature sensors, light sensors, pressure sensors, and ultrasonic sensors are some of the most frequently used sensors. Sensors must be chosen wisely to work for different types of applications in smart cities [32].

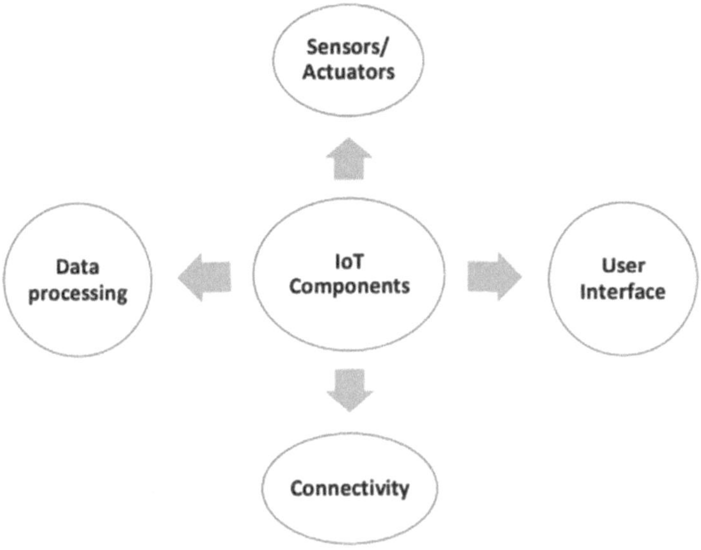

Figure 9.11 Components of IoT for smart eco-system.

- **Connectivity:** Sensors gather the required information which is then transferred to cloud storage for processing. A variety of connectivity technologies are available for this purpose, for example Wi-Fi, Cellular Networks, Bluetooth, NFC, etc. and to use these technologies various protocols are available like AMQP, MQTT, CoAP, DDS, and many more [33].
- **Data processing:** Data stored in the cloud is then processed by cloud analytic software which uses different tools to retrieve useful information from stored data.
- **User interface:** At last, the required information is provided to the user. This information is provided using various methods like emails, texts, notifications, or triggering alarms, etc.

9.10 POTENTIAL THREATS AND RISKS IN IOT WITH POSSIBLE SOLUTIONS

- **Internal security risk in IoT** [34]: To provide better connectivity in smart cities, the number of IoT devices is increasing every day. This increasing number of smart devices leads to more security issues for its users. These smart devices may suffer due to internal threats as these small networks are easier to hack. Some forms of internal threats are distributed denial of service attacks, violation by the operator of the security policy of the network, unauthorized access into the network,

malware attacks, and many more. There are two more types of internal threats to IoT – undirect and direct attack. In undirect attacks, smart devices are attacked by malware before access to the network because in general, malware is used to attack more and more devices rather than attacking the specific network.
In a direct attack, malware is embedded in the smart device. Since these smart devices are directly connected to the internal network, so the entire network gets corrupted. In a direct attack, the hacker mostly gets the control rights of smart devices in the network. A possible solution for such attacks can be creating a whitelist. A device identification system is used to identify the devices which are not on the whitelist. Access to the internal network for such devices is restricted. Moreover, for devices present in the whitelist, security measures are checked by the security management system to impose restrictions for sharing data to/from the network. The protection method using the whitelist is very effective in avoiding the connection with hazardous devices and hence securing the IoT network.

- **Transmission threats in IoT networks:** The physical environment is combined with the digital environment using IoT. IoT uses different sensors and communication technologies to identify the devices automatically and to exchange useful data among them. Due to the continuous evolution in the IoT framework, security mechanisms are still partially effective. The user interface, networking protocols, and passwords are three major aspects that can allow attacks in IoT networks. Generally, the firmware uses the default password of smart devices due to which hackers can easily access the devices. Weak passwords of smart devices are also major loopholes for the security of IoT networks. Since wireless IoT networks are open for all, hackers can easily use eavesdropping to steal the private information of users. Encryption must be used to protect the IoT networks from such attacks [35].

Another threat to IoT networks is the replay attack. Replay attacks are used to send messages repeatedly over the network to the targeted device. These attacks can be more harmful because packets sent in these attacks are legal and many firewalls do not block them. Such attacks can be prevented by applying a timestamp mechanism or random number mechanism in IoT networks.

9.11 OTHER CHALLENGES FOR INTEGRATING IOT IN SMART CITY DEVELOPMENT

- **Connectivity and sharing of huge data:** Smart devices produce large sets of data and this huge dataset is shared among devices, cloud infrastructure, and user applications which is very challenging. The number of network devices is going high every day and all devices use different tools and technology which increases the challenge more. These challenges can be addressed by making use of technologies with low energy consumption like LoRa, Sigfox, etc. [36].

- **Reliability, scalability, and availability:** The performance of an IoT network should not degrade after adding the new devices with different storing and processing capabilities. It must be scalable for heterogeneous devices and services. The reliable and scalable performance of IoT networks is challenging because one wrong decision can lead to disruption of the entire system. To address these challenges multidimensional scaling, automatic bootstrapping, and IoT pipelining of data can be used in combination [37].
- **Interoperability:** Interoperability is the biggest challenge faced by IoT networks due to the variety of technologies and platforms used by devices in the network [38]. To ensure the quality of service, industries should use different approaches for strategical, technological, and tactical interoperability. Intelligent devices can use advanced technologies to identify the devices of cross-domain in the network automatically and respond accordingly.
- **Power efficiency and quality of service:** Quality of service means the quality of communication for users. Several factors reduce the quality of communication in any network and some of these factors are packet loss, jitter, and latency. Quality of service can be controlled with the use of efficient and effective models. Smart devices face power constraints because these devices work on battery power only for remote applications. These challenges can be addressed by harvesting energy. Smart devices can utilize different forms of energy like wind energy, mechanical energy, solar energy, or thermoelectrical energy. Protocols like AI edge processing and fog edge processing can be created for efficient use of energy [39].
- **Regulatory and privacy issues:** IoT networks suffer from the unregulated flow of data. Some advanced technologies must be used to preserve the security and privacy of data in IoT networks.

9.12 INTEGRATING AN IOT SENSOR NODE WITH RENEWABLE ENERGY SOURCE [40]

A hardware experiment for Solar Energy Harvesting Internet of Things (SEH-IoT) scenario is shown in Figure 9.12. In this scenario, we monitor the room temperature wirelessly using a SEH-IoT node consisting of an energy harvester system. The complete SEH-IoT system is divided into two parts, that is, IoT system and an energy harvesting system [41].

- **Scientech 2311 IoT system:** It consists of a temperature sensor (LM35), IoT end node, a IoT USB gateway node, and an IoT monitoring software installed on a laptop PC. The temperature-sensing

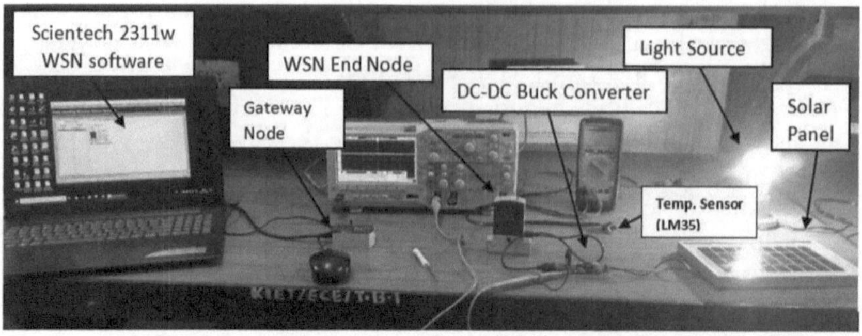

Figure 9.12 **Hardware experiment setup of SEH-IoT system.**

module (LM35) is connected to Input Output (I/O) port 1 of the end IoT node. The end IoT node measures the temperature and sends data wirelessly to the remote gateway. The measured data is sensed by the IoT node and sent to the remote gateway node. The gateway node is connected via USB cable to the computer system. At the computer system, a software Scientech-2311 w is installed which can show the visual representation of the sensor nodes topology and measured temperature (degree Celsius) as shown in Figure 9.12. In our experimental setup, the maximum distance between remote end IoT node and the gateway IoT node is less than 10 meters. The maximum distance between gateway node and sensor node can be up to 100 meters using ZigBee communication protocols [42].

- **Energy harvesting system:** A 5 watts solar panel and a commercial PWM controlled buck converter module is used as an energy harvesting system for Scientech 2311 W node. The output voltage from the solar panel is fed to the buck converter, which removes ripples and regulates the output voltage. This ripple-free and purified dc voltage (3.3 v) is used to charge the rechargeable battery of the IoT node. The LM2575 MOSFET is used for switching action in buck converter. It provides regulated dc output of 3.3 volts, 1 A to the IoT node with maximum 88% efficiency [42] (Table 9.2).

The measured room temperature is 0.301 × 100 = 30.1 degree Celsius as shown in the Scientech 2311 IoT monitoring software in Figure 9.13. The MAC address of the end IoT node is also shown with the actual date and time of the reporting of the IoT node. Figure 9.13 shows the output of the IoT sensor nodes with date, time, MAC address, and I/O channels data values. At present, the data shown represents the temperature with a sensor reading scaling factor of (x100). So, if we multiply the I/O channel data value 0.30 by 100 we get 30-degree Celsius temperature value.

Table 9.2 Experimental parameters

Hardware experiment parameters	*Number of components and details*
Scientech 2311 w IoT system	
IoT gateway node	1
IoT end node	1
Temperature sensor module (LM35)	1
Scientech 2311 w IoT monitoring software installed on a laptop PC	1
Energy harvesting system	
Solarcraft solar panel	5 w, 8 V, 0.65 A
Generic LM2576, 88% efficient, DC-DC Buck converter	3 V–40 V, 2 A
Measuring instruments	
Tektronix 200 MHz Digital Storage Oscilloscope (DSO)	1
Multimeter	1

9.13 CONCLUSION

In today's era, the huge development of multiple technologies has made the quality of life much better in smart cities. IoT technologies platforms have brought drastic growth in the development of smart cities. Renewable energy resources and balanced computing are required for the successful development of smart cities. Correct allocation of energy and power distribution are the key factors for providing continuous service to citizens of smart cities. The integration of IoT in renewable energy for smart cities can reshape the whole world. The use of an advanced eco-system of IoT can make the lives of citizens smarter and better. Although, IoT suffers from security threats but by using advanced technologies and integrating different security protocols and mechanisms privacy issues can be solved. We have discussed that an IoT-integrated management system for smart green energy can upgrade energy management in smart grids. This aids in limiting the wastage of energy that occurs due to power surges. We can reduce the dependency by utilizing local renewable energy resources. We also discussed that the use of green IoT can improve the efficient exploitation of energy and power consumption can be decreased. Green IoT can help in decreasing pollution and e-waste also. We conclude that for sustainable, eco-friendly smart cities integration of IoT is needed. IoT integration for smart cities can give better utilization of energy with more efficiency, less wastage, and reduced hazardous emissions. Hence, IoT technology plays a critical role in the construction of smart cities.

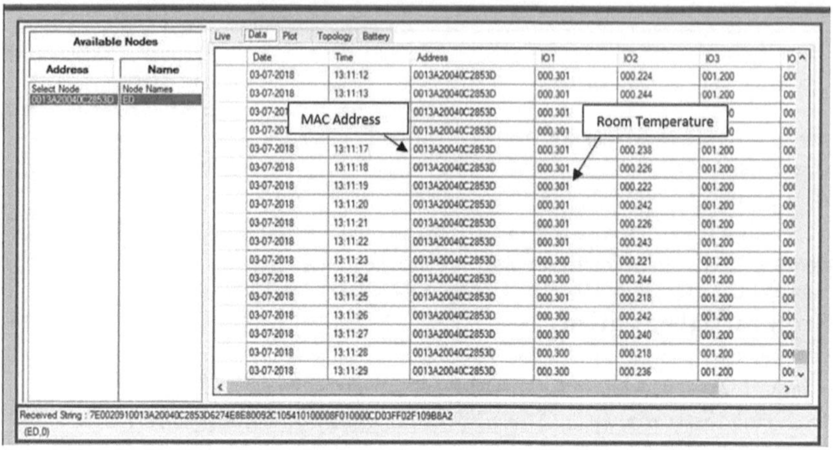

Figure 9.13 Temperature monitoring using SEH-IoT.

REFERENCES

[1] Faris Almalki, Saeed Alsamhi, Radhya Sahal, Jahan Hassan, Ammar Hawbani, Chen Weilin, Navin Rajput, Abdu Saif, Jeff Morgan, and John Breslin. (2021). Green IoT for eco-friendly and sustainable smart cities: Future directions and opportunities. *Mobile Networks and Applications*. https://doi.org/10.1007/s11036-021-01790-w

[2] Aparna Raj and Sujala Shetty. (2022). IoT eco-system, layered architectures, security and advancing technologies: A comprehensive survey. *Wireless Personal Communications*, 122. https://doi.org/10.1007/s11277-021-08958-3

[3] Deepak Kalra and Manas Pradhan. (2021). Enduring data analytics for reliable data management in handling smart city services. *Soft Computing*, 25. https://doi.org/10.1007/s00500-021-05892-1

[4] Zhihan Lv, Liang, Qiao, Amit Kumar Singh, and Qingjun Wang. (2021). AI-empowered IoT security for smart cities. *ACM Transactions on Internet Technology*, 21, 4, Article 99 (November 2021), 21 pages. https://doi.org/10.1145/3406115

[5] Meiyi Ma, Sarah M. Preum, Mohsin Y. Ahmed, William Tärneberg, Abdeltawab Hendawi, and John A. Stankovic. (2019). Data sets, modeling, and decision making in smart cities: A survey. *ACM Transactions on Cyber-Physical Systems*, 4, 2, Article 14 (April 2020), 28 pages. https://doi.org/10.1145/3355283

[6] Umer Majeed, Latif U. Khan, Ibrar, Yaqoob, S.M. Ahsan Kazmi, Khaled Salah, and Choong Seon Hong. (2021). Blockchain for IoT-based smart cities: Recent advances, requirements, and future challenges, *Journal of Network and Computer Applications*, 181, 103007. ISSN 1084-8045. https://doi.org/10.1016/j.jnca.2021.103007

[7] R. Verma (2022). Smart city healthcare cyber physical system: characteristics, technologies and challenges. *Wireless Personal Communication*, 122, 1413–1433. https://doi.org/10.1007/s11277-021-08955-6

[8] Waqar Ahmed, Hibba Ansari, Bahauddin Khan, Zahid Ullah, Shafaqat Ali, Chaudhry Mehmood, Muhammad Qureshi, I. Hussain, Muhammad Jawad, Muhammad Khan, Usman Shahid, A. Ullah, and Raheel Nawaz. (2020). Machine learning based energy management model for smart grid and renewable energy districts. *IEEE Access*. https://doi.org/10.1109/ACCESS.2020.3029943

[9] Ghulam Hafeez, Khurram Alimgeer, Zahid Wadud, Imran Khan, Muhammad Usman, Abdul Qazi, Farrukh Khan. (2020). An innovative optimization strategy for efficient energy management with day-ahead demand response signal and energy consumption forecasting in smart grid using artificial neural network. *IEEE Access*, 1–1. https://doi.org/10.1109/ACCESS.2020.2989316

[10] Wenyu Zhang, Zhenjiang Zhang, Sherali Zeadally, and Han-Cheih Chao. (2019). MASM: A multiple-algorithm service model for energy-delay optimization in edge artificial intelligence. *IEEE Transactions on Industrial Informatics*, 1–1. https://doi.org/10.1109/TII.2019.2897001

[11] G. Aceto, V. Persico, and A. Pescape (2019). A survey on information and communication technologies for Industry 4.0: State of the art, taxonomies, perspectives, and challenges. *IEEE Communications Surveys & Tutorials*, 21(4), 3467–3501. https://doi.org/10.1109/COMST.2019.2938259

[12] K. Bellman, C. Landauer, N. Dutt, L. Esterle, A. Herkersdorf, A. Jantsch, N. Taheri Nejad, P. R. Lewis, M. Platzner, and K. Tammemae (2020). Self-aware cyber-physical systems. *ACM Transactions on Cyber-Physical Systems*, 4(4), Article 38. https://doi.org/10.1145/3375716

[13] Yacine Atif, Jianguo Ding, and Manfred A. Jeusfeld. (2016). Internet of things approach to cloud-based smart car parking. *Procedia Computer Science*, 98, 193–198.

[14] Azevedo Guedes, Andre Luis. (2018). Smart cities: The main drivers for increasing the intelligence of cities. *Sustainability* 10(9), 3121.

[15] Michael Batty, Kay W. Axhausen, Fosca Giannotti, Alexei Pozdnoukhov, Armando Bazzani, Monica Wachowicz, Georgios Ouzounis, and Yuval Portugali. (2012). Smart cities of the future. *The European Physical Journal Special Topics*, 214(1), 481–518.

[16] E. F. Z. Santana, A. P. Chaves, M. A. Gerosa, et al. (2017). Software platforms for smart cities: Concepts, requirements, challenges, and a unified reference architecture. *ACM Computing Surveys*, 50(6), 1–37.

[17] Q. Zhu, S. W. Loke, R. Trujillo-Rasua, et al. (2019). Applications of distributed ledger technologies to the internet of things: A survey. *ACM Computing Surveys*, 52(6), 1–34.

[18] L. Gonzalez-Manzano, J. M. D. Fuentes, and A. Ribagorda. (2019). Leveraging user-related internet of things for continuous authentication: A survey. *ACM Computing Surveys*, 52(3), 1–38.

[19] R. Khatoun and S. Zeadally. (2016). Smart cities: Concepts, architectures, research opportunities. *Communications of the ACM*, 59(8), 46–57.

[20] J. Liu, H. Shen, H. S. Narman, et al. (2018). A survey of mobile crowdsensing techniques: A critical component for the internet of things. *ACM Transactions on Cyber-Physical Systems*, 2(3), 1–26.

[21] S. Kubler Kolbe, J. Robert, et al. (2019). Linked vocabulary recommendation tools for internet of things: A survey. *ACM Computing Surveys*, 51(6), 1–31.

[22] Y. Mehmood, F. Ahmad, I. Yaqoob, et al. (2017). Internet-of-things-based smart cities: Recent advances and challenges. *IEEE Communications Magazine*, 55(9), 16–24.
[23] M. Gharaibeh, A. Salahuddin, S. J. Hussini, et al. (2017). Smart cities: A survey on data management, security, and enabling technologies. *IEEE Communications Surveys and Tutorials*, 19(4), 2456–2501.
[24] A. Mozzaquatro, R. Jardim-Goncalves, and C. Agostinho. (2015). Towards a reference ontology for security in the internet of things. In *Proceedings of the IEEE International Workshop on Measurement & Networking*. IEEE, 2015, 289–296.
[25] W. Li, H. Song, and F. Zeng. (2017). Policy-based secure and trustworthy sensing for internet of things in smart cities. *IEEE Internet of Things Journal*, 5(2), 716–723.
[26] J. Lin, W. Yu, N. Zhang, et al. (2017). A survey on internet of things: Architecture, enabling technologies, security and privacy, and applications. *IEEE Internet of Things Journal*, 4(5), 1125–1142.
[27] M. Stolpe. (2016). The internet of things: Opportunities and challenges for distributed data analysis. *ACM SIGKDD Explorations Newsletter*, 18(1), 15–34.
[28] M. C. Schraefel, R. Gomer, A. Alan, et al. (2017). The internet of things: Interaction challenges to meaningful consent at scale. *Interactions* 24(6), 26–33.
[29] Xiangdong Zhang, Gunasekaran Manogaran, and BalaAnand Muthu. (2021). IoT enabled integrated system for green energy into smart cities. *Sustainable Energy Technologies and Assessments*, 46, 101208. ISSN 2213-1388. https://doi.org/10.1016/j.seta.2021.101208
[30] M. Ambrosin, P. Braca, M. Conti, et al. (2017). Odin: Obfuscation-based privacy-preserving consensus algorithm for decentralized information fusion in smart device networks. *ACM Transactions on Internet Technology*, 18(1), 1–22.
[31] C. Makhoul M. Guyeux, et al. (2016). Using an epidemiological approach to maximize data survival in the internet of things. *ACM Transactions on Internet Technology*, 16(1), 1–15.
[32] Y. Qian, D. Wu, W. Bao, et al. (2019). The internet of things for smart cities: Technologies and applications. *IEEE Network*, 33(2), 4–5.
[33] Y. Zhang, Z. Xiong, D. Niyato, et al. (2020). Information trading in internet of things for smart cities: A market-oriented analysis. *IEEE Network*, 34(1), 122–129.
[34] J. Hu, K. Yang, S. T. Marin, et al. (2018). Guest editorial special issue on Internet-of-Things for smart cities. *IEEE Internet of Things Journal*, 5(2), 468–472.
[35] K. F. Tsang, and V. Huang. (2019). Conference on sensors and internet of things standard for smart city and inauguration of IEEE P2668 internet of things maturity index [Chapter News]. *IEEE Industrial Electronics Magazine*, 13(4), 130–131.
[36] S. Saxena. (2021). Achieving cyber security through deep learning. Computer science and information technology. ESN publications, ISBN: 978-93-90781-10-2, 272–278.

[37] Raja Anwar, Tariq Abdullah, and Flavio Pastore. (2021). Firewall best practices for securing smart healthcare environment: A review. *Applied Sciences*, 11, 9183. https://doi.org/10.3390/app11199183
[38] Himanshu Sharma and Ahteshamul Haque. (2021). Artificial intelligence, machine learning & internet of medical things (IoMT) for COVID-19 & future pandemics: An exploratory study. *IEEE Smart Cities Newsletter.*
[39] Himanshu Sharma, Ahteshamul Haque, and Frede Blabjerg. (2021). Machine learning in WSN-IoT for smart cities. *Electronics Journal*, MDPI, 10(1012). ISSN 2079-9292.
[40] Himanshu Sharma, Ahteshamul Haque, and Zainul A. Jaffery. (2019). Maximization of wireless sensor networks lifetime using solar energy harvesting for smart agriculture monitoring. *Adhoc Networks Journal*, 94. ISSN 1570-8705.
[41] Himanshu Sharma, Ahteshamul Haque, and Zainul A. Jaffery. (2018). Solar energy harvesting wireless sensor network nodes: A survey. *Journal of Renewable and Sustainable Energy*, 10(2), 1–33. ISSN 1941-7012.
[42] Himanshu Sharma, Ahteshamul Haque, and Zainul A. Jaffery. (2018). Modelling and optimization of a solar energy harvesting system for wireless sensor network nodes. *Journal of Sensor and Actuator Networks, MDPI, USA*, 7(3), 1–19. ISSN 22242708.

Chapter 10

Power electronics and IoT for electric vehicles in smart cities

Himanshi Chaudhary, Manish Bhardwaj, and Shweta Singh
KIET Group of Institutions, Ghaziabad, India

Himanshu Sharma
Noida Institute of Engineering & Technology, Greater Noida, India

10.1 INTRODUCTION

The rapid urbanization and growing environmental concerns have led to the emergence of smart cities, where intelligent technologies are leveraged to enhance the quality of life and reduce the carbon footprint. EVs, with their zero-emission capabilities, have gained significant popularity as a clean and sustainable mode of transportation. However, the widespread adoption of EVs in smart cities necessitates addressing various challenges, including power management, charging infrastructure, and integration with the existing grid infrastructure [1]. This article aims to explore how power electronics and IoT can provide effective solutions to overcome these challenges and enable seamless integration of EVs into smart cities. Power electronics and internet of things (IoT) integration in electric vehicles (EVs) offers numerous benefits in terms of enhanced connectivity, efficient energy management, and improved user experience. This integration leverages the power of IoT to connect EVs with various devices, infrastructure, and services, while power electronics play a vital role in enabling seamless communication and control [2].

Electric vehicles (EVs) are playing a pivotal role in shaping the future of transportation, particularly in the context of smart cities. The integration of EVs in smart cities offers numerous benefits, including reduced greenhouse gas emissions, improved air quality, enhanced energy efficiency, and increased energy resilience. Here are some key aspects of electric vehicles in smart cities [3]:

- **Sustainable mobility**: EVs offer a sustainable alternative to traditional gasoline-powered vehicles, as they produce zero tailpipe emissions. By replacing internal combustion engine vehicles with EVs, smart cities can significantly reduce air pollution and combat climate change.
- **Energy efficiency**: EVs are more energy-efficient compared to conventional vehicles. Electric drivetrains convert a higher percentage of energy from the grid into propulsion, resulting in reduced energy

 DOI: 10.1201/9781032669809-10

consumption per kilometer traveled. This efficiency, coupled with advancements in regenerative braking and energy management systems, contributes to the overall energy optimization of smart city transportation.

10.2 POWER ELECTRONICS IN EVS

This section provides an overview of the role of power electronics in EVs. It discusses key components such as power converters, motor drives, and energy storage systems. Furthermore, it highlights the importance of bidirectional power flow for vehicle-to-grid (V2G) applications and the integration of renewable energy sources with EV charging infrastructure [4].

10.3 IOT-ENABLED SMART EV CHARGING STATIONS

IoT integration allows EVs to connect with smart charging infrastructure. IoT-enabled charging stations can communicate with EVs to determine optimal charging parameters, manage charging schedules based on electricity pricing or grid demand, and provide real-time status updates. Power electronics enable bidirectional power flow between the EV and the charging station, supporting features like load balancing, demand response, and vehicle-to-grid (V2G) capabilities [5].

The integration of IoT in EV charging infrastructure enhances the efficiency, reliability, and scalability of the charging process. This section discusses the IoT-enabled components and functionalities in EV charging stations, including smart meters, communication protocols, and cloud-based platforms. It also emphasizes the significance of data analytics and real-time monitoring for optimizing charging operations and managing grid load [6].

The term "connected car" refers to automobiles equipped with wireless connectivity that can communicate both internally and outside. Figure 10.1 shows three types of charging stations for electric vehicles for smart cities [7]. Next, Figure 10.2 shows a detailed block diagram of an electric vehicle charging station. As connected vehicles constitute the foundation of the internet of vehicles (IoV), extensive research and various industrial projects have laid the groundwork for this new era of connected vehicles.

Electric vehicle (EV):

- Represents the device that requires charging.

IoT devices:

- Sensors: Measure various parameters such as battery status, charging current, voltage, temperature, and occupancy status of charging stations.

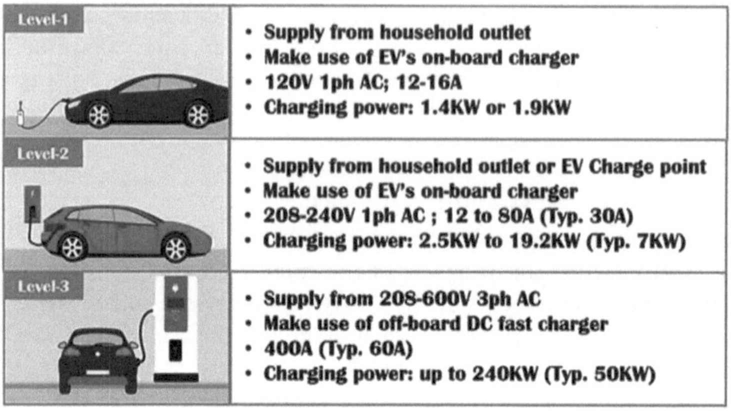

Figure 10.1 Three types of charging stations for electric vehicles.

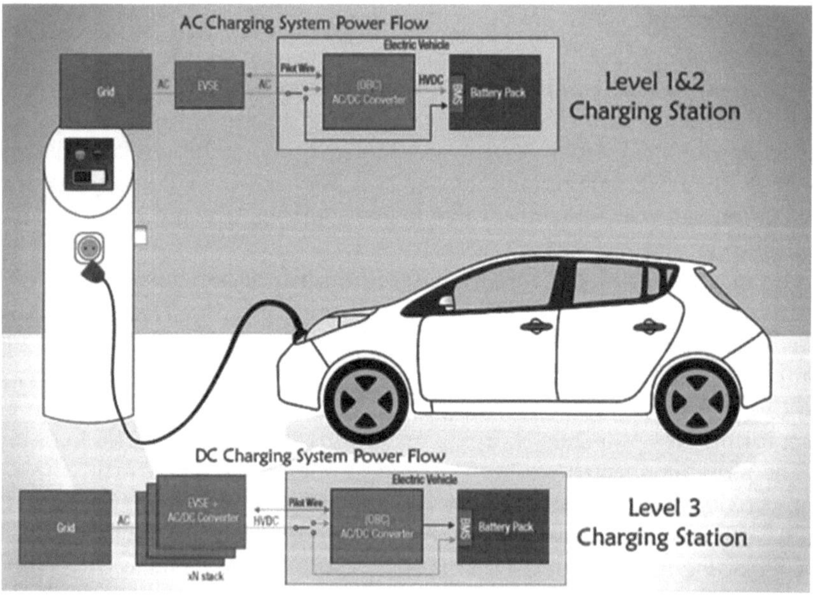

Figure 10.2 Detailed block diagram of a charging station.

- Actuators: Control charging stations, switch power supply, and manage charging processes.
- Communication modules: Enable connectivity and data exchange between the EV, charging station, and central management system.

Charging station:

- Charging point: Physical interface for connecting the EV to the charging infrastructure.

- Power management unit: Manages the power flow, controls charging rates, and ensures safe and efficient charging.
- IoT gateway: Facilitates communication between the charging station and the central management system.
- Metering and payment system: Monitors energy consumption, calculates charging costs, and manages payment transactions.

Cloud platform:
- Data storage: Stores real-time and historical data related to EV charging, energy consumption, and billing information.
- Data analytics: Performs analysis on collected data to extract insights, optimize charging operations, and predict charging demand.
- Charging management system: Provides centralized control and monitoring of charging stations, manages charging schedules, and optimizes energy management.
- User interface: Enables users to access charging information, make reservations, and monitor charging progress through web or mobile applications.

Electric grid:
- Power supply: Represents the electrical grid that supplies power to the charging stations.
- Demand response onterface: Facilitates bidirectional communication between the grid and charging stations, allowing load management and demand response mechanisms.
- Renewable energy integration: Enables the integration of renewable energy sources, such as solar or wind, to power the charging infrastructure.

EV communication:
- Vehicle-to-grid (V2G) communication: Establishes bidirectional communication between the EV and the charging station, enabling power flow control and V2G services.

In Figure 10.3 the block diagram represents the fundamental components of an IoT-based smart EV charging system, illustrating the flow of data and control between the EV, charging station, cloud platform, and the electric grid. The integration of IoT devices, communication modules, and cloud-based systems enables real-time monitoring, remote management, and optimization of EV charging operations, enhancing the efficiency, reliability, and sustainability of the charging infrastructure.

- **Vehicle connectivity**: IoT enables EVs to connect to the internet and interact with a wide range of devices and services. Power electronics facilitate the integration of IoT devices, sensors, and communication

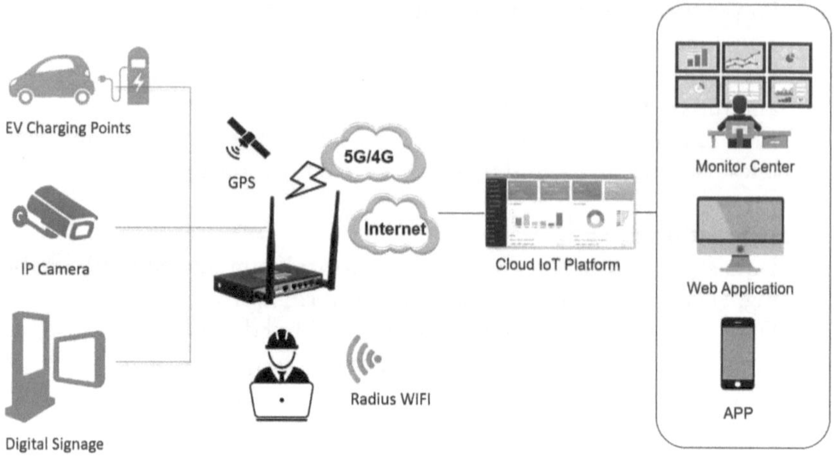

Figure 10.3 IoT-based smart EV charging system.

modules within the vehicle, enabling data exchange and communication with external networks. This connectivity enables real-time monitoring, remote diagnostics, over-the-air software updates, and integration with smart city infrastructure [8].

- **Enhanced user experience:** Power electronics and IoT integration offer various features that enhance the user experience of EVs. IoT connectivity enables features like remote vehicle monitoring, battery status monitoring, and vehicle tracking through mobile apps or web interfaces. Power electronics facilitate the integration of IoT devices with the vehicle's infotainment system, enabling personalized services, smart navigation, and integration with smart home systems.
- **Safety and security:** The integration of power electronics and IoT in EVs also addresses safety and security aspects. Power electronics components ensure the safe operation of the vehicle's electrical systems, including battery management and high-voltage circuits. IoT security measures, such as secure communication protocols and encryption, safeguard data transmission and protect against cyber threats [9].
- **Data analytics and predictive maintenance:** IoT integration enables the collection of vast amounts of data from EVs. Power electronics enable the processing and analysis of this data to derive valuable insights. Data analytics can be used for predictive maintenance, identifying potential faults or failures in power electronics components, and scheduling maintenance proactively. This helps in maximizing vehicle uptime and reducing maintenance costs.

The integration of power electronics and IoT in EVs opens up opportunities for advanced features, efficient energy management, and seamless connectivity with the broader smart city ecosystem [10]. It is expected to play

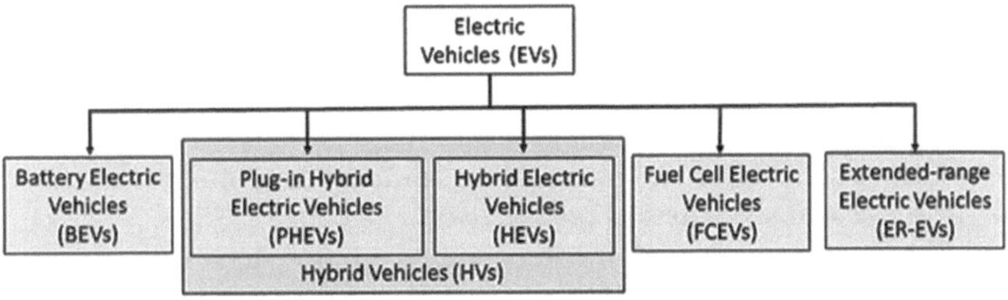

Figure 10.4 Types of electric vehicles.

a significant role in the development of intelligent transportation systems, improving the sustainability, efficiency, and overall performance of electric vehicles in the context of smart cities.

10.4 TYPES OF ELECTRIC VEHICLES

Figure 10.4 shows different types of electric vehicles. There are mainly three types of electric vehicles as [11]:

- Battery electric vehicle
- Hybrid vehicles
- Fuel cell electric vehicles

Hybrid EVs are nowadays most preferred choice for EV manufacturers and users. Hybrid EVs combine independent fossil fuels (oil, gas) powered engines and a rechargeable battery-based motor-driven system.

10.5 POWER MANAGEMENT IN EV CHARGING STATIONS USING IOT AND POWER ELECTRONICS

Efficient power management is crucial to ensure the optimal utilization of energy resources and grid stability. This section explores advanced power management strategies enabled by IoT and power electronics [12]. It discusses load balancing techniques, demand response mechanisms, and energy storage integration for mitigating peak load demand and minimizing the impact of EV charging on the grid. Efficient power management is essential to optimize charging operations and reduce stress on the power grid. This section discusses the key aspects of power management, including:

- **Power balancing:** Balancing the power load among charging stations to avoid overloading the grid and ensure efficient utilization of available power resources.

- **Demand response:** Enabling bidirectional communication between charging stations and the grid to respond to grid demands and adjust charging rates accordingly.
- **Renewable energy integration:** Integrating renewable energy sources, such as solar or wind, with charging stations to minimize reliance on the grid and promote sustainable charging practices.
- **Energy storage integration:** Utilizing energy storage systems to store excess energy during off-peak periods and release it during high-demand periods to minimize strain on the grid.
- **Intelligent energy management:** Power electronics and IoT integration enable intelligent energy management in EVs. IoT sensors and data analytics can collect and analyze real-time information about battery state-of-charge, vehicle usage patterns, traffic conditions, and charging infrastructure availability. Power electronics components, such as power converters and motor drives, can then optimize energy flow and usage based on this data. This integration ensures efficient utilization of energy resources, extends the vehicle's range, and enhances overall energy efficiency.

10.6 GRID INTEGRATION FOR EV CHARGING STATIONS

Seamless integration with the power grid is crucial for the efficient operation of EV charging stations [13]. This section focuses on grid integration considerations, including:

- **Grid connection and communication:** Enabling secure and reliable communication between charging stations and the power grid to facilitate real-time data exchange and grid monitoring.
- **Grid stability:** Implementing mechanisms to ensure stable grid operation, including voltage regulation, frequency control, and mitigation of power quality issues.
- **Grid services:** Leveraging EVs and charging stations as distributed energy resources to provide grid services, such as frequency regulation and peak load management through vehicle-to-grid (V2G) technology.
- **Grid infrastructure upgrades:** Identifying infrastructure upgrades, such as distribution transformers or grid capacity enhancements, to support the increased power demand from EV charging stations.

10.7 IOT INTEGRATION FOR POWER MANAGEMENT AND GRID INTERACTION

IoT plays a pivotal role in enabling intelligent power management and seamless grid interaction in EV charging stations [14]. This section highlights the key IoT-enabled functionalities, including:

- **Real-time monitoring:** Utilizing IoT sensors and connectivity to monitor charging station status, energy consumption, and power demand in real-time.
- **Data analytics:** Applying data analytics techniques to collected data to gain insights into charging patterns, grid conditions, and energy usage, facilitating informed decision-making.
- **Remote control and management:** Enabling remote access and control of charging stations, allowing operators to optimize charging schedules, adjust power levels, and troubleshoot issues.
- **Demand forecasting:** Utilizing historical data and predictive algorithms to forecast charging demands, enabling proactive management of power supply and grid integration.

10.8 BENEFITS AND CHALLENGES

This section presents the potential benefits of integrating power electronics and IoT in EVs for smart cities. It discusses the reduction of greenhouse gas emissions, improvement in energy efficiency, and the potential for creating a decentralized and resilient energy system. Additionally, it addresses the challenges related to interoperability, cybersecurity, standardization, and privacy that need to be addressed for the successful deployment of IoT-enabled EVs in smart cities.

10.9 CASE STUDY

This section provides real-world case studies showcasing the successful integration of power electronics and IoT in EVs within smart cities. It highlights the implementation of intelligent charging infrastructure, V2G systems, and smart grid integration in various cities around the world [15].

10.9.1 Smart EV fleet management with IoT

The IoT has revolutionized various industries, including transportation. This case study explores the implementation of IoT in EVs through the lens of smart fleet management. The case study highlights the benefits, challenges, and key considerations in adopting IoT solutions for EV fleet management. It focuses on real-time monitoring, predictive maintenance, route optimization, and energy management, demonstrating how IoT integration can enhance operational efficiency, reduce costs, and promote sustainable transportation. Here, we provide an overview of the increasing adoption of IoT in the transportation sector, particularly in the context of electric vehicles. It emphasizes the significance of smart fleet management and the potential benefits offered by IoT integration in EVs. This section presents an overview of smart fleet management and its relevance to EVs. It discusses the

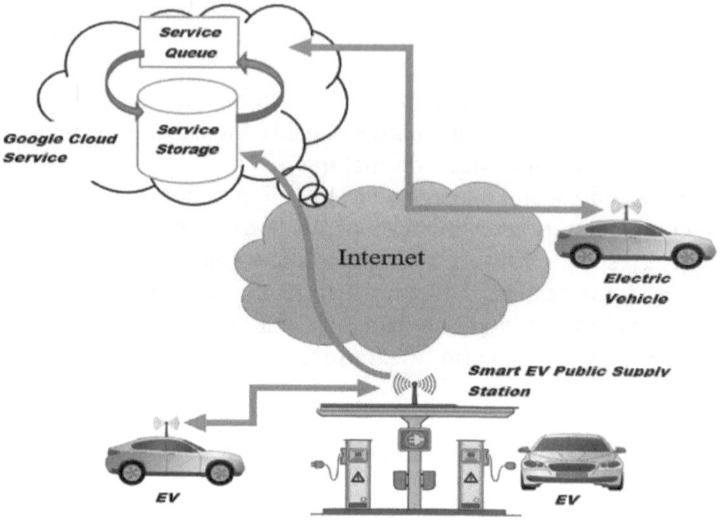

Figure 10.5 Basic block diagram of an internet cloud-based smart EV charging system.

key components and functionalities of IoT-based fleet management systems, including real-time monitoring, remote diagnostics, predictive maintenance, and data analytics (Figure 10.5).

10.9.2 Real-time monitoring and telematics

The case study delves into the real-time monitoring capabilities offered by IoT in EVs. It explores the integration of sensors, GPS, and communication modules to collect and transmit data on vehicle location, speed, battery status, and other relevant parameters. The benefits of real-time monitoring, such as improved fleet visibility, driver behavior analysis, and optimized routing, are discussed.

10.9.3 Predictive maintenance and remote diagnostics

This section highlights the role of IoT in enabling predictive maintenance and remote diagnostics for EV fleets. It explores how data collected from IoT sensors can be used to monitor vehicle health, detect anomalies, and schedule maintenance proactively. The case study demonstrates how predictive maintenance can reduce downtime, enhance vehicle reliability, and optimize maintenance costs.

10.9.4 Route optimization and traffic management

The case study examines how IoT integration facilitates route optimization and traffic management for EV fleets. It explores the utilization of real-time traffic data, weather information, and historical data to optimize route planning, reduce travel time, and minimize energy consumption. The benefits of efficient route planning, such as improved fleet productivity and reduced emissions, are highlighted.

10.9.5 Energy management and charging infrastructure

This section focuses on IoT-enabled energy management in EV fleets. It discusses the integration of IoT with charging infrastructure to optimize charging schedules, manage energy consumption, and balance power demand. The case study showcases how IoT integration helps in peak load management, demand response, and integration with renewable energy sources.

The case study presents an analysis of the benefits and challenges of implementing IoT in EV fleet management. It discusses the advantages, such as improved operational efficiency, reduced costs, enhanced sustainability, and increased customer satisfaction. It also addresses the challenges related to data security, interoperability, scalability, and infrastructure requirements. The case study concludes by summarizing the key findings and highlighting the significance of IoT integration in electric vehicles for smart fleet management. It emphasizes the transformative impact of IoT in optimizing operations, enhancing sustainability, and shaping the future of transportation.

10.10 CONCLUSION AND FUTURE DIRECTIONS

The article concludes by outlining potential future directions in the field of power electronics and IoT for EVs in smart cities. It emphasizes the need for continued research and development to address the challenges and maximize the benefits of this integration. The seamless integration of power electronics and IoT in EVs has the potential to revolutionize transportation systems, promote sustainability, and accelerate the transition to smart cities. Power management and grid integration in EV charging stations using IoT offer immense opportunities for optimizing charging operations, enhancing grid stability, and promoting sustainable energy practices. The integration of IoT technology enables real-time monitoring, data analytics, and remote control, empowering efficient energy management and seamless interaction with the power grid. By addressing challenges and leveraging the benefits, the future of EV charging infrastructure can pave the way for a sustainable and resilient energy ecosystem.

- EV charging with sustainable power sources: The EV charging stations can be additionally upgraded by incorporating it with environmentally friendly power sources, for example, sun-oriented and wind power. This won't just decrease the carbon impression yet additionally make the charging system more practical and financially savvy.
- EV charging with smart grid innovation: The task can be incorporated with savvy network innovation to enhance energy use and lessen the heap on the power lattice. This will empower the charging stations to be more effective and solid, and lessen the gamble of blackouts.
- EV charging with mobile applications: A portable application can be created to permit clients to find the closest charging station, really take a look at accessibility, and make installments. This will upgrade the client experience and urge more individuals to change to electric vehicles.
- EV charging with different businesses: The undertaking can team up with different enterprises, for example, vehicle producers, energy organizations, and metropolitan organizers to foster a comprehensive way to deal with maintainable metropolitan versatility. This will empower the undertaking to greatly affect decreasing fossil fuel by-products and advancing practical living.

REFERENCES

[1] Urooj, S.; Alrowais, F.; Teekaraman, Y. IoT based electric vehicle application using boosting algorithm for smart cities. *Energies J.*, 14, 2021. https://doi.org/10.3390/en14041072

[2] Alahi, M.E.E., Sukkuea, A.; Tina, F.W.; Nag, S.C. Integration of IoT-enabled technologies and artificial intelligence (AI) for smart city scenario: Recent advancements and future trends. *Sensors J.*, 2023, 23, https://doi.org/10.3390/s23115206

[3] Kabalci, Y.; Kabalci, E.; Padmanaban, S.; Holm-Nielsen, J.B.; Blaabjerg, F. Internet of things applications as energy internet in smart grids and smart environments. *Electronics* 2019, 8, 972.

[4] Farmanbar, M.; Parham, K.; Arild, O.; Rong, C. A widespread review of smart grids towards smart cities. *Energies* 2019, 12, 4484.

[5] Yao, L.; Chen, Y.Q.; Lim, W.H. Internet of Things for electric vehicle: An improved decentralized charging scheme. In *Proceedings of the 2015 IEEE International Conference on Data Science and Data Intensive Systems*, Sydney, Australia, 11–13 December 2015; pp. 651–658.

[6] Benedetto, M.; Ortenzi, F.; Lidozzi, A.; Solero, L. Design and implementation of reduced grid impact charging station for public transportation applications. *World Electr. Veh. J.* 2021, 12, 28.

[7] Sousa, R.A.; Melendez, A.A.N.; Monteiro, V.; Afonso, J.L.; Ferreira, J.C.; Afonso, J.A. Development of an IoT system with smart charging current control for electric vehicles. In *Proceedings of the IECON 2018-44th Annual Conference of the IEEE Industrial Electronics Society*, Washington, DC, 21–23 October 2018, pp. 4662–4667.

[8] Savari, G.F.; Krishnasamy, V.; Sathik, J.; Ali, Z.M. Abdel; Aleem, S.H.E. Internet of Things based real-time electric vehicle load forecasting and charging station recommendation. *ISA Trans.* 2020, 97, 431–447.
[9] Gao, D.; Zhang, Y.; Li, X. The internet of things for electric vehicles: Wide area charging-swap information perception, transmission and application. *Adv. Mater. Res.* 2013, 608, 1560–1565.
[10] Asaad, M.; Ahmad, F.; Alam, M.S.; Rafat, Y. IoT enabled electric vehicle's battery monitoring system. *EAI SGIOT* 2017, 8, 994–1005.
[11] Helmy, M.; Wahab, A.; Imanina, N.; Anuar, M.; Ambar, R.; Baharum, A.; Shanta, S.; Sulaiman, M.S.; Fauzi, S.S.M.; Hanafi, H.F. IoT-based battery monitoring system for electric vehicle. *Int. J. Eng. Technol.* 2018, 7, 505–510.
[12] Divyapriya, S.; Amudha, A.; Vijayakumar, R. Design and implementation of grid connected solar/wind/diesel generator powered charging station for electric vehicles with vehicle to grid technology using IoT. *Curr. Signal Transduct. Ther.* 2018, 13, 59–67.
[13] Muralikrishnan, P.; Kalaivani, M.; College, K.R. IOT based electric vehicle charging station using Arduino Uno. *Int. J. Sci. Technol.* 2020, 29, 4101–4106.
[14] Ayob, A.; Wan Mahmood, W.M.F.; Mohamed, A.; Wanik, M.Z.C.; Siam, M.F.M.; Sulaiman, S.; Azit, A.H.; Mohamed Ali, M.A. Review on electric vehicle, battery charger, charging station and standards. *Res. J. Appl. Sci. Eng. Technol.* 2014, 7, 364–372.
[15] Motlagh, N.H.; Mohammadrezaei, M.; Hunt, J.; Zakeri, B. Internet of things (IoT) and the energy sector. *Energies* 2020, 13, 494.

Chapter 11

Machine learning-based DNS traffic monitoring for securing IoT networks

Mehwish Weqar and Shabana Mehfuz
Jamia Millia Islamia, New Delhi, India

Dhawal Gupta
Group Business Director, Public Policy, Chase India

11.1 INTRODUCTION: IOT COMMUNICATION USING DNS

Internet of things (IoT) is a vast network of billions of connected devices. These devices have embedded automated systems, like sensors to sense the physical environment, communication hardware to send the collected data, and processors and actuators. These interconnected IoT networks have been gaining fast acceptance because these systems make our lives smarter and easier. They are present everywhere in the form of different smart applications like smart health care systems, smart traffic monitoring systems, smart heating and lightning systems in buildings, smart water and waste management systems, and so on, thus making a smart city.

In order to have seamless communication in IoT networks all the IoT devices should be able to get recognized, identified, and registered automatically in the network. For this purpose, a DNS naming scheme has been used, which provides unique names to IoT devices [1]. It is the most common and most important naming infrastructure used in the current IoT environment. It is an efficient and scalable naming scheme. As soon as an IoT device enters a network, a DNS naming scheme allows each IoT device to autoconfigure its human-readable DNS name and the corresponding IP address and get it registered in the required DNS server. The DNS server performs name resolution by translating the DNS name of the target IoT device or service into its corresponding IP address.

11.2 DNS TRAFFIC IN IOT NETWORKS

Whenever communication is needed to be performed in the IoT network which is either between device and device or device and service, the device

DOI: 10.1201/9781032669809-11

or the service, or whoever wants to start the communication performs the following steps:

(i) Look up the DNS name of the target IoT device or service in the DNS server.
(ii) DNS server performs DNS name resolution and returns the registered corresponding IP address of the target device or service.

After this, the requesting device (or service) gets the IP address of the target device and starts communication. This set of request and response DNS messages transfer generally takes place in the following cases:

(i) Before the start of device to device or device to service communication
(ii) When a device enters an IoT network
(iii) In case of software updates

Generally, an IoT network is a dynamic network of billions of connected devices and services, which most of the time transfer data among themselves. This, in turn, results in millions of DNS message transfers. The dynamic nature of the IoT network allows addition and removal of the IoT devices at a very fast pace and these devices keep on auto-registering themselves in DNS servers, hence resulting in a huge DNS traffic in the IoT network.

Most of the time IoT devices are resource-constrained in terms of computation power, battery backup and memory storage capacity and work in an unmanned and heterogeneous environment. All these limitations make the devices more vulnerable to cyber threats.

A study shows that within one year, cyber threats have increased four times. The cybercriminals have been continuously developing more dangerous and advanced malware that target these vulnerable IoT networks. They find DNS the most vulnerable and easy mechanism to attack [2].

11.3 TYPES OF DNS ATTACKS IN IOT NETWORKS

The cyberattacks that adversely affect the accessibility, attainability, availableness, obtainability, and stability of DNS service are considered as DNS attacks [3]. The main strategy of these attacks is to distort the functioning of DNS infrastructure [4]. DNS is the most used part of any network during communication and to date it remains the most unnoticed part of any organization or application at the time of security design. That is why it is considered as one of the most vulnerable mechanisms that can be attacked easily [5, 6].

Many types of DNS attacks occur in IoT networks, some of which are discussed and shown in Figure 11.1.

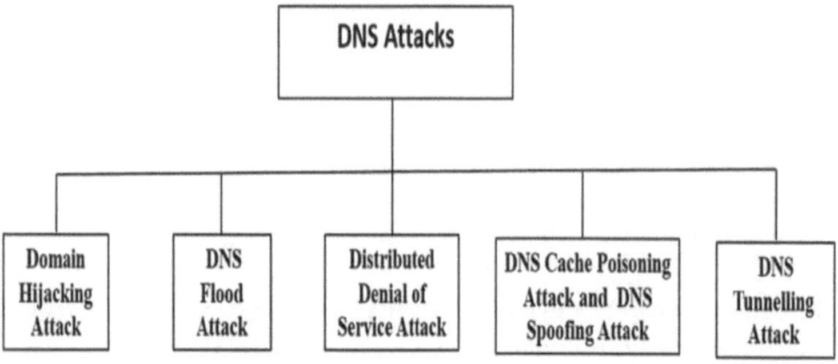

Figure 11.1 Types of DNS attacks.

11.3.1 Domain hijacking attack

These types of attacks involve changing the domain name registration in the DNS server and DNS registrar without the permission of the original owner. This results in diverting the DNS queries away from the original server to different server locations. The hijackers control the DNS records and the functioning of the DNS mechanism [7]. This results in many damages such as:

(a) **Monetary damages:** Those organizations which perform their function through online trading like many e-commerce companies solely depend on their websites, and for them losing their Domain means losing a large amount of money. That is why currently it is the biggest security challenge for online trading companies.
(b) **Trust damages:** The cyber criminals can use the hijacked domain to send virus installations to the linked devices. Thus, it results in damaging the trust, name, and fame of the hijacked domain.
(c) **Administrative damages:** The domain hijackers acquire full control of the hijacked domain so they can change the website to their benefit and can also steal data from the webpage like any confidential details of the contacting devices.

11.3.2 DNS flood attack

This attack is another form of denial of service attack, where the cyberattacker attacks the domain name resolution mechanism of the DNS server. In this attack the DNS server is flooded with numerous queries at the same time by the attacker, resulting in resource exhaustion of the server and saturation of network. Hence this situation leads to the DNS server unable to respond to further legitimate DNS queries [8].

11.3.3 Distributed denial of service attack

In this type of attack the cyber criminals target more than one DNS server at a time and utilize the weakness present in the DNS infrastructure and in turn make the server unresponsive to the valid DNS queries. They do this by first flooding the network with amplified DNS traffic and then resulting in an unfulfilled DNS request [9].

11.3.4 DNS cache poisoning attack or DNS spoofing

This is one of the most frequently occurring DNS attacks. In this type of attack the attacker targets the DNS cache and enters the wrong IP address corresponding to DNS entries [10]. Thus, this misinformation leads to directing the requests to malware websites. In this way the cybercriminals transfer all the DNS traffic to their corrupt servers which results in many damages like information leakage, malware, and spyware injection [11]. Hence, this damages the devices that are sending requests.

11.3.5 DNS tunneling attack

In this type of attack the cybercriminal sends the encoded data via DNS requests and responses to remote servers. Hence, they use DNS queries as a tunnel for transferring all the wrong messages like malware, stolen data, and so on. For this purpose, they use the compromised DNS servers. These attackers do so because they know that DNS is the least monitored mechanism in any organization, so they use this to bypass the security enablers of any particular organization. Through this they can channel data theft attacks easily and without being caught [12].

All these DNS security challenges result in many security issues in almost every IoT network [13].

11.4 WHY DNS TRAFFIC MONITORING IS REQUIRED?

The main aim of all the DNS security threats discussed in the previous section is to distort the normal functioning of the DNS mechanism and steal the confidential, personal, and financial data records of individuals or organizations and use them in numerous unlawful ways. This all, in turn, leads to confidentiality, integrity, and availability issues.

These types of security challenges are faced mostly by IoT networks. The reason behind this is, first, the resource restriction of IoT networks which leads to security limitations and, second, the numerous numbers of connected devices lead to heavy DNS traffic. That is why in the case of DNS attacks in IoT networks it is very difficult to figure out whether an attack has occurred or not, and if yes then where has this attack occurred, that is, which DNS server has been compromised.

So, to perform security checks, DNS traffic monitoring is necessary [14]. Many techniques have been proposed over a period of time to monitor DNS traffic [15]. Some of these include different types of firewalls, using intrusion detection systems and different DNS traffic analyzers. After performing a comparison analysis, it was found that the least cost-effective method is to use machine learning techniques for DNS traffic monitoring and analysis [16].

11.5 MACHINE LEARNING FOR DNS TRAFFIC MONITORING

The techniques of machine learning use data and algorithms to learn the behavior of the network like humans do and then predict the problem present in the network [17]. These techniques improve their accuracy over a period of time.

Generally, the basic function of these techniques is to continuously observe the working pattern of the network in real-time. Over a period of time, they collect the network traffic data and analyze the dataset, then notify as soon as they detect the anomalous behavior of the network which is due to the presence of some malware or threats.

The machine learning techniques are mainly divided into three categories shown in Figure 11.2.

(i) Supervised learning mechanism: In this method, the algorithms train themselves by the provided labeled data called the training dataset, over some time, and then they can identify the anomalies present. In the case of traffic monitoring, the training dataset is the data of legitimate traffic, which is then used to identify the compromised traffic. Figure 11.3 shows the block diagram of this technique.

(ii) Unsupervised learning mechanism: In this method there is no categorized data set to train the algorithm, so the algorithm has to learn and

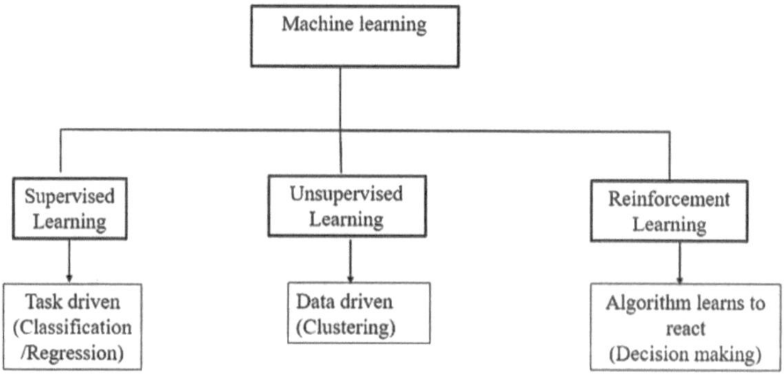

Figure 11.2 Machine learning types.

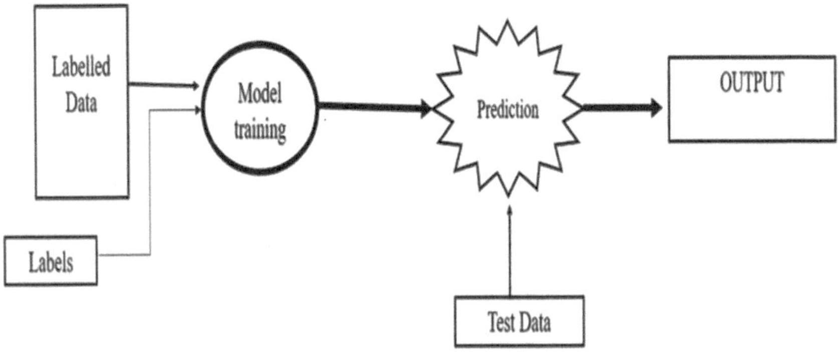

Figure 11.3 Supervised machine learning.

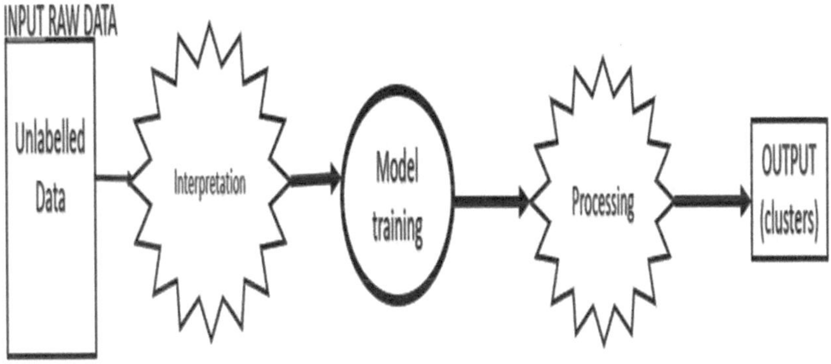

Figure 11.4 Unsupervised machine learning.

improve itself over the period. Then, it uses a different set of steps to identify the anomalous data present. Thus, in the case of network traffic monitoring the main objective is to analyze the behavior of traffic data and detect any suspicious data present. Figure 11.4 shows the block diagram of this technique.

(iii) Reinforcement learning mechanism: This is an iterative process. In this method, learning is performed based on the feedback data from the previous computations. Thus, the more feedback received, the better will be the performance.

It is evident from all the facts stated in the previous section that the IoT networks are prone to many security issues because of the presence of limitations like resources-constrained connected devices, and heterogenous and unmanned working environments. Therefore, to mitigate these security issues many types of machine learning techniques have been proposed and applied in analyzing the various aspects of the IoT network, one of which is

DNS traffic monitoring [18]. Also, the recently proposed machine learning techniques for traffic monitoring discussed in the next section are based on either supervised or unsupervised paradigms.

11.6 RECENTLY PROPOSED DNS TRAFFIC MONITORING APPROACHES USING MACHINE LEARNING TECHNIQUES

To minimize the evolving security threats in IoT networks, many machine learning-based DNS traffic monitoring approaches have been proposed in recent times. This section discusses some of the important proposed techniques.

11.6.1 Proposed technique: Classifying IoT devices in smart environments using network traffic characteristics

In the current scenario where the IoT network is dynamic in nature and communicating with a vast number of resource-constrained devices, then in such a situation identification of malware or compromised devices is a very difficult task. Thus, in order to make this identification easier and more accurate, a technique has been proposed [19] which is based on the supervised machine learning mechanism. Here a testbed of 29 IoT devices was created imitating a smart environment. Then, network traffic data was collected over some months. This collected data was of two types: (i) device-specific data like server port numbers, Domain names requested, and TLS handshakes data; (ii) traffic-specific data, like traffic flow volume, active time duration of device, sleep time duration of device, and frequency of traffic. A combination of these parameters is then fed into the proposed model which is a machine learning-based multi-stage algorithm which is used in IoT device classification. Figure 11.5 shows the working architecture of the

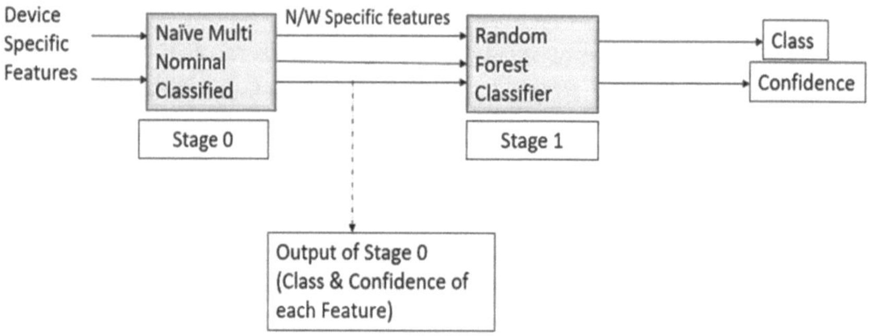

Figure 11.5 Multi-stage machine learning-based classification architecture.

proposed machine learning technique; here one classifier is placed at stage zero which takes device-specific parameters as input and gives class and confidence of the given parameters as intermediate output which along with network traffic-specific parameters are given to stage one classifier as input. The final output is class and confidence of each IoT device.

The accuracy of prediction of this analyzer depends on the number of instances of parameters in each input. The cost of computation of each attribute and the merit of each attribute are also calculated in this analysis. It was analyzed that most device-specific parameters have high computation costs and medium merits. It was analyzed that DNS-based parameters have a high cost of computation and medium merits (confidence). Figures 11.6 and 11.7 show a comparison of some important parameters.

The experimental results show they exhibit maximum accuracy in the prediction of IoT devices. The drawback of this technique is that to have a high accuracy of prediction, it requires a large training dataset with high confidence (merit). Hence cost and time are two demerits of this technique which can be resolved in future with some enhancements.

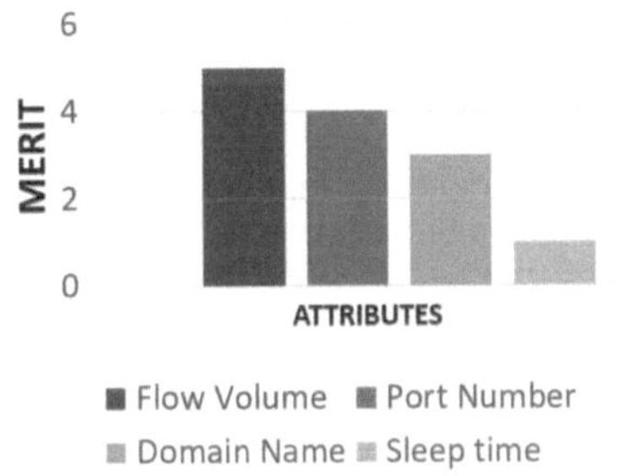

Figure 11.6 Comparative analysis of parameters considering merit.

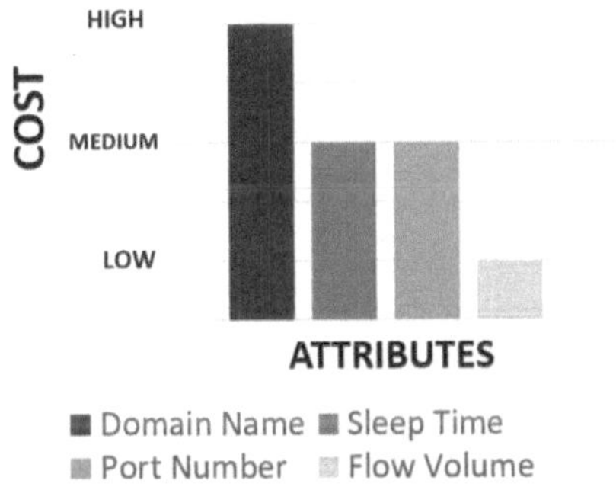

Figure 11.7 Comparative analysis of parameters considering cost.

11.6.2 Proposed technique: Detecting volumetric attacks on IoT devices via SDN-based monitoring of MUD activity

IoT networks are being targeted by more improvised and advanced cyber-attacks, which in turn, increases the requirement for more cost-effective, less time-consuming and accurate threat monitoring and controlling mechanisms. All previously proposed machine learning-based network traffic monitoring schemes involved high computation costs to create training datasets. Therefore, the infrastructure proposed in [20] aims to reduce the previous drawbacks.

Denial of service attacks and DNS flooding attacks are two basic types of volumetric attacks on IoT networks that hinder the normal working of connected IoT devices. The main aim of the prototype infrastructure developed [20] is to detect volumetric attacks. This proposed system contains two working parts: (i) the training part: which makes the system learn the MUD (manufacturer User Description) profile of the connected IoT devices [21]; and (ii) the monitoring (checking) part which analyzes the traffic and detects any anomalous behavior in the DNS network traffic pattern. Figure 11.8 shows the working of the proposed architecture. The traffic passing through SDN goes to the MUD profile module which first trains the network with connected devices by taking MUD details from the network and stores in the server after that. A module is also present which contains details of the attack pattern in the Intrusion detection module. After training, monitoring of traffic is performed using traffic data and comparing its similarity with normal traffic. Discovery of a deviation leads to the functioning of an anomaly detection module.

Figure 11.9 shows how two-stage machine learning algorithms used for anomaly detection works. In Figure 11.9, stage 1 takes device-specific

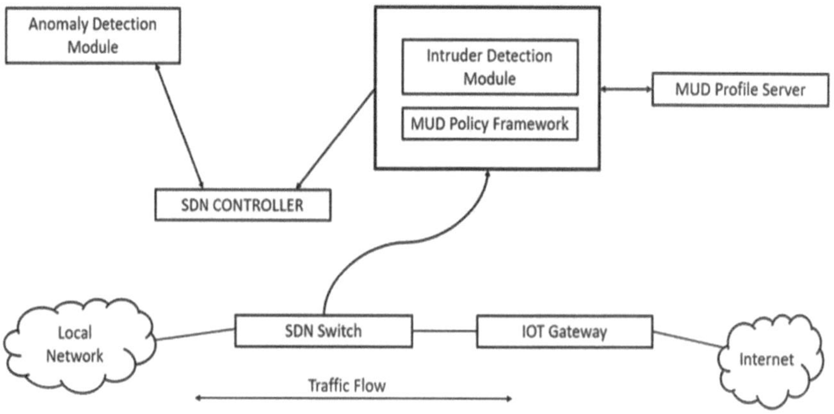

Figure 11.8 MUD profile training-based threat detection prototype infrastructure.

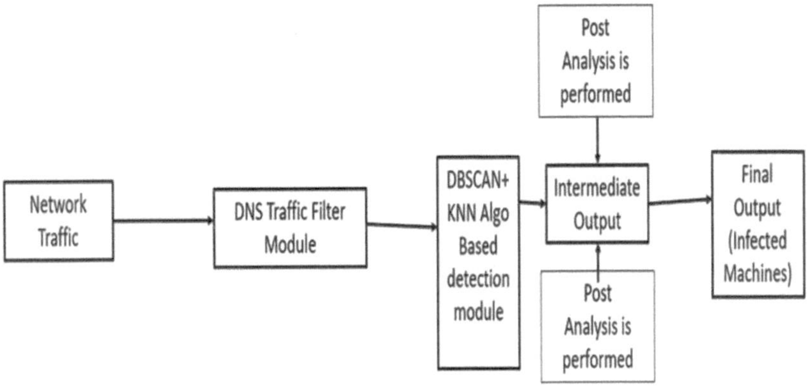

Figure 11.9 Two-stage anomaly detection scheme.

parameters as input and as soon as a malicious behavior is detected then an alarm is generated which activates stage 2 of the algorithm. This stage takes flow-specific parameters as input and then checks the traffic flow for the abnormal behavior.

The accuracy of detecting a threat in the traffic is high due to the implementation of two-stage machine learning algorithms. Some drawbacks of this technique are:

(i) computation cost and memory requirement for calculating device-specific parameters is high
(ii) threats generated from local network are not detected accurately
(iii) for high accuracy, large training dataset is required which contains all normal traffic patterns from the MUD profile for each connected IoT device, which is a complex and time-consuming task

11.6.3 Proposed technique: Internet of Things traffic characterization using flow and packet analysis

With the rapid growth in IoT applications, the security threat domain is also evolving at a fast pace. So, in order to render this resource-constrained IoT network optimum security, a threat detection mechanism is required which has the capability to identify a larger number of attacks. The methodology proposed [22] focuses on training a network with two available public datasets, provided by previous techniques. One dataset is of normal traffic which makes the system learn normal network behavior, another dataset is of malicious traffic, which trains the system to differentiate between normal and abnormal traffic. These datasets are further divided into two categories of parameters which are packet-level attributes and flow level attributes. The main objective of this technique is to test and prove how packet level and flow level analysis improves the accuracy of prediction. The main advantage of this technique is its simplicity. This mechanism can be further improved in future to predict the time of the attack, the attack participants, and ports involved in the attack. Some drawbacks of this technique are lack of scalability and dynamicity.

11.6.4 Proposed technique: Detecting abnormal DNS traffic using unsupervised machine learning

With the increase in the development of IoT-based smart applications, the DNS threat spectrum is also improving itself at a very fast pace. To avoid being caught, many DNS attacks use a DNS tunneling technique, often called Advance Persistent Treat (APT). They hide themselves in legitimate DNS traffic, so many intrusion detection systems are not able to detect them. Nguyen et al. [23] first performed a comparative analysis on a performance basis of four present unsupervised learning algorithms, which are K-means, Gaussian Mixture Model (GMM), Density-Based Spatial Clustering of Applications with Noise (DBSCAN), and checked their ability in detecting threats present in DNS traffic. It was deduced after testing with datasets that DBSCAN performs well in detecting APT. A novel hybrid approach has been presented which is a combination of DBSCAN machine learning algorithm and K Nearest Neighbor (KNN) machine learning algorithm. Figure 11.10

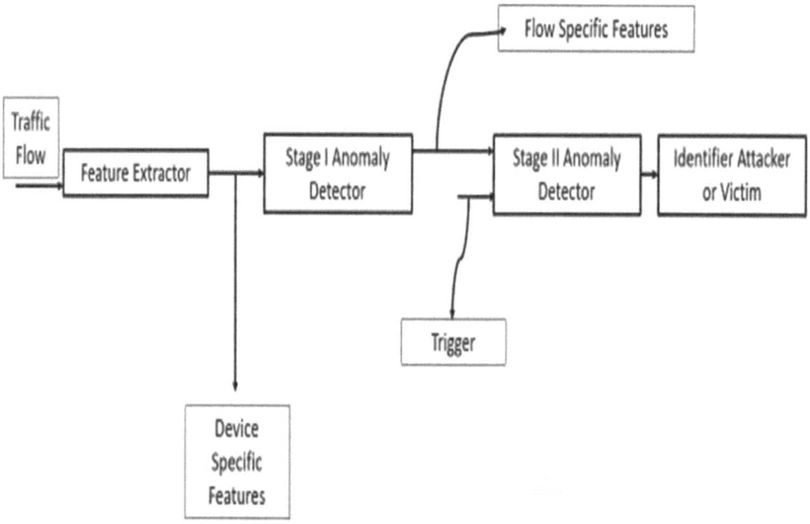

Figure 11.10 Hybrid architecture DNS attack detection mechanism (combination of DBSCAN and K Nearest Neighbor (KNN) algos).

shows the working of the proposed architecture. Here the network traffic is given to the first module where DNS traffic is filtered and passed on to the next module where the DBSCAN algorithm works to find out any traffic deviation from normal traffic behavior. After this, intermediate results have been generated which are analyzed further using the KNN algorithm to find out DNS attacks.

This approach gives a 100% attack detection rate with less error percentage. This technique tries to overcome the drawbacks of previous techniques like requirement of large training dataset and extra computation cost required in labeling of class. But this technique also has some limitations, for example, it requires post analysis of intermediate results which incur extra computation cost. Until now it has been tested on DNS protocol only.

11.7 COMPARATIVE ANALYSIS

In order to have a seamless communication between the connected IoT devices and services in a network, it is necessary to have a secure and error-free connection. To achieve this, many evolutionary schemes have been proposed over the period of time that involve DNS traffic monitoring as the main focus to detect, predict, and mitigate sophisticated IoT network attacks. All of the techniques discussed in the previous sections are based on a machine learning mechanism, which allows detecting the more advanced, intelligent, and ever evolving attacks [24].

After comparing and analyzing them, it was found that all the recently proposed schemes exhibit a high accuracy rate in threat prediction. But achieving this high accuracy rate is not so easy and it involves some extra work in terms of computation power, memory capacity, algorithm learning cost and more time.

Table 11.1 presents the comparative analysis of the recently proposed relevant DNS traffic monitoring techniques for securing IoT networks against various threats which have been presented in Section 11.3. The table provides the advantages and novelty present in each technique. Table 11.1 also illustrates the research gaps and limitations present in these techniques and suggests possible future improvements.

Some proposals, like in Sivanathan et al. [19] use Joy, a software package which is based on a cross-platform library – Libpcap. This is used to monitor live network traffic and capture the IP packets and performs feature extraction from the obtained data. Then, this data containing extracted features is converted and stored in JSON file format. These files can be easily analyzed further using different analysis tools to find out malicious traffic present.

Another popular tool named Weka is used [19] to perform data mining on gathered network traffic data. It applies different machine learning algorithms to the dataset. Then it performs different computation tasks for DNS

Table 11.1 Comparative analysis of machine learning-based DNS traffic monitoring techniques for securing IoT networks

Research paper	*Proposed work*	*Novelty of the proposed work*	*Research gaps*	*Future scope of enhancement*
1. Classifying IoT devices in smart environments using network traffic characteristics [19]	1. Create an IoT network setup. 2. Collect and synthesize traffic traces from the infrastructure for 6 months. 3. Choose some network traffic characteristic s as attributes to monitor. 4. Develop a multi-stage machine learning-based classification algorithm and demonstrate its ability to identify specific IoT devices.	1. Uniquely identify each IoT device. 2. Tests the proposed multi-stage machine learning-based classification algorithm in real-time on a created testbed. 3. 99% accuracy achieved.	1. Cost of impleme ntation is high. 2. More training dataset required for better accuracy takes more time to setup.	Regulation is required between cost of attribute collection and classification accuracy.
2. Detecting volumetric attacks on iot devices via sdn-based monitoring of MUD activity [20]	1. Developed a prototype infrastructure. 2. Train the system to learn the MUD-compliant functioning for every IoT device in the network. 3. Use coarse-grained and fine-grained activity patterns to train. 4. Network behavioral analysis of IoT devices is performed to identify volumetric attacks.	1. Development of a machine learning-based, self-learning protype. 2. Uses MUD (Manufacturer Usage Description) profile of IoT devices to check any volumetric attack in the system.	1. Attack detection depends on training dataset. 2. Tested on fixed number of devices. 3. Not suitable for dynamic and scalable IoT networks.	Deploy the generated prototype model in multiple applications and check the accuracy in DNS attack detection.

3. Internet of things traffic characterization using flow and packet analysis [22]	1. Study the traffic data set of the IoT network. 2. Differentiate the normal traffic from the malicious traffic using existing public dataset. 3. Perform Flow level analysis and packet level analysis on the real-time traffic data and detect identifiers of attack.	Security, prevention and detection of cyberattacks.	Method can be faster but less precise when it comes to classifying each device.	More accurate analysis can predict, when and what is going to be attacked and who and how will attack. Can predict frequency and periodicity of attack.
4. Detecting abnormal DNS traffic using unsupervised machine learning [23]	1. Perform comparative analysis of four present unsupervised learning algorithms in detecting malicious DNS traffic. 2. Propose a hybrid technique, by combining DBSCAN and K Nearest Neighbor (KNN) algorithms. 3. Perform real-time analysis of test to check the accuracy of prediction.	1. Combining the two algorithms & KNN gives 100% DNS attack detection rate and less error. 2. More suitable to detect DNS tunneling attacks.	1. Only tested on DNS traffic. 2. Require pre-processed DNS traffic data as Input. 3. More time is consumed in filtering on DNS basis.	Extend to monitor other protocols like HTTP and SMTP.

traffic characterization and results in identifying malicious traffic behavior. This tool is also used in calculating the computation power of the chosen parameters at each step of the proposed technique.

It can be analyzed and deduced from the Table 11.1 that most of the threat-predicting, traffic monitoring techniques present in literature exhibit some advantages and improvements as compared to previously proposed techniques in terms of accuracy of prediction but it is quite evident that most of them incur extra computation cost.

The IoT networks are resource-constrained and low-cost devices, hence involving extra computation is not a good idea as this leverages extra burden on the IoT network and in turn reduces the overall efficacy of the IoT application.

Hence, it is evident from the detailed analysis and performance evaluation that the currently proposed techniques need improvements in terms of:

(i) computation power
(ii) memory consumption
(iii) cost

11.8 CONCLUSION AND FUTURE DIRECTIONS

With the evolution of smart and intelligent environments, involving more and more smart applications like those of smart homes, smart grids, smart health, smart traffic management systems, and so on, results in connecting a tremendous number of IoT devices. This huge number of devices results in a massive amount of DNS message transfer during device to device or device to service communication. Cyber criminals target this DNS traffic to include their malware traffic, by either hijacking a device/system or DNS server or by encoding their malware traffic in normal DNS queries. Due to the resource limitations in these networks most of the time these criminals become successful in their aim which is mainly data theft.

Thus, to enable these IoT applications, to function systematically and offer smooth and threat-free services in such vulnerable conditions, many DNS traffic monitoring and threat detection schemes have been proposed which are based on different machine learning algorithms. However, after performing the relative investigation in detail, it was concluded that there is a need to reduce incurred computation costs, which can be done through some enhancements like:

I. Pre-training a system before deploying.
II. Adding a module of more computation power capable of performing classification and accurate labeling of the dataset.
III. Combining the good features of more than one machine learning algorithms.

Thus, in order to allow complete security to IoT networks, a more accurate, intelligent, easy-to-deploy, and lightweight security prediction mechanism is required to deal with ever-developing cyber threats.

REFERENCES

[1] C. Hesselman et al., "The DNS in IoT: Opportunities, Risks, and Challenges," in *IEEE Internet Computing*, vol. 24, no. 4, 1 July–Aug. 2020, pp. 23–32.

[2] netrusion, https://www.netsurion.com

[3] Adam Ali, Zare Hudaib and Esra'a Ali Zare Hudaib, "63 DNS Advanced Attacks and Analysis," in *International Journal of Computer Science and Security (IJCSS)*, vol. 8, no. 2, 2014, pp. 63–74.

[4] G. Schmid, "Thirty Years of DNS Insecurity: Current Issues and Perspectives," in *IEEE Communications Surveys & Tutorials*, vol. 23, no. 4, Fourthquarter 2021, pp. 2429–2459, doi: 10.1109/COMST.2021.3105741

[5] securityTrails, https://securitytrails.com

[6] Tae Kim and Douglas Reeves, "A Survey of Domain Name System Vulnerabilities and Attacks," in *Journal of Surveillance, Security and Safety*, 2020, doi: 10.20517/jsss.2020.14

[7] M. Lyu, H. Habibi Gharakheili and V. Sivaraman, "A Survey on DNS Encryption: Current Development, Malware Misuse, and Inference Techniques," *arXiv e-prints*, 2022.

[8] M. Jazzar and M. Hamad, "An Analysis Study of IoT and DoS Attack Perspective," in Agarwal, B., Rahman, A., Patnaik, S., Poonia, R. C. (eds) *Proceedings of International Conference on Intelligent Cyber-Physical Systems. Algorithms for Intelligent Systems*. Springer, Singapore, 2022, pp 127–142, doi:10.1007/978-981-16-7136-4_11

[9] I. Georgiev and K. Nikolova, "An Approach of DNS Protection against DDoS Attacks," in *2017 13th International Conference on Advanced Technologies, Systems and Services in Telecommunications (TELSIKS)*, 2017, pp. 140–143, doi: 10.1109/TELSKS.2017.8246248

[10] Keyu Man, Zhiyun Qian, Zhongjie Wang, Xiaofeng Zheng, Youjun Huang and Haixin Duan, "DNS Cache Poisoning Attack Reloaded: Revolutions with Side Channels," in *Proceedings of the 2020 ACM SIGSAC Conference on Computer and Communications Security*, 2020, pp. 1337–1350.

[11] M. A. Hussain, H. Jin, Z. A. Hussien, Z. A. Abduljabbar, S. H. Abbdal and A. Ibrahim, "DNS Protection against Spoofing and Poisoning Attacks," in *2016 3rd International Conference on Information Science and Control Engineering (ICISCE)*, 2016, pp. 1308–1312, doi: 10.1109/ICISCE.2016.279

[12] B. Rajendran, Sanjay and D. P. Shetty, "DNS Amplification & DNS Tunnelling Attacks Simulation, Detection and Mitigation Approaches," in *2020 International Conference on Inventive Computation Technologies (ICICT)*, 2020, pp. 230–236, doi: 10.1109/ICICT48043.2020.9112413

[13] Lei Fang, Hongbin Wu, Kexiang Qian, Wenhui Wang and Longxi Han, "A Comprehensive Analysis of DDoS attacks based on DNS," in *Journal of Physics: Conference Series*, vol. 2024, 2021, 012027, doi: 10.1088/1742-6596/2024/1/012027

[14] Sunil Kumar Singh and Pradeep Kumar Roy, "Malicious Traffic Detection of DNS over HTTPS using Ensemble Machine Learning," in *International Journal of Computing and Digital Systems*, vol. 11, no. 1, 2022, pp. 189–197.

[15] S. Torabi, A. Boukhtouta, C. Assi and M. Debbabi, "Detecting Internet Abuse by Analyzing Passive DNS Traffic: A Survey of Implemented Systems," in *IEEE Communications Surveys & Tutorials*, vol. 20, no. 4, 2018, pp. 3389–3415, doi: 10.1109/COMST.2018.2849614

[16] Ilhan Firat Kilincer, Fatih Ertam and Abdulkadir Sengur, "Machine Learning Methods for Cyber Security Intrusion Detection: Datasets and Comparative Study," in *Computer Networks*, vol. 188, 107840, 2021 ISSN 1389-1286, doi: 10.1016/j.comnet.2021.107840

[17] H. Nguyen-An, T. Silverston, T. Yamazaki and T. Miyoshi, "IoT Traffic: Modeling and Measurement Experiments," *IoT*, vol. 2, no. 1, pp. 140–162, Feb. 2021, doi: 10.3390/iot2010008

[18] K. Xu, F. Wang, S. Jimenez, A. Lamontagne, J. Cummings and M. Hoikka, "Characterizing DNS Behaviours of Internet of Things in Edge Networks," in *IEEE Internet of Things Journal*, vol. 7, no. 9, pp. 7991–7998, Sept. 2020, doi: 10.1109/JIOT.2020.2999327

[19] A. Sivanathan et al., "Classifying IoT Devices in Smart Environments Using Network Traffic Characteristics," in *IEEE Transactions on Mobile Computing*, vol. 18, no. 8, pp. 1745–1759, 1 Aug. 2019.

[20] Ayyoob Hamza, Hassan Habibi Gharakheili, Theophilus A. Benson and Vijay Sivaraman. "Detecting Volumetric Attacks on loT Devices via SDN-Based Monitoring of MUD Activity," in *Proceedings of the 2019 ACM Symposium on SDN Research*, 2019, pp. 36–48.

[21] Ayyoob Hamza, Hassan Habibi Gharakheili and Vijay Sivaraman, "Combining MUD Policies with SDN for IoT Intrusion Detection," in *Proceedings of the 2018 Workshop on IoT Security and Privacy*, 2018, pp. 1–7.

[22] M. Preda, I. Bica and V.-V. Patriciu, "Internet of Things Traffic Characterization Using Flow and Packet Analysis," in *2020 12th International Conference on Electronics, Computers and Artificial Intelligence (ECAI)*, 2020, pp. 1–7.

[23] Thi Q. Nguyen, R. Laborde, A. Benzekri and B. Qu'hen. "Detecting Abnormal DNS Traffic Using Unsupervised Machine Learning". *4th Cyber Security in Networking Conference: Cyber Security in Networking (CSNet 2020)*, IEEE Communications Society, Oct 2020, Lausanne, Switzerland, pp. 1–8.

[24] Y. Liu, J. Wang, J. Li, S. Niu and H. Song, "Machine Learning for the Detection and Identification of Internet of Things Devices: A Survey," in *IEEE Internet of Things Journal*, vol. 9, no. 1, pp. 298–320, 1 May 2020, doi: 10.1109/JIOT.2021.3099028

Chapter 12

Machine learning in power electronics for smart cities

Ahteshamul Haque
Jamia Millia Islamia, New Delhi, India

K. V. S. Bharat
Silicon Austria Labs, Graz, Austria

Fatima Shabir Zehgeer and Azra Malik
Jamia Millia Islamia, New Delhi, India

12.1 INTRODUCTION

The increasing urbanization calls for more advancement in the various domains of technology. The concept of the smart city focuses primarily on providing a quality lifestyle to all the dwellers with optimum resource utilization and enhanced sustainability. Power electronics plays a key role in green energy, electrification of transportation, smart grids, smart energy management, and the optimization, smart fault detection, and diagnostics, and to various other important elements of smart cities. Machine learning has revolutionized the world and its impact on power electronics has helped tremendously in the development of smart cities. The integration of renewable energy resources, especially the solar energy, has been one of the important components of the smart city.

The ever-increasing developments taking place in various fields such as data science, big data analysis [1], digital twin [2], and internet of things (IoT) has laid the strong foundation for the smart city development with a huge availability for power electronic systems. The voluminous data has enabled the immense application of Artificial Intelligence (AI), especially the machine learning in power electronics [3]. For smart city development, it becomes necessary to implement various machine learning algorithms that have the ability to analyze and learn from such a huge data available in smart cities [4].

The functions of AI for the application in power electronics may be essentially optimization, classification, regression, or data structure exploration tasks. Among the various categories of AI, machine learning has the highest contribution (45.8%) for the application in power electronics, especially for control operations [3]. Machine learning in power electronics may be classified as supervised learning, unsupervised learning, or the reinforcement learning. The supervised learning tasks which essentially develop an implicit

DOI: 10.1201/9781032669809-12

relation between input and output may be classification tasks such as in the fault diagnosis [5] or regression as in the remaining useful life (RUL) prediction [6]. Neural networks, kernel methods, and adaptive neural fuzzy inference systems are some of the supervised machine learning methods. For the classification and regression, neural networks have been very effective and have been successfully implemented for the various essential components of a smart city [7]. Unsupervised learning-based machine learning methods essentially employing data clustering and data compression have been used for data preprocessing as in anomaly detection [8]. A more advanced subset of the machine is deep learning and it has also been extensively implemented for various applications. The machine learning in power electronics for various smart city applications has been shown in Figure 12.1. Machine learning has helped a lot in increasing the accuracy and efficiency of the PV systems. Machine learning techniques have proved to enhance the reliability, efficiency, and performance of power electronic systems in smart cities. It is applied for the optimal sizing of the PV system [9] and in the forecasting techniques for solar irradiance [10, 11] and the solar output power [12].

The forecasting of solar irradiance using artificial neural network (ANN) [13], Naive Bayes classifier [14], k-nearest neighborhood (k-NN) [15], and Gaussian process regression [16] has also been proposed. An extensive review of the application of machine learning in smart cities for energy management has been done [17]. Machine learning has been implemented in controlling the power electronic converters, for tracking the maximum power point [18], and in fault monitoring for the filters [19]. Besides, it has immensely been applied for the maintenance of power electronic converters such as fault detection and diagnostics [5], condition monitoring [20], and RUL prediction, and machine learning has successfully been utilized for the enhancement of performance, power quality, efficiency, and reliability of the power electronic systems in the smart cities.

12.2 MACHINE LEARNING FOR CONTROLLING OF POWER ELECTRONIC CONVERTERS

Machine learning is applied to the optimization and regression tasks to control the power electronic converters. Based on the mode of the operation, the control of power electronic converters may be grid-connected control or the stand-alone control [21]. As the complexity of the system increases, the conventional control methods become infeasible, time-consuming, and demand more information about the system control. Also, the conventional controller has less adaptability, is a static operation, and thus renders it useless for time-variant systems. The system parameters change with the changing environmental conditions, resulting in the limited robustness and deteriorated performance of the conventional controllers. The AI-based control methods which include the metaheuristic methods, fuzzy methods, and machine learning methods can be used to mitigate such issues. The metaheuristic methods

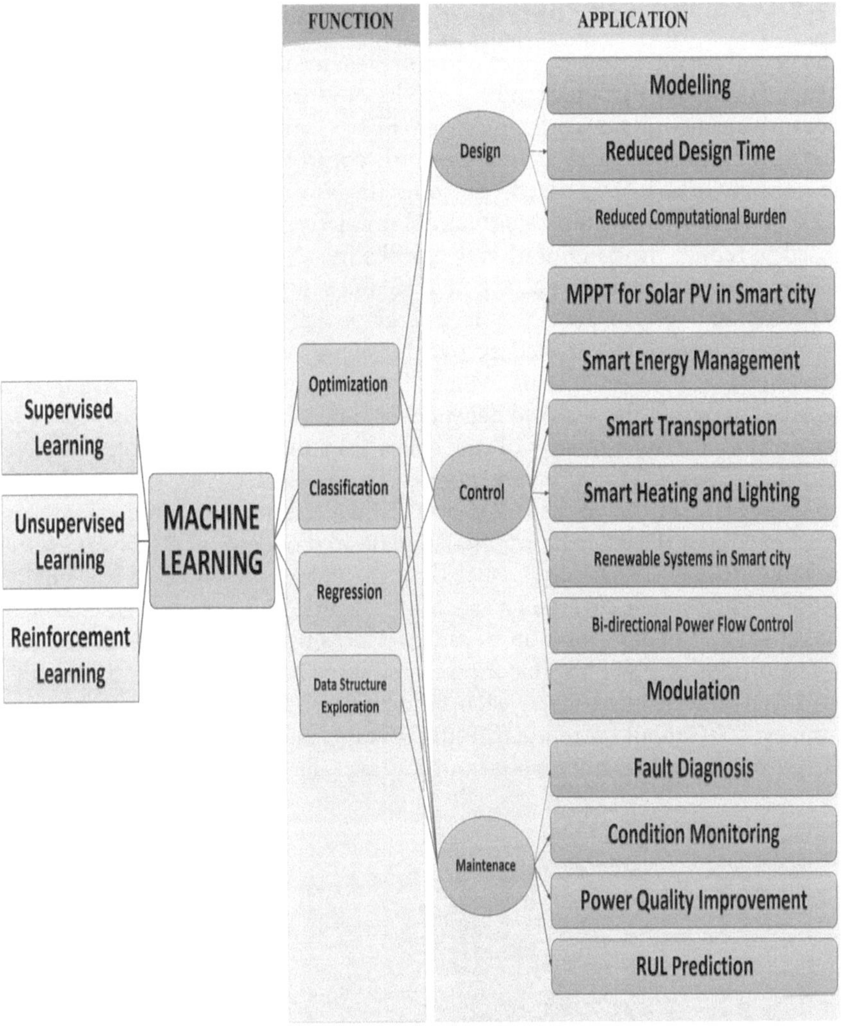

Figure 12.1 Application of machine learning in power electronics for smart cities.

like Genetic Algorithms, ant colony optimization, and so on are found to be much more superior to the conventional methods. However, the metaheuristic methods are generally applied to the static optimization tasks while the machine learning method is applied to the dynamic optimization. Machine learning methods are more accurate and flexible than the fuzzy logic-based methods. Machine learning has emerged as the most powerful AI method. The performance can be improved with the fuzzy logic integrated with the machine learning. The various machine learning-based controllers include the neural network-based controller (along with its combination with the fuzzy logic), FFNN (feedforward neural network), adaptive neural fuzzy inference system (ANFIS), reinforcement learning (RL), and so on [3].

12.2.1 ANN-based control of an inverter

The neural network-based controllers offer many advantages such as robustness, adaptability, dynamic, and so on. The NN-based controllers include the feedforward neural network controller (FFNN), radial basis neural network (RBNN) controller, fuzzy neural network controller (FNN), and recurrent neural network (RNN)-based controller. In Novak and Dragicevic's study [22], an ANN-based finite – set model predictive controller has been proposed to reduce the computational complexity.

Generation of clean, green, sustainable, and renewable energy is the essential component of a smart city. Solar energy systems which include solar photovoltaic systems are installed across the smart cities. This system is a key distributed generation system, which has minimized the transmission and distribution losses and reduced dependency on depleting fossil fuels. Solar PV systems are also integrated with the smart grid and used for electric vehicle charging. However, the power generated by the solar energy systems is DC which needs to be converted to AC power. Inverters are thus required for this conversion. For the efficient, optimal, and desired operation, the inverters are to be controlled. In Demirtas et al [23], ANN-based control of the single-phase voltage source inverters as shown in Figure 12.2 is proposed in which the design and implementation of the inverter is trained and ANN-driven.

The proposed ANN has one input layer, three hidden layers and one output layer with 10 neurons at each hidden layer. The input layer has three inputs viz. DC input current, DC input voltage, and AC output current. The AC output voltage is obtained at output layer. First, the network is trained

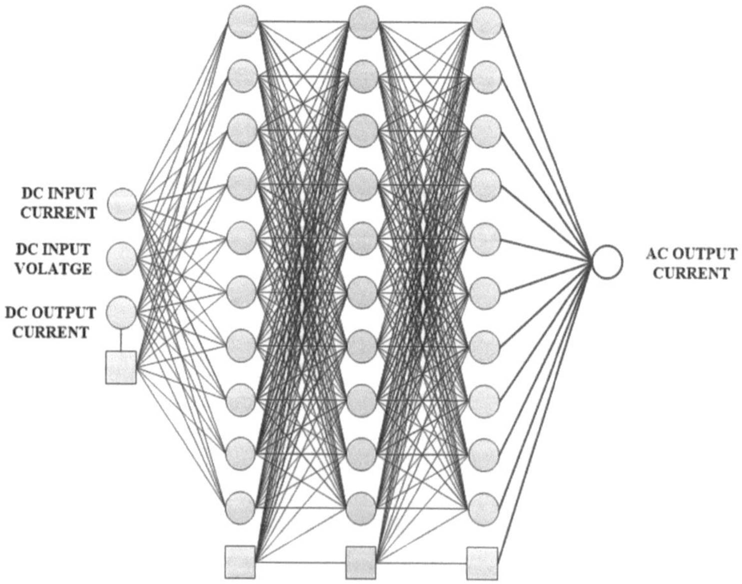

Figure 12.2 ANN architecture for controlling single-phase inverter.

in order to determine the activation function and the training methods to be used. Then for the ANN learning, the training data is shown to the network followed by the input of the test data. Based on this training, ANN then calculates the output (duty period) in order to maintain the desired output voltage for the varying load. With the ANN, more accurate and efficient results are obtained.

12.2.2 ANFIS-based controller

For a smart grid in a smart city, bidirectional flow of power from grid to vehicle and from vehicle to grid is desirable. To achieve the bidirectionality, the power electronic converters play a crucial role as the interface between the vehicle the and smart grid. When the load changes are non-linear and mathematical calculations are complex, controllers based on ANFIS are feasible. In Islam et al. [24] an ANFIS is proposed for the control of the bidirectional power flow. For achieving the required control, two ANFIS controllers are used. The DC-DC converter controls the battery current which is to be maintained at the reference value. This reference battery current is controlled by the first ANFIS as shown in Figure 12.3.

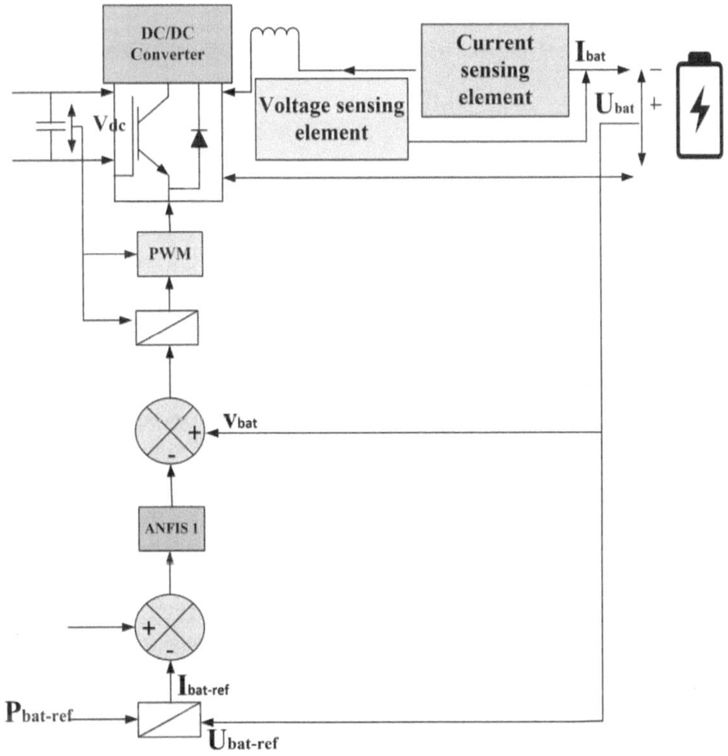

Figure 12.3 DC-DC converter with ANFIS 1 controller.

The second ANFIS controller maintains the constant DC bus voltage irrespective of what the power flow direction is. With a proposed five-step control strategy, the ANFIS controller controls the charging of the battery and the power injection by the PEV into the smart grid. The resultant charging of the battery is smooth and the distortion is reduced. Also, the power flow to the grid is smooth with reduced harmonics. For the design of the ANFIS controller, a five-step optimization algorithm produces the optimal data set values for the ANFIS controller. The structure for the ANFIS is operated in the neural network stage and the fuzzy stage. The ordering of the information and the discovery of the designs is done by the neural network stage while the other stage is the fuzzy based. The ANFIS comprises of five layers and this system is based on fuzzy if-then rules for the input and output parameters of the neural network.

Another ANFIS-based controller has been proposed by Karuppusamy, Natarajan, and Vijeyakumar [25] to control the multilevel inverter for the solar photovoltaic system. Based on the input (the grid voltage and the difference voltage) and the output (control voltage), the neural networks generate the fuzzy rules. As per the different inputs, ANFIS results in the control voltage which is then used to control the IGBTs of the multilevel inverter.

12.2.3 ANN-based control of DC-DC boost converter

In Mohammadzadeh et al. [26] two control techniques are used together to control the DC-DC converter for the application in the EVs. The two techniques are output feedback control (FBC) and the model predictive control (MPC) and machine learning-based control method has been incorporated for the performance optimization of the DC-DC boost converter.

FBC is robust with a simple design and it results in better management of power loss in the output voltage. The MPC works better for the dynamic state and results in improved settling time with the better voltage undershoot or overshoot. Using the two different neural networks both the techniques are separately used to train the ANNs which are then used for boost converter control. For the neural network training, the inputs are the inductor current and the voltage of the capacitor. The output target is the switching state which is given to the Pulse Width Modulator for the generation of the gate signal which is the input to the gate driver circuit of the main MOSFET switch in the DC-DC converter. With the proposed control, the steady state response of the converter was enhanced with the decrease in steady state error and optimized dynamic behavior of the DC-DC converter was obtained. The ANN-based control of the DC-DC boost converter is shown in Figure 12.4.

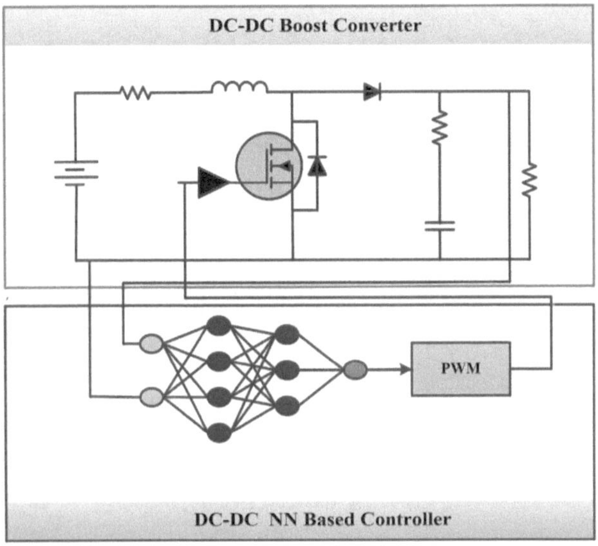

Figure 12.4 ANN-based control of DC-DC boost converter.

12.2.4 Reinforcement learning-based control of DC-DC converters

The development of the smart cities is based on the concept of sustainability and all the associated technologies should help in mitigating the climate change. Thus, the integration of the renewable energy resources which solar and wind energy is the chief element of the smart city. However, the for the regulated and desired DC voltage the DC-DC converters with the appropriate controller are required.

In RL, there exists a control strategy for the controllers that is oriented toward the predefined goals [27]. The controller interacts with the simulation model or the physical system to learn the control strategy which helps to achieve the goal and the targets. For an interleaved DC-DC converter, a novel integral reinforcement learning (IRL) -based control has been pro posed [28]. Compared with the conventional controller, the proposed IRL controller doesn't rely on the process of updating the disturbances which result in the reduced complexity of the computation process. With the asumption that parameters like L, C, R, D are bounded, the interleaved buck converter is modeled. C denotes the output capacitance, R is the load resistence, D is the duty ratio of the converter, and L denotes the phase inductance. With the output voltage V_o being close to the reference value V_{ref}, the control method objective is to reduce the steady state error to zero. Also, the DC-DC converter system should be stable within the given bounded limits and the the gain of the controller should be adjusted autonomously with the change in the operating conditions. The flowchart of the control

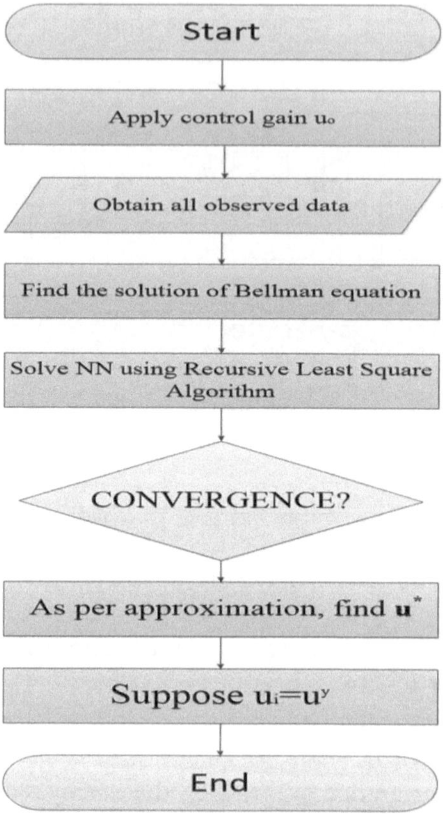

Figure 12.5 Flowchart for RL-based control of DC-DC converter.

method is shown in Figure 12.5. For implementing the prosed control method, the various observations are obtained after the initial stabilizing policy gain u_o is applied. Then Bellman equation given in (12.1) is solved.

$$e^{-\alpha T}V_i\big(x(t+T)\big)-V_i\big(x(t)\big)=\int_t^{t+T}e^{-\alpha(\tau-t)}\big(-xTQx-u_i^TRu_i+\gamma^2d_i^Td_i\big)d\tau + \int_t^{t+T}e^{-\alpha(\tau-T)}\big(-2u_{i+1}^TR(u-u_i)\big)d\tau \tag{12.1}$$

Where $x(t)$, $u(t)$ are the collected observations, V_i is the cost function, u_i is the optimal control policy at the ith iteration, d is the disturbance, Q represents the positive semidefinite Hermitian matrix, and R denotes a positive scalar for the weight determination on the control inputs. With the reinforcement learning algorithms, the approximation of the weights of the neural network is done. Finally, the updated optimal control policy is obtained with the convergence of the reinforcement learning system algorithm.

12.2.5 RL-based controller for tracking MPP

Wind energy contributes significantly in smart city development. It is the cleaner, greener source of energy and promotes the sustainability, resilience, and economic growth of a smart city. Most of the electricity generated from the wind is from the variable speed wind energy conversion systems (WECSs). The speed of the shaft in the WECS is to be controlled so that the maximum power points are tracked and consequently for varying wind speeds, the power generated is maximum. Based on the interaction with the environment through actions and rewards, the RL learns the system behavior and has been extensively used for MPPT control. In Qie et al. [29] the MPPT method based on a novel RL is proposed to control the variable speed WECS. The proposed online method is based on two processes:

1. Online learning Process
2. Online Application Process

In the online learning process, the controller uses the model-free Q learning algorithm to learn about the maximum power points based on the experience. The learned values are stored in a Q-table and are used to find the optimal relationship between the rotor speed and the electrical power. In the online application process, based on the learned relationship from the learning process, the control of the variable speed WECS is executed. When compared with conventional methods like Perturb and Observe (P&O), this proposed RL method results in the generation of higher electrical power. Also, the fast MPPT control can be obtained with the fast convergence to MPP during the variable wind speed.

In Wei et al. [30] the RL has been used together with ANN for the WECS. With the ANN the efficiency of learning is improved and the online learning time is reduced as ANN provides the generalized learning with the predictions even when it has not been previously experienced. The integration of the ANN with RL results in the elimination of the look-up table which results in considerable computation cost savings and makes the MPPT algorithm implementation and design easier.

Another application of the RL is in the MPPT control of the photovoltaic systems. Various RL-based control methods have been proposed for maximum power point tracking. Another effort to optimize the performance of the PV system was done by Kofinas et al. [31] in which again the RL technique for MPPT control of the buck-type DC-DC converter shown in Figure 12.6 has been proposed. The short circuit current and the open circuit voltage are the two parameters for the proposed RL MPPT control

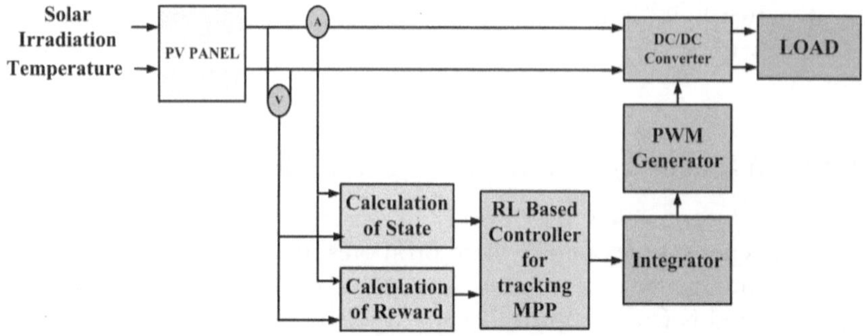

Figure 12.6 RLMPPT control of DC-DC converter.

method. Without the prior information about the PV source and its dynamics, the RL method tracks the MPP by learning the system from the response of the PV source. The Markov Decision Process (MDP), which is a sequential model of the source behavior, has to be defined for the implementation of the RL. In Chou et al. [18] the tracking of MPP with the Q learning algorithm-based RL method is proposed.

12.2.6 ML for controlling power flow in multi-active -bridge converter

The multi-active bridge (MAB) converters have their role in smart energy management, grid stability, electric vehicle (EV) charging, and energy storage systems, thus these converters have a potential significance for the smart energy in the smart city.

An MAB converter connects the source and the load by providing a single conversion stage. The MAB converter has the multi-winding transformer and is based on the phase angle the power is injected into or extracted from the windings. Thus, the multiway power flow has to be controlled. The conventional iterative algorithms like Newton Raphson have not been used but for the multi-power flow, it may not converge. Machine Learning-based techniques like neural networks have replaced the traditional numerical methods for the multi-input multi-output power flow control of the MAB converters. The feedforward neural network (FFNN) -based machine learning technique has been proposed by Liao et al. [32] for the power flow control in 6 port MAB. The inputs to the input neural layer are the power flows which are to be targeted and at the output layer, phase angles that can result in the targeted power flow are the outputs. The sigmoid function is used as the activation function at the hidden layer.

The FFNN-based control results in more accurate results as compared to the traditional control methods. The FNN-based architecture for the power flow control of the MAB converter is shown in Figure 12.7.

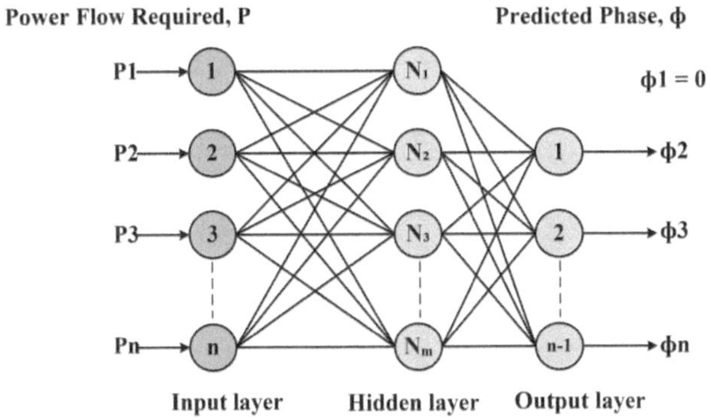

Figure 12.7 The FNN-based architecture for the power flow control of the MAB converter.

12.2.7 Other ML techniques for the control applications

In smart cities, for smart street lighting, indoor and outdoor lighting, traffic signals, and information displays use of the LED (Light Emitting Diodes) technology is quite popular. An ANN-based multi-color control LED system is proposed [33]. A buck type (single inductor multiple output) single input multiple output (SIMO) LED driver which has a series connected channel is used to drive the LED system. The ANN control is linked with the spectral peak control. The high color rendering index data points are provided by the spectral control to ANN control and these data points are then used to update and train the neural network. Even in the presence of the ambient light the color control of the LED system is achieved with the trained neural network. Also, without any additional hardware, the performance is improved.

In a smart city, induction heating has been encouraged due to its efficient heating, smart energy and waste management, medical and industrial applications, and in smart home applications. Lucia et al. [34] have proposed a deep learning-based model of predictive control for the resonant converters used for the induction heating. To supply the ac voltage to the induction coil, resonant power converters such as a series resonant half bridge converter is used. For the desired performance, the MPC is used for the prediction of future behavior prediction along with the computation of the control inputs. The deep learning networks which have several hidden layers learn the optimal control policy given by the MPC so that using the neural network is only required for the evaluation for the online controller.

Ahmed and Al-Othman [35] have proposed an effective SVM-based MPPT control for the photovoltaic system. For an inverter connected to a grid, NN with the recurrent units has been proposed [36] while Phan, Lai, and Lin [37] proposed deep learning-based MPPT control.

Table 12.1 summarizes the machine learning in power electronics for control applications.

Table 12.1 Summary of machine learning in power electronics for various control functions

Machine learning method	*Learning method type*	*Method and variant*	*Application*	*References*	*Advantages and disadvantages*
Supervised learning method	Neural network method	(Artificial neural network) ANN	Control of the single-phase inverter to be used for solar energy conversion systems.	[23]	• With the changing loads, the output is maintained at constant value. • Efficiency and accuracy are better.
			Two ANN (trained using MPC and FBC) used for control of DC-DC converter	[26]	• Better static and dynamic performance • Dependent on specific application and converter.
			FFNN for controlling power flow in MAB converter.	[32]	• More accurate results as compared with traditional methods. • Precise converter model is not required. • Doesn't have embedded capability for expert knowledge.
			FFNN for multi-color control LED system for smart street lightening.	[33]	• High CRI color control is attained. • Without any additional hardware, the performance is improved.
		(Adaptive neuro-fuzzy inference system) ANFIS	Controlling bidirectional power flow in DC-DC converter.	[24]	• The power flow to the grid is smooth with reduced harmonics. • Compared with the fuzzy rule methods, without expert knowledge the fuzzy rule is automatically generated.
			Controlling multilevel inverter for solar PV system.	[25]	
	Kernel method	Sparse kernel method-based support vector machine (SVM)	MPPT controller.	[35]	• Economic, robust and effective for varying insulations.

Unsupervised learning methods	k − means clustering algorithm	Optimization of buck regulator.	[38]	• Sensitive to outliners. • Execution is simple.
	Principal componentanalysis	Compression of data in Cuk converter.	[39]	• Improved and flexible framework.
Reinforcement learning (RL) method	Integral reinforcement learning	Control of interleaved DC-DC converter.	[28]	• Less computational complexity. • Real-time implementation.
	Q learning algorithm-based RL	Control of maximum power point tracking for WECS.	[29]	• Lower computational costs. • No prior system knowledge is required. • Higher amount of the electrical power is generated.
	ANN-based reinforcement learning	Control of maximum power point tracking for variable speed drive WECS.	[30]	• The maximum power obtained is smooth with less fluctuations.
	RL MPPT	MMPT control of buck type DC-DC converter.	[31]	• Better output and accuracy than conventional methods. • Optimal performance under different operating states.

12.3 MACHINE LEARNING FOR MONITORING OF POWER ELECTRONICS CONVERTERS

For smart city applications, the power electronic converters are prone to harsh and severe working conditions, due to which it becomes necessary to monitor the devices for reliable operations. To improve reliability, various maintenance activities which include condition monitoring, prediction of the RUL, fault diagnosis, and so on are used. All the maintenance activities are carried out as per the IEEE standards [40]. With the proper health and condition monitoring, both at the component and system level, the desired operation of the power electronic system is ensured. Condition monitoring and assessment of the health of the power electronic units is one of the main maintenance functions. In the condition monitoring, with the parameter identification, information about the critical components is obtained. The identification of the system parameters may be based on model or model-free depending upon whether the prior information about the system dynamics is required or not [41]. The machine learning techniques are applied in the model-free method of the parameter identification to develop relationships between inputs and output to be monitored. The relation between the input current, ripple voltage, and the output capacitance are developed with the training of the feedforward neural network [3]. The model-free method for parameter identification incurs less cost and thus finds suitable industrial applications in the smart city. Condition monitoring also includes the processing of data and feature mining to refine the raw data for specific applications. With noise reduction the data is cleaned, and the clustering of the similar data types is done, which results in improved performance and enhanced accuracy for the various applications like fault diagnosis [40]. Soliman et al. [19] monitored the health of the electrolytic capacitors comprehensively. It is very important to monitor the operation of the power converters since they are susceptible to faults There are numerous machine learning methods in the literature for monitoring applications.

12.3.1 Machine learning for condition monitoring in back-to-back converters

About 60% of the failures in power electronic devices is due to the fault in the electrolytic capacitors [42]. Thus, it becomes inevitably important to condition monitor (CM) the electrolytic capacitors for the health assessment of the power electronic converters. Some of the machine learning methods have been proposed [43] for the condition monitoring of the electrolytic capacitors used in back-to-back converters. First, the capacitor ripple voltage is measured and using the four-level wavelet decomposition, the capacitor degradation is determined and the various features such as skewness, root mean square (RMS), crest factor (CF), and energy entropy (EE) are extracted. Then, machine classifiers such as k-nearest classifier, Support vector machine classifier (SVM), and Naive Bayes classifier are used for the CM.

K-nearest neighbor classifier (KKN): this supervised learning classifier is based on the decisions that are made considering the neighborhood points and the distances measured by the Euclidian distance. The overall accuracy of the KKN is about 89.6% [43]. Based on the maximum distance between the different classes, SVM classifier optimally classifies the data points using a hyperplane which is found by minimizing the target function given in eq. (12.2).

$$\min J(\Omega, b, \phi) = \frac{1}{2}\|\Omega\|^2 + c\sum_{i=1}^{n}\phi_i \text{ with} \forall_i \begin{cases} y_i(\Omega.\delta(x_i) + b) \geq 1 - \phi_i \\ \phi_i \geq 0 \end{cases} \tag{12.2}$$

Where Ω denotes the weight vector, b stands for the bias, ϕ_i denotes the slack variable, and for mapping inputs with the feature space function $\delta(.)$ is used. Based on the probability theory and the Bayesian rule [44], the Bayesian classifier which is also known as the Naive Bayes classifier is proposed [19]. With the utilization of this machine learning classifier, the model becomes complex [45] and the classifier has the accuracy of about 72% while the accuracy of SVM is 92% which results in the better performance [43].

12.3.2 ML-based monitoring methods for DC-DC converters

Bindi et al. [46] proposed the monitoring of the zeta converter which is a fourth order DC-DC converter used for the PV conversion. The multi-layer multi-valued neuron (MLMVN) based classifier is used for the health monitoring of the zeta converter. The zeta converter integrated with the smart grid has been shown in Figure 12.8.

The multi-layer neural network has neurons that are multi-valued. The inputs and the weights are complex-valued. The proposed classifier has three neural network layers with the discrete activation function [47]. The activation function is used for the decomposition of the complex plane into k number of sectors which are equal. At the bottom of the sector the output is placed. The complex inputs are given by (X_1, X_2 ... X_n) and the complex weights are denoted by (W_1, W_2 ... W_n). The weighted sum of the inputs given by $z = X_1W_1 + X_2W_2 + \ldots X_nW_n$ is also placed at the lower end of the sector. The discrete activation function is represented in Eq. (12.3).

$$P(z) = Y = \varepsilon_k^j = e^{i2\pi j/k} \; if \; 2\pi j / k \leq \arg(z) < 2\pi(j+1)/k \tag{12.3}$$

Where the sector at which z is placed is denoted by j. The sector division-based CM has been shown in Figure 12.9.

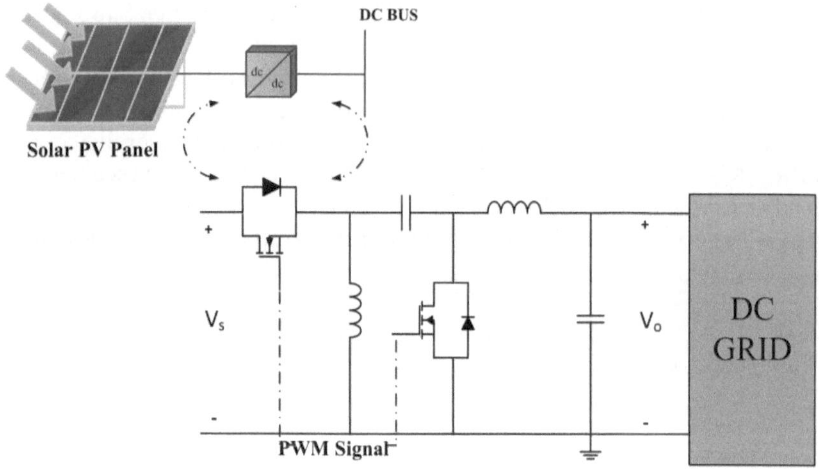

Figure 12.8 DC-DC converter integrated with solar PV panel in smart grid.

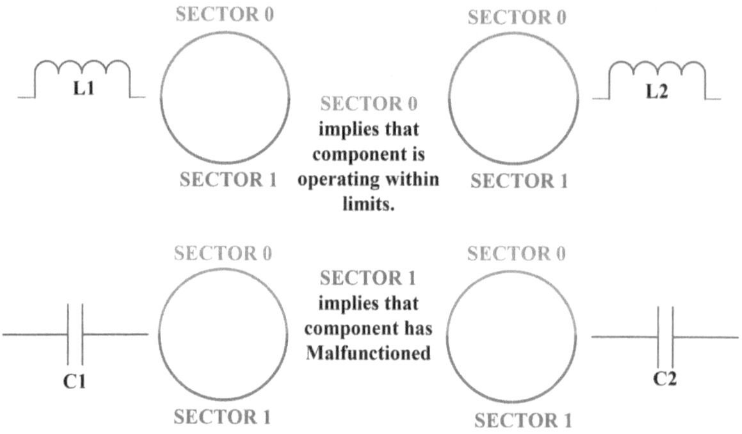

Figure 12.9 Sector-based condition monitoring.

12.3.3 A multi-layer perceptron neural network (MLPNN)-based condition monitoring

In smart cities, the installation of the PV system is done for renewable energy integration. For, a reliable and efficient PV system, condition monitoring for maintenance plays an important role. In Haque et al. [48], multi-layer perceptron neural network is used for the monitoring of the photovoltaic modules. The thermal images of the PV module are obtained and based on these images the faulty PV module is subjected to a flash test for no more than 50 ms. The flash tester gives the IV characteristics of the module. After the preprocessing of the signal, the discrete wavelet transform (DWT) is applied to the output voltage and its various features like power, energy, SNR, and total harmonic distortion are extracted. These features are fed to the multi-perceptron neural network (MLPNN) machine learning classifier which classifies this data for the procurement of the trained data. Using the MLPP, the fault type is displayed based on the training of input data for multiple outputs [49]. The trained data is then used for the fault classification based on the nature, location, and impact of the fault on the power system.

MLPNN is a hybrid system that is used in training the given set of data for multiple outputs to display the type of fault. The block diagram for the monitoring and fault diagnostics is shown in Figure 12.10.

Bharath, Haque, and Khan [20] also used the MLPNN classifier for the monitoring of the PV modules. During the normal and the faulty conditions, different features are extracted from the PV output using the discrete wavelet transform and the principal component analysis (PCA) is applied to reduce the dimension of the feature set so that the optimal and most useful features are accommodated in the reduced dimension feature set. Then the features obtained are given to the multi-perceptron-based neural network which is a feedforward network. Various algorithms like the back propagation algorithms are used for the training of multi-layer perceptron. The multi-layer perceptron (MLP) architecture comprises the input layer with six neurons, 12 neurons in the hidden layer and for the four classes of the

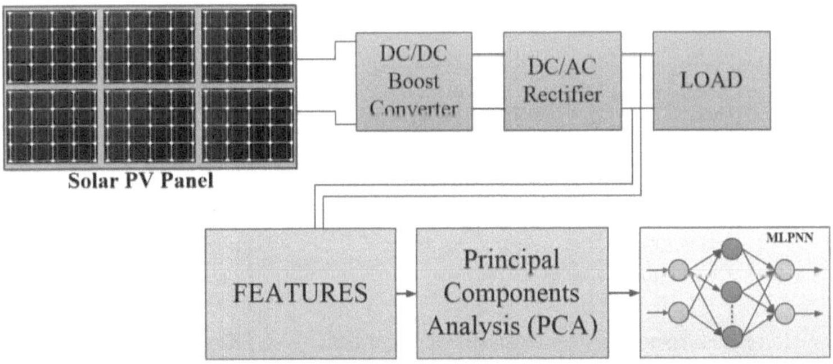

Figure 12.10 Machine learning-based conditioning monitoring of solar PV system.

fault the output layer has four neurons. The four binary outputs represent the different fault type [50] as shown in Table 12.2.

The summary of various machine learning-based conditioning systems is shown in Table 12.3.

Table 12.2 Fault classification based on the output of MLPNN

MLP binary output	*(1000)b*	*(0100)b*	*(00100)b*	*(0001)b*
Indicated fault	MF	FF	CF	IF
Fault description	Module has failed	No fault is present	Fault in the converter	Inverter has failed

Table 12.3 Summary of machine learning-based conditioning monitoring system

Reference	*Machine learning classifier*	*Application*	*Advantages/disadvantages*
[43, 44]	K-nearest neighbor classifier SVM classifier Naive Bayes classifier	Condition monitoring for the electrolytic Capacitors in back-back converters.	• With the Naive Bayes classifier model becomes complex and efficiency is 72%. • Better performance with SVM classifier with 92% efficiency. • Efficiency of KNN classifier is 89.6%.
[46]	Multi-layer multi-valued neuron-based classifier	Condition monitoring of zeta converter, a fourth order DC-DC converter.	• As compare to SVM, rate of classification is higher. • Can operate in varying environmental conditions.
[19]	ANN-based classifier	Monitoring of capacitors.	• Error is within acceptable limits. • No supplementary hardware is required.
[48]	Multi-layer perceptron neural network-based classifier	Monitoring of the PV modules.	• Detects fault in less than 9 seconds with 100% efficiency when fault data is given.

12.4 MACHINE LEARNING IN FAULT DIAGNOSTICS

From the binary output of the neural network, any malfunction is detected and then the detailed investigation about the failure is done through the proper fault diagnosis. The machine learning methods for fault detection and subsequent fault diagnosis may be supervised or unsupervised [3]. In Akram and Lotfifard [51], real-time-based open circuit and short circuit fault diagnosis using probabilistic neural network (PNN) which is a radial basis neural network has been proposed. Based on the collected data such as current, voltage, and temperature, the fault is classified using the four-layer-based PNN. An ANN-based diagnostic tool is proposed [52] for an inverter operating in changing load conditions in a microgrid. For the implementation of the smart city concept, rapid fault diagnosis is important. A machine learning-based method is proposed [53] for the fault diagnosis of the inverter used for the electric propulsion system for water transportation. It has been estimated that almost 50% of the failures occurring in the PECs are due to the fault in DC filters [54]. The fault diagnosis of the capacitor in the DC filters using ANFIS is proposed [55] in which the DC filter voltage is one of the three inputs to the ANFIS and the fault due to the capacitor aging is identified. The industries in the smart cities require devices with high-power ratings and multilevel inverters are quite suited for such applications. In Wang et al. [56], the fault diagnosis of the multilevel cascaded inverter based on the PCA and the multi-case relevance vector machine (mRVA) is proposed. The diagnosis time of the proposed method is less as compared to the traditional methods. One of the chief features of a smart city is that an uninterrupted power supply should be provided to industrial and the domestic customers. Multilevel inverters present a potential solution for meeting the high-power requirements of modern industries and EV charging in a smart city. In Chowdhury et al. [57], a machine learning method has been proposed for the open circuit fault diagnosis of the five-level cascaded H-bridge converter. For the fault identification, a DWT is used in which db10, a mother wavelet which is a scaling function and resembles the fault signal, is used to decompose the fault signal into 12 levels, and energy extraction from all the 12 levels is done. The extracted features are fed as input to the feedforward neural network which analyses the signal and detects the fault. For the fault diagnosis in converters used in wind energy systems, the Adaboost-SVM-based machine learning method is proposed in [58]. Other machine learning methods for fault diagnosis are explained in the following sub-sections.

12.4.1 Random Forest (RF) methods based on supervised learning used for fault diagnosis

To maintain high efficiency and enhance the conversion of solar energy to electrical energy, monitoring along with fault diagnosis of the solar PV system is necessary. In Chen et al. [59], for the fault diagnosis of the PV system, machine learning-based Random Forest method is proposed. In the proposed RF algorithm, two of the pre-processed features (voltage of the array and the current of the string) are extracted. The data obtained is divided randomly into the test data and the T training data sets. The m features are selected from the training sets and the decision tree is grown using the CART algorithm. The data which has not been used for training is used for validation and the predicted data obtained after validation is tested for the accuracy of the major votes acquired from all DTs. The RF method is found to be highly accurate with the removal of overfitting limitations encountered in another machine learning-based fault diagnosis such as in Belaout et al. [60]. The flowchart for the fault diagnosis based on RF method is shown in Figure 12.11.

12.4.2 Fault diagnosis using deep learning-based convolutional neural network

Fault detection is a challenging task and thus several methods has been proposed. Aziz et al. [61] have proposed a novel idea for fault diagnosis based on the deep convolutional network. The PV data is collected and converted

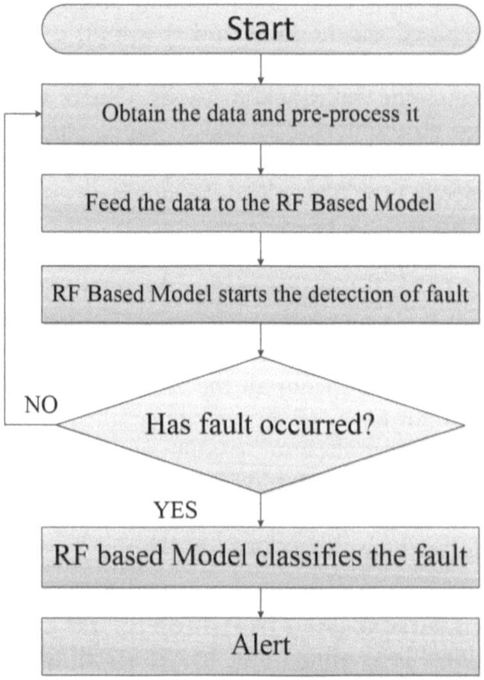

Figure 12.11 The flowchart for the fault diagnosis based on RF method.

into the 2D visual data called Scalogram with the use of Continuous wavelet transforms [62]. Then the Scalogram is fed to the AlexaNet deep convolution neural network (CNN) which has been pre-trained and has fine-tuned three layers at the end. The features are extracted, and from the Softmax layer, based on probability, the data is classified into the fault classes which are obtained from the output layer. Compared with the other machine learning methods, the proposed method is highly accurate with 73.53% accuracy for the detection of fault [61].

12.4.3 CNN-based fault diagnosis of inverter

A novel method for fault diagnosis is the CNN-based global average pooling (GAP) in which the convolution neural network is modified to have five layers is proposed [63]. As shown in Figure 12.12, 1D fault data is given as the input to the input layer of the CNN-GAP model in which 1D data is converted to the 2D maps representing the features. The features are then extracted from the maps in the feature extraction layer. The fourth layer is the GAP layer which operates the output of the preceding layer for the appropriate dimension and size. Then at last the output of the global average pooling layer is given to the fifth layer called the Softmax layer and the output of this layer gives the relevant diagnosis of the fault for an inverter. Another CNN-based method for fault diagnosis of the inverter used in electric propulsion for the hybrid electric ships has been proposed [53].

Kurukuru et al. [64] have proposed the fault diagnosis of the inverter based on the Radial basis function network with the Gaussian kernel. Table 12.4 summarizes the machine learning-based fault diagnosis techniques for power electronic converters (PECs).

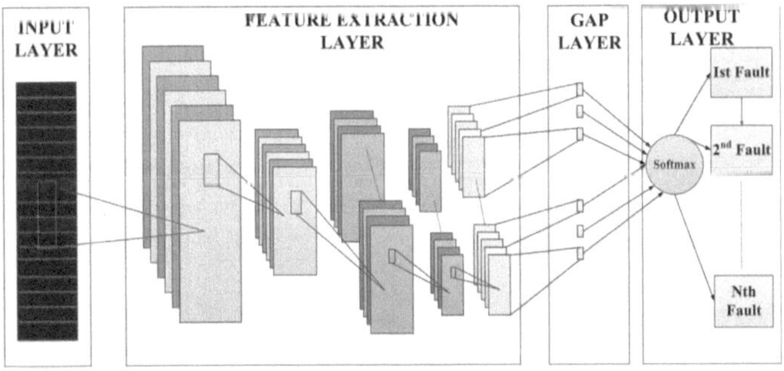

Figure 12.12 CNN-based GAP network architecture.

Table 12.4 Summary of the ML-based fault diagnosis for PECS

References	*Machine learning technique*	*Application*	*Advantages/disadvantages*
[52]	ANN	Fault diagnosis of inverter in microgrid.	• Lower cost. • No supplementary hardware is required.
[63]	CNN-based global average pooling	Fault diagnosis of inverter and IGBT.	• Highly accurate (99.5%). • Fast fault diagnosis.
[55]	ANFIS	Fault diagnosis of capacitors in filter.	• Fast and robust. • Accurately detects and classifies fault due to aging.
[56]	PCA and multi-case relevance vector machine (PCA-mRVM)	Fault diagnosis of multilevel inverter.	• Diagnosis time is less (0.47s). • Output is probabilistic and not binary.
[57]	ANN with wavelet decomposition	Fault diagnosis of 5 level inverter.	• During faulty condition diagnostic signal may get distorted. • Classification accuracy is very high.
[59]	Random Forest based on supervised learning	Fault diagnosis in PV panel.	• Highly accurate with removal of overfitting limitation.
[63]	Deep learning-based CNN	Fault diagnosis of PV module.	• 73.53% accuracy. • Works accurately in noisy environment.

12.5 OTHER MACHINE LEARNING APPLICATIONS

Machine learning in power electronics has been majorly applied for control and the maintenance functions. However, the machine learning methods have also been applied to reduce the design time of the power electronic systems and the computational burden. The NN are trained to learn about the input-output relationship and without the requirement of the real circuit topology, the NN can imitate the circuit behavior. In Chiozzi et al. [65], the MLP topology-based NN is utilized for building the electrothermal model of a power MOSFET. The ANN has three inputs viz. gate-source voltage V_{GS}, V_{GS} (drain-to-source voltage), and the junction temperature (T_j) and the drain current I_D as output. With the sigmoid function used as the activation function, the relation between the inputs and output is formulated as:

$$Y = f(x) \Leftrightarrow I_D = F(V_{GS}, V_{DS}, T_j) \tag{12.4}$$

Then, a non-linear RC network in which resistors are temperature dependent is modeled. Finally, the two generated models are used in conjunction with each other which results in an appropriate model with improved convergence.

Another example of the application of machine learning in the optimal, automated reliable design is proposed [66] in which two artificial neural networks are used. The first ANN maps the four inputs to the two outputs, the two inputs viz. ambient temperature and solar irradiance (annual) are related to the operating conditions and the other two inputs (switching frequency and the DC voltage) are the design parameters. With the first ANN used as the surrogate model for the converter, the inputs are mapped to the thermal profile of the device (junction temperatures). Using the first ANN and the yearly operating conditions, the second ANN is trained to obtain the yearly consumption (LC). In Zhang et al. [67], exogenous input-based non-linear auto regressive network is used to develop an efficient thermal model of a power electronic converter. The proposed method has a higher training speed as compared to RNN. The other relevant applications of machine learning in power electronics for smart cities are discussed in the following sub-sections.

12.5.1 ML for improving the power quality

For improvement of power quality, gradient descent least square regression (GDLSR) control algorithm along with the MPPT control for the grid-connected PV system has been proposed [68]. The proposed GDLSR which uses the Laplacian kernel function [69] is suitable for application in a high frequency system as it is the hybrid of the least square regression and the gradient descent vector. The active and the reactive components of the grid current and the weights are obtained from the neural networks of the two GDLSR control blocks. During the extraction, the active and reactive components obtained from the NN have reduced harmonics, noise, and distortions. Thus, under various operating conditions of the load and grid such as the grid over/under voltage or when the load is non-linear, various power quality issues are mitigated. With the implementation of a neural network based on GDLSR, correction in power factor, reduction in harmonics, compensated reactive power and improved utilization factor for the PV system is achieved. In Sekar et al. [70] a deep learning-based noise tolerant combined model of the CNN and the long short-term memory (LSTM) is proposed for the detection and classification of the power quality disturbances.

12.5.2 ML for prognosis health management (PHM) and RUL prediction

In order to enhance the system's reliability, encourage reusability, and reduce the maintenance costs and computational effort, the PHM helps in monitoring the system's health, identification of the defects, and in estimating the RUL. The approach for RUL prediction may be model-based or data-based [71]. The approach based on data uses the information received from the condition monitoring system, to predict the residual lifetime of the working unit. Various machine learning methods such as SVM, DL, and RVM are applied for the data-driven methods of RUL prediction. Deep learning methods such as CNN and RNN are also incorporated for predicting RUL. Capacitors, MOSFETs, and IGBTs are the critical components of a power system whose monitoring becomes necessary. In Wang et al. [72] a hybrid model based on the three deep learning methods viz. a linear regression model ARIMA (Autoregressive Integrated Moving Average), Bidirectional long short-term memory (Bi-LSTM), and BO (Bayesian Optimization) is used to estimate the RUL with the prediction of the losses and early stabilizing time of the capacitors when subjected to varying degree of the electrical stresses. Also, when the equivalent series capacitance of a capacitor is twice the initial value, the capacitor is categorized as degraded. And for an electrolytic capacitor, the loss of an electrolyte is the major cause of its failure [73]. In Rigamonti et al. [74], using a prognostic model based on a particle filter, the RUL of the capacitor and its level of degradation under the varying conditions is predicted using a temperature independent index which is a ratio between the ESR at the given temperature and the initial ESR value at the same temperature.

A Modified Grey Wolf Optimizer-Support Vector Machine algorithm is proposed [75] to estimate the RUL of a super-buck converter during the degradation of the vital components viz. inductor, capacitor, and resistance of the load. Also, Chaturvedi et al. [71] have proposed the deep learning-based multivariable LSTM model to predict the RUL for a fourth order DC-DC converter (super-buck converter) under the multiple component degradation. For the RUL prediction of MOSFETs, a prognostic model based on particle filter and an upgraded and faster version of RNN called ESN (Echo-State Network) which uses the on-state resistance of the MOSFET as health index, is proposed [76]. Figure 12.13 shows the basic principle of the RUL prediction in MOSFETs using an ESN model.

One of the most popularly used semiconductor devices especially in renewable energy systems is IGBT [77]. For the estimation of an accurate and more certain RUL for the IGBT, a Gaussian process regression-based Bayesian interface is proposed [78] for the RUL estimation. An ANN and linear regression-based prognostic model for accessing the residual lifetime of electrolytic capacitors has been proposed [73]. The six factors which affect the performance of an electrolytic capacitor are fed to the input layer of ANN consisting of six neurons. The ripple current, ESR, temperature along with frequency, vibration, and humidity are the six inputs to ANN while the number of hidden layers is ten. With the proper training of ANN, the output predicts the remaining life of the residual capacitor. For a solar PV system, fault prognostics using a Gaussian mixture model are proposed [79]. First, using the t – distributed stochastic neighbor embedding (t – SNE), the high dimensional input data is pre-processed for the conversion to a feature space of two dimensions. For fault detection, a fast-clustering technique looks for a center inverter and its status is chosen as the reference. Finally, with the Gaussian mixture model, the divergence between the humps

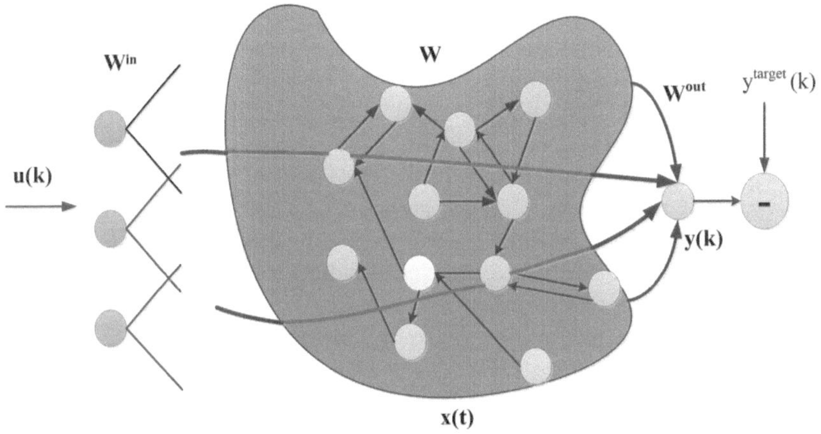

Figure 12.13 Prediction of RUL for MOSFETs.

of center inverter and the inverter under concern is noticed which predicts the fault occurrence of the inverter.

12.5.3 ML-based techniques for anti-islanding and enhancing the low voltage ride through (LVRT) capability

For the development of a smart micro grid in a smart city, the installation of distributed generation such as solar power and wind power is encouraged due to their positive impact on the environment along with the capability to reduce the pressure on the existing grid, especially during the peak hours [80]. An anti-islanding technique is an important protection approach for the identification of any system abnormality. The conventional anti-islanding local techniques may be active [81] in which an external disturbance is added to the system or passive [82] in which the abnormality is detected based on the comparison with a predefined threshold, or it may be a hybrid of both active and passive islanding. However, with the advent of machine learning techniques, much more accurate and faster anti-islanding techniques have evolved which are greatly required for the detection of discrepancies and quick and accurate disconnection of DGs.

In Khan et al. [83], the wavelet-based multi-feature NN is proposed for the islanding detection of a solar PV system. The features of the input current and voltage signal such as entropy and energy are extracted by the decomposition of the signal using the discrete wavelet transform and after the pre-preprocessing, these features are fed to the eight-input layer-based multi-layer perceptron neural network. The sigmoid function is used as an activation function and with this machine learning technique, the islanding detection time is less than 0.2s while the ANFIS-based islanding detection as proposed [84] does the detection around 0.4s. The testing efficiency of the machine learning classifier proposed [83] is around 98.1% and it is more as compared to the other passive islanding detection techniques. For example, ANN-based wavelet transform [85], DWT with decision tree [83], and DWT with Ridgelet probabilistic NN [86] have the efficiencies of about 95%, 94%, and 93.3% respectively.

When the fault is detected, abrupt disconnection of DGs is not advisable as it may lead to the instability of the grid [87]. For grid stability and the recovery of the voltage, reactive power should be injected into the grid [21] and this is done with the low voltage/fault ride through operations. Though FACT devices or the modified controller can be used for the low voltage ride through [88], various machine learning techniques have been proposed which have enhanced the LVRT capability [89], in which reinforcement learning-based two controllers have been designed to enhance the LVRT capability for a hybrid wind power system. One of the controllers is designed based on Q learning and the other one is designed based on dynamic fuzzy Q learning. A comprehensive review about the applications of machine learning for various operations in smart cities has been done [90], Table 12.5 summarizes the application of the machine learning in design, RUL prediction, and other applications in power electronics for a smart city.

Table 12.5 Summary of the machine learning methods in design, RUL prediction and other applications

References	*Machine learning method*	*Application*	*Advantages*
[65]	Multi-layer perceptron-based NN	To build electrothermal model of power MOSFET.	• An appropriate model. • Improved convergence.
[68]	NN-based gradient descent least square regression (GDLSR) with Laplacian function	Improving power quality of grid.	• Power factor correction. • Reduction in harmonics. • Compensated reactive power.
[72]	Deep learning-based hybrid method	Estimate RUL of capacitors.	• Increase reliability. • Encourage reusability.
[76]	Echo-State network (ESN)	RUL prediction of MOSFET.	• Faster as compared to recurrent neural network.
[83]	Wavelet-based multi-feature NN	Anti-islanding technique for solar PV.	• Shorter detection time (0.2s). • High testing efficiency (98.1%).
[89]	Reinforcement learning-based on Q learning and fuzzy Q learning	Enhance LVRT capability of wind energy system.	• Grid stability. • Recovery of voltage.

12.6 SUMMARY

This chapter presents extensive details and reviews about how the functionality and the development of a smart city can be improved with the applications of machine learning tools in power electronics. Smart energy management, smart transportation, smart lighting and heating, smart power generation, and smart grids are the some of the essential components of a smart city and power electronics play a crucial role in all these smart functions. AI and especially machine learning has been applied to power electronics for the implementation of smart cities all around the globe. In the recent times, all the worldwide stakeholders have been focusing on how climate change can be mitigated and reducing the carbon footprint has been declared as a global goal alongside the developments taking place for smart cities. This has called for the electrification of transportation in which power electronic converters like the DC-DC converters and rectifiers are the interface between the vehicle and the grid. For a smart grid, a vehicle should also be able to energize the grid whenever it is required. For vehicle-grid (V2G) power flow bidirectional converters are required. The power generation from distributed generation via renewables is an important feature of a smart city and the power electronic converters-based renewable energy resources, especially the solar PV systems and the wind energy, have been at a forefront for distributed generation for a micro grid in a smart city. For the efficient and desired execution of all such smart city applications, machine learning has been used in the design, control, and maintenance of the power electronic converter. Machine learning is used for condition monitoring, fault diagnosis, and RUL prediction of power electronic converters, capacitors, MOSFETs, and so on. Also, the power quality and LVRT capability have been improved with the application of machine learning techniques in power electronics. Thus, machine learning techniques in the power electronics have been extensively exploited so as to fulfill all the necessary requirements for smart city development.

REFERENCES

[1] Y. Zhang, T. Huang, and E. F. Bompard, "Big Data Analytics in Smart Grids: A Review," *Energy Informatics*, vol. 1, no. 1, p. 8, Dec. 2018, doi: 10.1186/s42162-018-0007-5

[2] K. Y. H. Lim, P. Zheng, and C.-H. Chen, "A State-of-the-Art Survey of Digital Twin: Techniques, Engineering Product Lifecycle Management and Business Innovation Perspectives," *J. Intell. Manuf.*, vol. 31, no. 6, pp. 1313–1337, Aug. 2020, doi: 10.1007/s10845-019-01512-w

[3] S. Zhao, F. Blaabjerg, and H. Wang, "An Overview of Artificial Intelligence Applications for Power Electronics," *IEEE Trans. Power Electron.*, vol. 36, no. 4, pp. 4633–4658, Apr. 2021, doi: 10.1109/TPEL.2020.3024914

[4] C. V. Mahamuni, Z. Sayyed, and A. Mishra, "Machine Learning for Smart Cities: A Survey," in *2022 IEEE International Power and Renewable Energy Conference (IPRECON)*, IEEE, Dec. 2022, pp. 1–8, doi: 10.1109/IPRECON55716.2022.10059521

[5] I. Bandyopadhyay, P. Purkait, and C. Koley, "Performance of a Classifier Based on Time-Domain Features for Incipient Fault Detection in Inverter Drives," *IEEE Trans. Ind. Informatics*, vol. 15, no. 1, pp. 3–14, Jan. 2019, doi: 10.1109/TII.2018.2854885

[6] L. Wang, J. Yue, Y. Su, F. Lu, and Q. Sun, "A Novel Remaining Useful Life Prediction Approach for Superbuck Converter Circuits Based on Modified Grey Wolf Optimizer-Support Vector Regression," *Energies*, vol. 10, no. 4, p. 459, Apr. 2017, doi: 10.3390/en10040459

[7] G. Pahuja and T. N. Nagabhushan, "A Comparative Study of Existing Machine Learning Approaches for Parkinson's Disease Detection," *IETE J. Res.*, vol. 67, no. 1, pp. 4–14, Jan. 2021, doi: 10.1080/03772063.2018.1531730

[8] S. S. Moosavi, A. Kazemi, and H. Akbari, "A Comparison of Various Open-Circuit Fault Detection Methods in the IGBT-Based DC/AC Inverter Used in Electric Vehicle," *Eng. Fail. Anal.*, vol. 96, pp. 223–235, Feb. 2019, [Online]. Available: https://linkinghub.elsevier.com/retrieve/pii/S1350630717313237

[9] S. Kolawole, "Application of Neural Networks for Predicitng the Optimal Sizing Parameters of Stand-Alone Photovoltaic Systems," *SOP Trans. Appl. Phys.*, vol. 2014, no. 1, pp. 12–16, Mar. 2014, doi: 10.15764/APHY.2014.01003

[10] Ü. Ağbulut, A. E. Gürel, and Y. Biçen, "Prediction of Daily Global Solar Radiation Using Different Machine Learning Algorithms: Evaluation and Comparison," *Renew. Sustain. Energy Rev.*, vol. 135, p. 110114, Jan. 2021, doi: 10.1016/j.rser.2020.110114

[11] G. Mendonça de Paiva, S. Pires Pimentel, B. Pinheiro Alvarenga, E. Gonçalves Marra, M. Mussetta, and S. Leva, "Multiple Site Intraday Solar Irradiance Forecasting by Machine Learning Algorithms: MGGP and MLP Neural Networks," *Energies*, vol. 13, no. 11, p. 3005, Jun. 2020, doi: 10.3390/en13113005

[12] F. Grimaccia, S. Leva, M. Mussetta, and E. Ogliari, "ANN Sizing Procedure for the Day-Ahead Output Power Forecast of a PV Plant," *Appl. Sci.*, vol. 7, no. 6, p. 622, Jun. 2017, doi: 10.3390/app7060622

[13] M. Ding, L. Wang, and R. Bi, "An ANN-Based Approach for Forecasting the Power Output of Photovoltaic System," *Procedia Environ. Sci.*, vol. 11, pp. 1308–1315, 2011, doi: 10.1016/j.proenv.2011.12.196

[14] Y. Kwon, A. Kwasinski, and A. Kwasinski, "Solar Irradiance Forecast Using Naïve Bayes Classifier Based on Publicly Available Weather Forecasting Variables," *Energies*, vol. 12, no. 8, p. 1529, Apr. 2019, doi: 10.3390/en12081529

[15] C.-R. Chen and U. Kartini, "k-Nearest Neighbor Neural Network Models for Very Short-Term Global Solar Irradiance Forecasting Based on Meteorological Data," *Energies*, vol. 10, no. 2, p. 186, Feb. 2017, doi: 10.3390/en10020186

[16] F. Lubbe, J. Maritz, and T. Harms, "Evaluating the Potential of Gaussian Process Regression for Solar Radiation Forecasting: A Case Study," *Energies*, vol. 13, no. 20, p. 5509, Oct. 2020, doi: 10.3390/en13205509

[17] L. Cheng and T. Yu, "A New Generation of AI: A Review and Perspective on Machine Learning Technologies Applied to Smart Energy and Electric Power Systems," *Int. J. Energy Res.*, vol. 43, no. 6, pp. 1928–1973, May 2019, doi: 10.1002/er.4333

[18] K.-Y. Chou, S.-T. Yang, C.-S. Yang, and Y.-P. Chen, "Maximum Power Point Tracking of Photovoltaic System Based on Reinforcement Learning," in *2019 IEEE International Conference on Consumer Electronics - Taiwan (ICCE-TW)*, IEEE, May 2019, pp. 1–2, doi: 10.1109/ICCE-TW46550.2019.8991860

[19] H. Soliman, H. Wang, B. Gadalla, and F. Blaabjerg, "Condition Monitoring for DC-Link Capacitors Based on Artificial Neural Network Algorithm," in *2015 IEEE 5th International Conference on Power Engineering, Energy and Electrical Drives (POWERENG)*, IEEE, May 2015, pp. 587–591, doi: 10.1109/PowerEng.2015.7266382

[20] K. V. S. Bharath, A. Haque, and M. A. Khan, "Condition Monitoring of Photovoltaic Systems Using Machine Leaming Techniques," in *2018 2nd IEEE International Conference on Power Electronics, Intelligent Control and Energy Systems (ICPEICES)*, IEEE, Oct. 2018, pp. 870–875, doi: 10.1109/ICPEICES.2018.8897413

[21] V. S. B. Kurukuru, A. Haque, M. A. Khan, S. Sahoo, A. Malik, and F. Blaabjerg, "A Review on Artificial Intelligence Applications for Grid-Connected Solar Photovoltaic Systems," *Energies*, vol. 14, no. 15, p. 4690, Aug. 2021, doi: 10.3390/en14154690

[22] M. Novak and T. Dragicevic, "Supervised Imitation Learning of Finite-Set Model Predictive Control Systems for Power Electronics," *IEEE Trans. Ind. Electron.*, vol. 68, no. 2, pp. 1717–1723, Feb. 2021, doi: 10.1109/TIE.2020.2969116

[23] M. Demirtas, I. Cetinbas, S. Serefoglu, and O. Kaplan, "ANN Controlled Single Phase Inverter for Solar Energy Systems," in *2014 16th International Power Electronics and Motion Control Conference and Exposition*, IEEE, Sep. 2014, pp. 768–772, doi: 10.1109/EPEPEMC.2014.6980590

[24] M. A. Islam et al., "Modeling and Performance Evaluation of ANFIS Controller-Based Bidirectional Power Management Scheme in Plug-In Electric Vehicles Integrated With Electric Grid," *IEEE Access*, vol. 9, pp. 166762–166780, 2021, doi: 10.1109/ACCESS.2021.3135190

[25] P. Karuppusamy, A. M. Natarajan, and K. N. Vijeyakumar, "An Adaptive Neuro-Fuzzy Model to Multilevel Inverter for Grid Connected Photovoltaic System," *J. Circuits, Syst. Comput.*, vol. 24, no. 05, p. 1550066, Jun. 2015, doi: 10.1142/S0218126615500668

[26] M. Mohammadzadeh, E. Akbari, A. A. Salameh, M. Ghadamyari, S. Pirouzi, and T. Senjyu, "Application of Mixture of Experts in Machine Learning-Based Controlling of DC-DC Power Electronics Converter," *IEEE Access*, vol. 10, pp. 117157–117169, 2022, doi: 10.1109/ACCESS.2022.3218667

[27] M. Glavic, R. Fonteneau, and D. Ernst, "Reinforcement Learning for Electric Power System Decision and Control: Past Considerations and Perspectives," *IFAC-PapersOnLine*, vol. 50, no. 1, pp. 6918–6927, Jul. 2017, doi: 10.1016/j.ifacol.2017.08.1217

[28] T. Qie, X. Zhang, C. Xiang, Y. Yu, H. H. C. Iu, and T. Fernando, "A New Robust Integral Reinforcement Learning Based Control Algorithm for Interleaved DC/DC Boost Converter," *IEEE Trans. Ind. Electron.*, vol. 70, no. 4, pp. 3729–3739, 2023, doi: 10.1109/TIE.2022.3179558

[29] C. Wei, Z. Zhang, W. Qiao, and L. Qu, "Reinforcement-Learning-Based Intelligent Maximum Power Point Tracking Control for Wind Energy Conversion Systems," *IEEE Trans. Ind. Electron.*, vol. 62, no. 10, pp. 6360–6370, Oct. 2015, doi: 10.1109/TIE.2015.2420792

[30] C. Wei, Z. Zhang, W. Qiao, and L. Qu, “An Adaptive Network-Based Reinforcement Learning Method for MPPT Control of PMSG Wind Energy Conversion Systems,” *IEEE Trans. Power Electron.*, vol. 31, no. 11, pp. 7837–7848, Nov. 2016, doi: 10.1109/TPEL.2016.2514370

[31] P. Kofinas, S. Doltsinis, A. I. Dounis, and G. A. Vouros, “A Reinforcement Learning Approach for MPPT Control Method of Photovoltaic Sources,” *Renew. Energy*, vol. 108, pp. 461–473, Aug. 2017, doi: 10.1016/j.renene.2017.03.008

[32] M. Liao, H. Li, P. Wang, Y. Chen, and M. Chen, “Machine Learning Methods for Power Flow Control of Multi-Active-Bridge Converters,” in *2021 IEEE 22nd Workshop on Control and Modelling of Power Electronics (COMPEL)*, IEEE, Nov. 2021, pp. 1–7, doi: 10.1109/COMPEL52922.2021.9646055

[33] X. Zhan, W. Wang, and H. Chung, “A Neural-Network-Based Color Control Method for Multi-Color LED Systems,” *IEEE Trans. Power Electron.*, vol. 34, no. 8, pp. 7900–7913, Aug. 2019, doi: 10.1109/TPEL.2018.2880876

[34] S. Lucia, D. Navarro, B. Karg, H. Sarnago, and O. Lucia, “Deep Learning-Based Model Predictive Control for Resonant Power Converters,” *IEEE Trans. Ind. Informatics*, vol. 17, no. 1, pp. 409–420, Jan. 2021, doi: 10.1109/TII.2020.2969729

[35] N. A. Ahmed and A. K. Al-Othman, “Photovoltaic System with Voltage-Based Maximum Power Point Tracking Using Support Vector Machine,” in *2010 5th IEEE Conference on Industrial Electronics and Applications*, IEEE, Jun. 2010, pp. 2264–2269, doi: 10.1109/ICIEA.2010.5516664

[36] X. Fu and S. Li, “Control of Single-Phase Grid-Connected Converters with LCL Filters Using Recurrent Neural Network and Conventional Control Methods,” *IEEE Trans. Power Electron.*, pp. 1–1, 2015, doi: 10.1109/TPEL.2015.2490200

[37] B. C. Phan, Y.-C. Lai, and C. E. Lin, “A Deep Reinforcement Learning-Based MPPT Control for PV Systems under Partial Shading Condition,” *Sensors*, vol. 20, no. 11, p. 3039, May 2020, doi: 10.3390/s20113039

[38] J. Zhang, H. S.-H. Chung, and W.-L. Lo, “Clustering-Based Adaptive Crossover and Mutation Probabilities for Genetic Algorithms,” *IEEE Trans. Evol. Comput.*, vol. 11, no. 3, pp. 326–335, Jun. 2007, doi: 10.1109/TEVC.2006.880727

[39] N. Femia, G. Spagnuolo, and V. Tucci, “State-Space Models and Order Reduction for DC-DC Switching Converters in Discontinuous Modes,” *IEEE Trans. Power Electron.*, vol. 10, no. 6, pp. 640–650, Nov. 1995, doi: 10.1109/63.471283

[40] S. Committee and I. Reliability, *IEEE Standard Framework for Prognostics and Health Management of Electronic Systems IEEE Standard Framework for Prognostics and Health Management of Electronic Systems*. 2017.

[41] H. Soliman, P. Davari, H. Wang, and F. Blaabjerg, “Capacitance Estimation Algorithm Based on DC-Link Voltage Harmonics Using Artificial Neural Network in Three-Phase Motor Drive Systems,” in *2017 IEEE Energy Conversion Congress and Exposition (ECCE)*, IEEE, Oct. 2017, pp. 5795–5802, doi: 10.1109/ECCE.2017.8096961

[42] A. M. Imam, T. G. Habetler, R. G. Harley, and D. M. Divan, “LMS Based Condition Monitoring of Electrolytic Capacitor,” in *31st Annual Conference of IEEE Industrial Electronics Society, 2005. IECON 2005.*, IEEE, 2005, p. 6, doi: 10.1109/IECON.2005.1569015

[43] S. Rajendran, D. Jena, M. Diaz, and V. S. Kirthika Devi, "Machine Learning Based Condition Monitoring of a DC-Link Capacitor in a Back-to-Back Converter," in *2022 IEEE International Conference on Automation/XXV Congress of the Chilean Association of Automatic Control (ICA-ACCA)*, IEEE, Oct. 2022, pp. 1–5, doi: 10.1109/ICA-ACCA56767.2022.10006052
[44] T. Kim and J.-S. Lee, "Exponential Loss Minimization for Learning Weighted Naive Bayes Classifiers," *IEEE Access*, vol. 10, pp. 22724–22736, 2022, doi: 10.1109/ACCESS.2022.3155231
[45] Q. Xue, Y. Zhu, and J. Wang, "Joint Distribution Estimation and Naïve Bayes Classification under Local Differential Privacy," *IEEE Trans. Emerg. Top. Comput.*, vol. 9, no. 4, pp. 2053–2063, Oct. 2021, doi: 10.1109/TETC.2019.2959581
[46] M. Bindi et al., "Machine Learning-Based Monitoring of DC-DC Converters in Photovoltaic Applications," *Algorithms*, vol. 15, no. 3, p. 74, Feb. 2022, doi: 10.3390/a15030074
[47] I. Aizenberg, "The Multi-Valued Neuron," 2011, pp. 55–94, doi: 10.1007/978-3-642-20353-4_2
[48] A. Haque, K. V. S. Bharath, M. A. Khan, I. Khan, and Z. A. Jaffery, "Fault Diagnosis of Photovoltaic Modules," *Energy Sci. Eng.*, vol. 7, no. 3, pp. 622–644, Jun. 2019, doi: 10.1002/ese3.255
[49] D. Zhu, J. Bieger, G. Garcia Molina, and R. M. Aarts, "A Survey of Stimulation Methods Used in SSVEP-Based BCIs," *Comput. Intell. Neurosci.*, vol. 2010, pp. 1–12, 2010, doi: 10.1155/2010/702357
[50] P. K. Ray, B. K. Panigrahi, P. K. Rout, A. Mohanty, and H. Dubey, "Detection of Faults in a Power System Using Wavelet Transform and Independent Component Analysis," *Comput. Commun. Electr. Technol. - Proc. Int. Conf. Adv. Comput. Commun. Electr. Technol. ACCET 2016*, no. October, pp. 227–231, 2017, doi: 10.1201/9781315400624-45
[51] M. N. Akram and S. Lotfifard, "Modeling and Health Monitoring of DC Side of Photovoltaic Array," *IEEE Trans. Sustain. Energy*, vol. 6, no. 4, pp. 1245–1253, Oct. 2015, doi: 10.1109/TSTE.2015.2425791
[52] Z. Huang, Z. Wang, and H. Zhang, "A Diagnosis Algorithm for Multiple Open-Circuited Faults of Microgrid Inverters Based on Main Fault Component Analysis," *IEEE Trans. Energy Convers.*, vol. 33, no. 3, pp. 925–937, Sep. 2018, doi: 10.1109/TEC.2018.2822481
[53] G. Yan, Y. Hu, and Q. Shi, "A Convolutional Neural Network-Based Method of Inverter Fault Diagnosis in a Ship's DC Electrical System," *Polish Marit. Res.*, vol. 29, no. 4, pp. 105–114, Dec. 2022, doi: 10.2478/pomr-2022-0048
[54] A. M. R. Amaral and A. J. M. Cardoso, "Using Input Current and Output Voltage Ripple to Estimate the Output Filter Condition of Switch Mode DC/DC Converters," in *2009 IEEE International Symposium on Diagnostics for Electric Machines, Power Electronics and Drives*, IEEE, Aug. 2009, pp. 1–6, doi: 10.1109/DEMPED.2009.5292791
[55] T. Kamel, Y. Biletskiy, and L. Chang, "Capacitor Aging Detection for the DC Filters in the Power Electronic Converters Using ANFIS Algorithm," in *2015 IEEE 28th Canadian Conference on Electrical and Computer Engineering (CCECE)*, IEEE, May 2015, pp. 663–668, doi: 10.1109/CCECE.2015.7129353
[56] T. Wang, H. Xu, J. Han, E. Elbouchikhi, and M. E. H. Benbouzid, "Cascaded H-Bridge Multilevel Inverter System Fault Diagnosis Using a PCA and Multiclass

Relevance Vector Machine Approach," *IEEE Trans. Power Electron.*, vol. 30, no. 12, pp. 7006–7018, Dec. 2015, doi: 10.1109/TPEL.2015.2393373
[57] D. Chowdhury, M. Bhattacharya, D. Khan, S. Saha, and A. Dasgupta, "Wavelet Decomposition Based Fault Detection in Cascaded H-Bridge Multilevel Inverter Using Artificial Neural Network," in *2017 2nd IEEE International Conference on Recent Trends in Electronics, Information & Communication Technology (RTEICT)*, IEEE, May 2017, pp. 1931–1935, doi: 10.1109/RTEICT.2017.8256934
[58] X. X. Zheng and P. Peng, "Fault Diagnosis of Wind Power Converters Based on Compressed Sensing Theory and Weight Constrained AdaBoost-SVM," *J. Power Electron.*, vol. 19, no. 2, pp. 443–453, 2019, doi: 10.6113/JPE.2019.19.2.443
[59] Z. Chen et al., "Random Forest Based Intelligent Fault Diagnosis for PV Arrays Using Array Voltage and String Currents," *Energy Convers. Manag.*, vol. 178, pp. 250–264, Dec. 2018, doi: 10.1016/j.enconman.2018.10.040
[60] A. Belaout, F. Krim, A. Mellit, B. Talbi, and A. Arabi, "Multiclass Adaptive Neuro-Fuzzy Classifier and Feature Selection Techniques for Photovoltaic Array Fault Detection and Classification," *Renew. Energy*, vol. 127, pp. 548–558, Nov. 2018, doi: 10.1016/j.renene.2018.05.008
[61] F. Aziz, A. Ul Haq, S. Ahmad, Y. Mahmoud, M. Jalal, and U. Ali, "A Novel Convolutional Neural Network-Based Approach for Fault Classification in Photovoltaic Arrays," *IEEE Access*, vol. 8, pp. 41889–41904, 2020, doi: 10.1109/ACCESS.2020.2977116
[62] S. Guo, Q. Lv, B. Liu, Y. Lin, and R. Li, *Deep Convolutional Neural Networks for Electrocardiogram Classification*, vol. 536. Springer, Singapore, 2019, doi: 10.1007/978-981-13-6837-0_5
[63] W. Gong, H. Chen, Z. Zhang, M. Zhang, and H. Gao, "A Data-Driven-Based Fault Diagnosis Approach for Electrical Power DC-DC Inverter by Using Modified Convolutional Neural Network With Global Average Pooling and 2-D Feature Image," *IEEE Access*, vol. 8, pp. 73677–73697, 2020, doi: 10.1109/ACCESS.2020.2988323
[64] V. S. B. Kurukuru, F. Blaabjerg, M. A. Khan, and A. Haque, "A Novel Fault Classification Approach for Photovoltaic Systems," *Energies*, vol. 13, no. 2, p. 308, Jan. 2020, doi: 10.3390/en13020308
[65] D. Chiozzi, M. Bernardoni, N. Delmonte, and P. Cova, "A Neural Network Based Approach to Simulate Electrothermal Device Interaction in SPICE Environment," *IEEE Trans. Power Electron.*, vol. 34, no. 5, pp. 4703–4710, May 2019, doi: 10.1109/TPEL.2018.2863186
[66] T. Dragicevic, P. Wheeler, and F. Blaabjerg, "Artificial Intelligence Aided Automated Design for Reliability of Power Electronic Systems," *IEEE Trans. Power Electron.*, vol. 34, no. 8, pp. 7161–7171, Aug. 2019, doi: 10.1109/TPEL.2018.2883947
[67] Y. Zhang, Z. Wang, H. Wang, and F. Blaabjerg, "Artificial Intelligence-Aided Thermal Model Considering Cross-Coupling Effects," *IEEE Trans. Power Electron.*, vol. 35, no. 10, pp. 9998–10002, Oct. 2020, doi: 10.1109/TPEL.2020.2980240
[68] N. Kumar, B. Singh, and B. K. Panigrahi, "Framework of Gradient Descent Least Squares Regression-Based NN Structure for Power Quality Improvement in PV-Integrated Low-Voltage Weak Grid System," *IEEE Trans. Ind. Electron.*, vol. 66, no. 12, pp. 9724–9733, Dec. 2019, doi: 10.1109/TIE.2018.2886765

[69] M. R. Hajiaboli, M. O. Ahmad, and C. Wang, "An Edge-Adapting Laplacian Kernel for Nonlinear Diffusion Filters," *IEEE Trans. Image Process.*, vol. 21, no. 4, pp. 1561–1572, Apr. 2012, doi: 10.1109/TIP.2011.2172803

[70] K. Sekar, K. Kanagarathinam, S. Subramanian, E. Venugopal, and C. Udayakumar, "An Improved Power Quality Disturbance Detection Using Deep Learning Approach," *Math. Probl. Eng.*, vol. 2022, pp. 1–12, May 2022, doi: 10.1155/2022/7020979

[71] A. Chaturvedi, M. Sarma, S. K. Chaturvedi, and J. Bernstein, "Performance Assessment and RUL Prediction of Power Converters under the Multiple Components Degradation," *Microelectron. Reliab.*, vol. 144, p. 114958, May 2023, doi: 10.1016/j.microrel.2023.114958

[72] Z. Wang, J. Qu, X. Fang, H. Li, T. Zhong, and H. Ren, "Prediction of Early Stabilization Time of Electrolytic Capacitor Based on ARIMA-Bi_LSTM Hybrid Model," *Neurocomputing*, vol. 403, pp. 63–79, Aug. 2020, doi: 10.1016/j.neucom.2020.03.054

[73] C. Bhargava, V. K. Banga, and Y. Singh, "An Intelligent Prognostic Model for Electrolytic Capacitors Health Monitoring: A Design of Experiments Approach," *Adv. Mech. Eng.*, vol. 10, no. 10, p. 168781401878117, Oct. 2018, doi: 10.1177/1687814018781170

[74] M. Rigamonti, P. Baraldi, E. Zio, D. Astigarraga, and A. Galarza, "Particle Filter-Based Prognostics for an Electrolytic Capacitor Working in Variable Operating Conditions," *IEEE Trans. Power Electron.*, vol. 31, no. 2, pp. 1567–1575, Feb. 2016, doi: 10.1109/TPEL.2015.2418198

[75] L. Wang, J. Yue, Y. Su, F. Lu, and Q. Sun, "A Novel Remaining Useful Life Prediction Approach for Superbuck Converter Circuits Based on Modified Grey Wolf Optimizer-Support Vector Regression," *Energies*, vol. 10, no. 4, p. 459, Apr. 2017, doi: 10.3390/en10040459

[76] Z. Li, Z. Zheng, and R. Outbib, "A Prognostic Methodology for Power MOSFETs under Thermal Stress Using Echo State Network and Particle Filter," *Microelectron. Reliab.*, vol. 88–90, pp. 350–354, Sep. 2018, doi: 10.1016/j.microrel.2018.07.137

[77] Shaoyong Yang, A. Bryant, P. Mawby, Dawei Xiang, Li Ran, and P. Tavner, "An Industry-Based Survey of Reliability in Power Electronic Converters," in *2009 IEEE Energy Conversion Congress and Exposition*, IEEE, Sep. 2009, pp. 3151–3157, doi: 10.1109/ECCE.2009.5316356

[78] S. H. Ali, M. Heydarzadeh, S. Dusmez, X. Li, A. S. Kamath, and B. Akin, "Lifetime Estimation of Discrete IGBT Devices Based on Gaussian Process," *IEEE Trans. Ind. Appl.*, vol. 54, no. 1, pp. 395–403, Jan. 2018, doi: 10.1109/TIA.2017.2753722

[79] Z. He, X. Zhang, C. Liu, and T. Han, "Fault Prognostics for Photovoltaic Inverter Based on Fast Clustering Algorithm and Gaussian Mixture Model," *Energies*, vol. 13, no. 18, p. 4901, Sep. 2020, doi: 10.3390/en13184901

[80] Y. Liao et al., "Voltage and Var Control to Enable High Penetration of Distributed Photovoltaic Systems," in *2012 North American Power Symposium (NAPS)*, IEEE, Sep. 2012, pp. 1–6, doi: 10.1109/NAPS.2012.6336328

[81] W.-J. Chiang, H.-L. Jou, J.-C. Wu, K.-D. Wu, and Y.-T. Feng, "Active Islanding Detection Method for the Grid-Connected Photovoltaic Generation System," *Electr. Power Syst. Res.*, vol. 80, no. 4, pp. 372–379, Apr. 2010, doi: 10.1016/j.epsr.2009.09.018

[82] K. N. E. K. Ahmad, N. A. Rahim, J. Selvaraj, A. Rivai, and K. Chaniago, "An Effective Passive Islanding Detection Method for PV Single-Phase Grid-Connected Inverter," *Sol. Energy*, vol. 97, pp. 155–167, Nov. 2013, doi: 10.1016/j.solener.2013.08.011

[83] M. A. Khan, V. S. Bharath Kurukuru, A. Haque, and S. Mekhilef, "Islanding Classification Mechanism for Grid-Connected Photovoltaic Systems," *IEEE J. Emerg. Sel. Top. Power Electron.*, vol. 9, no. 2, pp. 1966–1975, Apr. 2021, doi: 10.1109/JESTPE.2020.2986262

[84] D. Mlakic, H. R. Baghaee, and S. Nikolovski, "A Novel ANFIS-Based Islanding Detection for Inverter-Interfaced Microgrids," *IEEE Trans. Smart Grid*, vol. 10, no. 4, pp. 4411–4424, Jul. 2019, doi: 10.1109/TSG.2018.2859360

[85] Y. Fayyad and A. Osman, "Neuro-Wavelet Based Islanding Detection Technique," in *2010 IEEE Electrical Power & Energy Conference*, IEEE, Aug. 2010, pp. 1–6, doi: 10.1109/EPEC.2010.5697180

[86] M. Ahmadipour, H. Hizam, M. L. Othman, and M. A. Radzi, "Islanding Detection Method Using Ridgelet Probabilistic Neural Network in Distributed Generation," *Neurocomputing*, vol. 329, pp. 188–209, Feb. 2019, doi: 10.1016/j.neucom.2018.10.053

[87] A. Q. Al-Shetwi, M. Z. Sujod, F. Blaabjerg, and Y. Yang, "Fault Ride-through Control of Grid-Connected Photovoltaic Power Plants: A Review," *Sol. Energy*, vol. 180, pp. 340–350, Mar. 2019, doi: 10.1016/j.solener.2019.01.032

[88] A. Z. Fatama, M. A. Khan, V. S. B. Kurukuru, A. Haque, and F. Blaabjerg, "Coordinated Reactive Power Strategy Using Static Synchronous Compensator for Photovoltaic Inverters," *Int. Trans. Electr. Energy Syst.*, vol. 30, no. 6, Jun. 2020, doi: 10.1002/2050-7038.12393

[89] L. Zhou, A. Swain, and A. Ukil, "Reinforcement Learning Controllers for Enhancement of Low Voltage Ride through Capability in Hybrid Power Systems," *IEEE Trans. Ind. Informatics*, vol. 16, no. 8, pp. 5023–5031, Aug. 2020, doi: 10.1109/TII.2019.2956509

[90] S. S. Band et al., "When Smart Cities Get Smarter via Machine Learning: An In-Depth Literature Review," *IEEE Access*, vol. 10, pp. 60985–61015, 2022, doi: 10.1109/ACCESS.2022.3181718

Chapter 13

Machine learning in renewable energy systems for smart cities

Ahteshamul Haque and Azra Malik
Jamia Millia Islamia, New Delhi, India

13.1 INTRODUCTION

A "Smart City" includes holistic ways of improving individuals' lives through economic growth, developed roads, technological advancements, smart transport developments, eco-friendly electrical infrastructure, high-quality air-purifiers, and so on. It does not have any standard definition; rather it is defined basis the requirements specific to the particular area [1]. The energy sector is crucial for expanding economic growth and it has the potential to pave the way for building more sustainable economy [2]. With reference to enhancing electrical power system infrastructure, one of the ways includes replacing conventional sources of energy like coal, natural gas, oil, and so on with renewable sources like solar, wind, biogas/biomass, fuel cells, tidal to name a few. The conventional energy sources once used up cannot be replenished, therefore they are non-renewable [3]. They lead to huge CO_2 and other greenhouse gas emissions globally and are hazardous to the environment [4]. Consequently, it becomes necessary to limit these emissions and enforce policies that help in improving the quality of life. The Kyoto Protocol engages the countries to adopt measures to curtail greenhouse gas emissions according to defined targets. These goals should align with the overall target of net zero emissions by the middle of the century and therefore, countries need to strive hard to take measures including increased investment in renewables, curbing deforestation, encouraging electric vehicle usage and so on. Whereas, renewable sources of energy are environment friendly, clean, and are available in abundance. A report by the International Energy Agency (IEA) reveals that solar photovoltaic (PV) comprised greater than half of overall renewable power growth in the year 2021, accompanied by wind and hydropower. It is forecasted that in the next five years, that is, by 2026, renewable capacity is set to account for approximately 95% of increase in total world capacity. Worldwide, renewable energy capacity is predicted to increase by 60% up to 2026, crossing 4,800 GW. Specifically, in the Indian context, renewable energy has great potential with solar PV leading the market. The government has also set an ambitious target to achieve 500 GW of renewable capacity by 2030. As reported by Global Wind Energy Council (GWEC), total available worldwide capacity of wind power is 743 GW, which can eliminate up to 1.1 billion tons of CO_2 worldwide. However, this existing rate of deployment of wind power will not be

DOI: 10.1201/9781032669809-13

sufficient to attain carbon neutrality even by the mid-century. Therefore, it is necessary to scale up wind power at the required rate. In addition, it is required that issues and challenges arising due to enormous penetration of renewable power to the grid like power and voltage fluctuations and so on should be mitigated. Further, along with enhancing the solar and wind capacity, policies must be stressed to focus on renewable energy utilization in industries, buildings, transport, and so on. There is a significant gap between the current growth rate and the trajectory essential to meet the net zero target by 2050. Annual growth rate of renewable capacity should increase by 80%, for that government should meet implementation and policy regulation challenges along with increasing their ambition.

Machine learning is a branch of artificial intelligence (AI). AI is the science of creating smart and intelligent machines, that provides the capability of utilizing computers to comprehend and recognize human intelligence for performing various tasks [5]. It has a very huge collection of applications with multiple variants, and further advancements in this field will help in causing a paradigm shift in the technical industry [6]. Machine learning, a subfield under AI, highlights the use of data, and methods to emulate human learning. It does not need to be explicitly programmed, and further focuses on slow and gradual improvement in accuracy and efficiency. Machine learning is further branched in to methods like supervised learning, deep learning, reinforcement learning, and unsupervised learning [7]. Machine learning has a vast scope in renewable energy systems in the case of smart grids. Power electronics converters have a significant part to perform in the renewable energy systems since they deal with the control and transfer of electrical energy [8]. They are capable of handling high power capacities with considerable accuracy in extensive power ranges. Machine learning can be adopted at the generation end (at the PV generator or wind generator side), or it can be utilized at the power converter side for multiple applications [9]. Figure 13.1 shows the utilization of machine learning in applications like forecasting of power from both PV and wind turbines, optimization for power module heatsink, control for maximum power point tracking (MPPT) for both solar and wind power applications, maintenance and fault diagnostics, and so on in renewable energy systems.

An optimization-based review presenting metaheuristic methods for the purpose of optimal power quality, and control design is presented [10]. Similar tasks utilizing neural networks are proposed with a proper network structure design, training procedure, and other relevant considerations [11]. This work covers various applications for power electronics as well as non-power electronics. Techniques like metaheuristic methods and fuzzy logic have been discussed for control design applications [12]. Various examples are illustrated based on the given techniques. An intensive and detailed discussion regarding utilizing the metaheuristic methods for MPP tracking in PV system is provided [13]. Further, AI techniques including machine

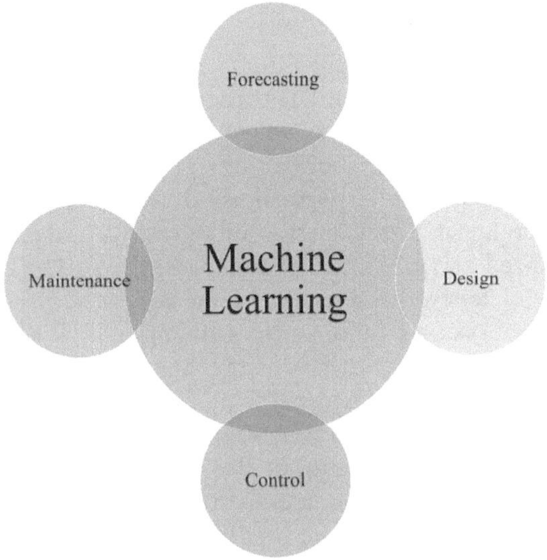

Figure 13.1 Different areas in RESs with regards to machine learning applications.

learning-based techniques are reviewed for PV systems focusing on many PV applications [14]. Similarly, machine learning can be utilized for maintenance in PV systems including condition monitoring, fault diagnosis, and so on. In PV systems, the two most vulnerable components are switching devices and capacitors. A review utilizing machine learning-related techniques for condition monitoring of capacitors in solar systems is presented [15]. It works on providing the root-cause analysis of failures in various types of capacitors specifically for PV systems. Reliability management utilizing machine learning methods for energy systems is detailed [16]. It also discusses maintenance in PV systems along with relevant machine learning algorithms. Many fault detection methods using machine learning-based classification techniques focusing on inverter switch faults in PV systems have also been discussed [17, 18]. Similar work for wind energy systems have also been presented in literature. Overall, for all four applications, mainly PV and wind turbines can be utilized successfully.

13.2 DESCRIPTION OF MACHINE LEARNING TECHNIQUES

Machine learning has the capability to automatically derive the patterns and regularities by learning from the experience attained through the collected data or through the collaborations using trial and error [19].

13.2.1 Supervised learning

The objective of supervised learning is to form valuable associations between the respective inputs and outputs. The training data usually comprises input and output pairs and as the name suggests, there is supervisory guidance available due to the relevant historical data [16]. It finds its importance in pattern recognition algorithms, particularly those which are challenging to establish. In general, it involves the task of regression and classification and these can be utilized according to the application [20]. The majority of the tasks in the case of renewable energy systems adopt the classification function. The classification task requires the training dataset with appropriate labels and the output deals with distinct groups to be labeled according to the considered algorithm [20]. An example of classification is the fault classification for inverter switches in renewable energy systems. On the other hand, regression functionality requires the outcome of the input and output sets comprising more than one consistent entity. For instance, one of the applications of regression is the prediction of the remaining useful life (RUL) of an inverter switch. Here, the output is a continuous entity and it requires the model to be trained such that it can evaluate new data objects, different from the training data objects.

A general procedure for a classification method based on machine learning is provided in Figure 13.2. First, it requires data acquisition for the desired task in particularly raw form. The input data is pre-processed involving outliers and noise removal from the data [21]. It may also involve scaling and standardization of the considered data. The next step is feature extraction, which inculcates quantifying and evaluating the relevant features corresponding to the pre-processed data. It also determines the continuum of the data object collections and associated features. Further, feature assortment is the course of segregating the desired features, which provides more accurate classifier behavior. Followed by the feature selection phase is the classifier training phase using the features obtained in the earlier step. Last, a testing system is required to validate the accuracy attained through classifier training. It further differentiates various classes from each other and processes the true test data points. Supervised learning techniques can be subdivided into techniques like artificial neural networks (ANN), k-nearest neighbor (kNN), random forest (RF), decision tree (DT), and so on, discussed below.

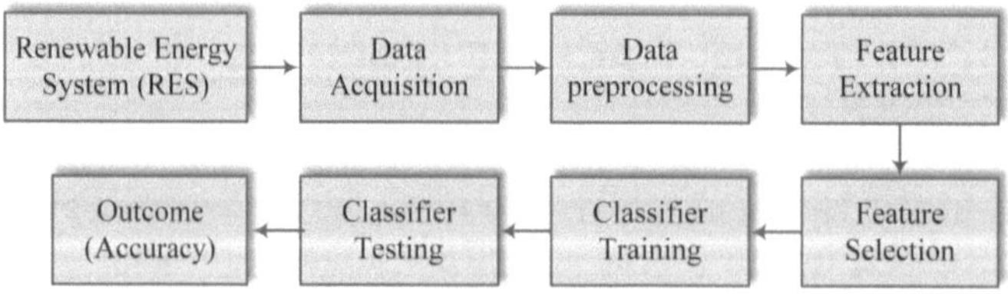

Figure 13.2 A general procedure for classification using machine learning.

13.2.1.1 Artificial neural network (ANN)

ANNs are mathematical structures similar to the biological neural network arrangement in a human brain. It is an attempt to emulate the human brain network such that the machine acquires the capability to think, understand, and decide like humans do [10]. The network model structure provides a supervisory process involving in-depth interaction between layers for conversion of the given input to output [22]. They are capable of performing more than one task simultaneously. The classification technique for ANN can be understood mathematically through the addition of bias and input weight product as given below

$$\text{Sum}\left(\text{Product of input and weights with bias}\right) = \sum_{j=1}^{n}\left(w_j + x_j\right) + \text{bias} \tag{13.1}$$

Further, output can be represented through activation layer

$$\text{Output} = f\left(x\right) = \begin{cases} 1, & \text{if } wx + b \geq 0 \\ 0, & \text{if } wx + b < 0 \end{cases} \tag{13.2}$$

ANNs can be of different structures and configurations depending on the considered application, like feedforward backpropagation, Radial basis function (RBF), Recurrent Neural networks (RNNs), and so on. A feed-forward neural network consists of information flow in a single direction, which begins at the input layer, progresses through the hidden layers, and culminates at the output layer. The data flow is ensured in only one direction, making no feedback loop. They are also called multilayer perceptron (MLP) networks [23]. This approximates the overall objective of the network through a function $y = f(x)$, mapping the input x to one of the output category y. Through training, it learns the parameter values, which provide the optimum function estimate. On the other hand, RNNs are the networks which can connect different layers through the formation of feedback loops or cycles. It has a defined, specific direction of data flow, which may include backpropagation of information to achieve the desired objective. RNN functions by defining different states, and transitioning from one state to the next state. The objective of RNN is achieved by defining a function consisting of an input and the preceding state, which then provides the new state. Utilizing these states can define the optimal output of RNN.

RBF is a category under ANN, which adopts an arbitrary function determined through the values located from the origin. Typically, this arbitrary function is RBF, which estimates the output through the linear arrangement of RBF of inputs and other network parameters with the input layer connected fully and directly to the next layer devoid of weights, as shown in

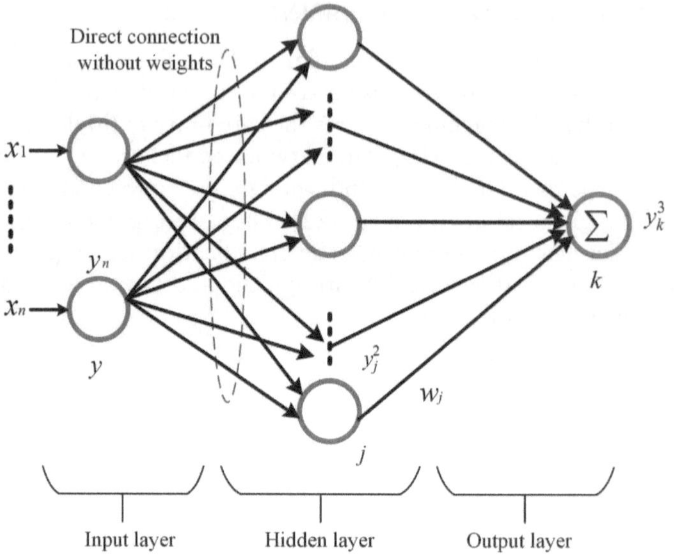

Figure 13.3 A typical RBF network structure.

Figure 13.3. The operation of RBF takes place by summing the size and center of the data objects with the respective weights to achieve the concluding output. It can be represented through a Gaussian distribution, for a scalar input, as shown below

$$f(x) = \exp\left(\frac{-(x-c)^2}{r^2}\right) \tag{13.3}$$

Where r is the radius, and c is the center parameters. The Gaussian distribution described above reduces with the increment in the distance from c.

13.2.1.2 k-nearest neighbors (kNN)

kNN is observed to be suitable for both regression and classification tasks. For classification, first it finds the closest neighbor by applying methods such as Euclidean distance Manhattan distance, Hamming distance, and so on. This distance is estimated by calculating the distance of the input from known data objects [24]. Among the given methods, Euclidean distance method is the most popular and is widely applied for kNN techniques. This calculation is helpful in finding and selecting k-nearest neighbors, which further classifies the input using the similarities obtained amongst the data points [25]. The kNN algorithm works by assuming that there is a likelihood between the new data point and already available data. It further puts the new data point in the category that has the most likelihood among the available categories or classes. The value of k is significant and is selected based on the dataset size and other important parameters.

13.2.1.3 Support vector machines (SVM)

Support vector machines or SVM is another machine learning tool useful for several classification applications. The main objective of SVM is to create a mapping of points in a space that maximizes the margin between examples of different target categories. This margin maximization is crucial for ensuring that SVM functions as an effective classifier with lower generalization errors [26]. The goal is to identify a hyperplane or a set of hyperplanes within an N-dimensional space. Support vectors refer to data points that are in close proximity to a given hyperplane and play a significant role in maximizing the margin of the classifier by adjusting the position and orientation of the hyperplane [27]. However, it is important to note that in some cases, the points in this space may not be linearly separable due to the arrangement of the data. To address this issue, SVM leverages kernel functions, also known as the "kernel trick," to transform the dataset from a lower-dimensional space to a higher dimensional space. This transformation preserves the information while enabling the creation of a linear classifier in the higher dimensional space. Multiple kernel functions, denoted as K kernels, are assigned to each point to aid in determining the optimal hyperplane within the transformed feature space. By utilizing a sufficient number of kernel functions, precise separation of data points becomes achievable. However, it is important to be cautious of overfitting, which occurs when the model becomes too tailored to the training data and performs poorly on unseen data.

13.2.1.4 Bayesian networks

A Bayesian network (BN) model, which includes the Bayes classifier, is a graphical model that consists of nodes representing variables along with their respective probability distributions [28]. These nodes are connected by directed arcs, which indicate the conditional dependencies between the variables and are associated with conditional distributions. The probability distribution for each node can be derived from various sources such as expert opinions, empirical data, statistical models, simulations, articles, or reports. A BN allows for the representation of variable interactions and provides a means to quantify and characterize complex outcomes [29]. It serves as a supervised model for both regression and classification problems and is particularly well-suited for handling incomplete datasets. The structure of a BN model can be constructed based on expert opinion and may also incorporate variables that reflect managerial decisions.

13.2.1.5 Random forests

The random forest algorithm operates by generating multiple decision trees and obtaining predictions through averaging the outputs of these trees. Each decision tree is constructed using a subset of the training dataset, and the remaining data is utilized to estimate the error of the decision tree [30]. When determining the splitting of each node, a random selection of independent variables is made [30]. Another variant of the random forest algorithm is the Extremely Randomized Trees (ERT), which involves slight modifications to the random forest approach.

13.2.1.6 Decision trees (DT)

The objective of this algorithm is to create a model that can predict a quantitative variable using a group of independent variables. It utilizes a recursive partitioning approach, where trees consist of decision nodes and leaves. In DT regression, the splitting of nodes into two or more branches is determined by assessing the reduction in standard deviation [31]. The initial decision node, known as the root node, is divided based on the most significant independent variable. Subsequently, nodes are split using the variable with the lowest sum of squared estimate of errors (SSE) as the decision node. Classification of the dataset is based on the chosen variable values. The process continues until a pre-defined termination criterion is met. The final nodes are referred to as leaf nodes and provide predictions for the dependent variable. These predictions are calculated as the mean of the values associated with the leaves.

13.2.2 Unsupervised learning

Unsupervised learning, in contrast to supervised learning, does not require output data during the learning process as it works with unlabeled datasets. In power electronics applications, unsupervised learning tasks are mainly focused on data clustering and data compression. Data clustering involves identifying patterns within a dataset and dividing it into distinct groups or clusters based on their similarities [32]. By doing so, data points within the same cluster exhibit similar characteristics and differ from those in other clusters. An example of data clustering in power electronics is the identification of discrete health states from continuous degradation data, commonly used in condition monitoring of power electronic converters. Data compression aims to reduce the number of features in a dataset by eliminating excessive information. Principal Component Analysis (PCA) is a popular technique used to achieve this goal [33]. It provides a reduced representation of the dataset with fewer features while maintaining the overall integrity of the data. Unsupervised learning algorithms, such as data clustering and data compression, are typically employed as data-preprocessing steps before subsequent analytics. Although optional, these preprocessing steps offer benefits such as reducing computational burden and improving the accuracy of analytics.

13.2.3 Deep learning

Deep learning (DL) algorithms employ ANNs for various applications. ANNs consist of interconnected nodes or neurons, inspired by biological neuronal connections. The connections are organized into three types of layers: input layer, hidden layers, and output layer. The input layer receives the original predictor variables, while the output layer generates predictions based on the inputs [34]. The hidden layers, which can be multiple in DL, contain non-observable neurons responsible for computation. Every node in a layer is connected to nodes in the subsequent layer, with each connection having a weight that combines the inputs. The weighted value is transformed by an activation function, such as a sigmoid or rectified linear unit, and passed as input to the next layer's nodes [35]. This process continues until the output layer, where the final prediction is produced. The objective of an ANN is to fit the weights to reduce an error function, often a quadratic function. To achieve this, ANNs use the backpropagation algorithm, which utilizes the gradient descent method [36]. The algorithm calculates the partial derivatives of the layers to determine the optimal weight for each node.

Different types of ANNs have been used with regards to DL:

Multilayer perceptron neural networks (MLP): Classical neural networks with one or multiple hidden layers.

Convolutional neural networks (CNN): Primarily applied to images, CNNs alternate between convolutional and pooling layers. Convolutional layers extract features using kernels or filters while pooling layers reduce feature size to decrease computational requirements. One of the representations of CNN can be observed in Figure 13.4.

Recurrent neural networks (RNN): Suitable for time series or sequential data, RNNs incorporate internal memory, allowing neurons to receive their previous output as input. This short-term memory is crucial for time series forecasting.

Long-short-term memory neural networks (LSTM): An extension of RNNs that address long-term dependencies. LSTMs employ a cell state that carries information and is updated through forget, input, and output gates.

Gated recurrent unit (GRU): A simplified version of LSTM that combines the forget and input gates, achieving similar results.

Encoder-Decoder neural networks (EDNN): Utilized for sequence-to-sequence prediction, EDNNs consist of an encoder, intermediate vector, and decoder. Recurrent units, such as LSTM or GRU, process input sequences to encapsulate information in the intermediate vector, which the decoder uses for generating predictions.

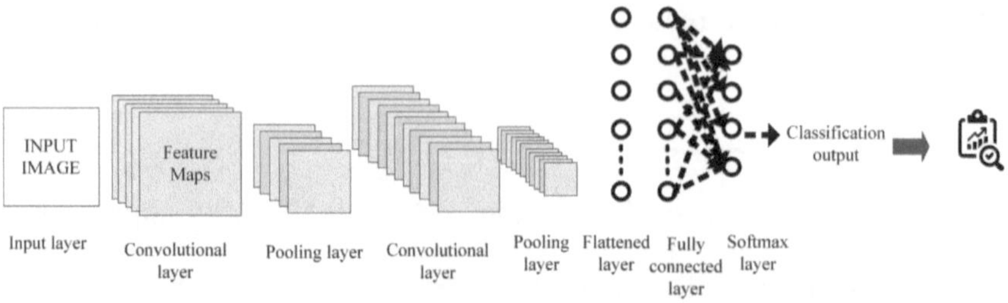

Figure 13.4 General architecture of CNN based classification technique.

13.2.4 Reinforcement learning

Reinforcement Learning (RL) is very different from other techniques as it does not rely on a training dataset. Instead, RL focuses on finding an optimal action strategy that maximizes the reward for a specific task, similar to a dynamic programming or optimization problem. This objective-driven strategy is developed through a trial-and-error process by interacting with systems or simulation models [37]. RL gradually collects the experience and slowly acquires a specific technique that maximizes an already set objective. Theoretical foundations of RL lie in the concept of a Markov decision process (MDP) [38]. The training of RL aims to create a Q-table, which represents an action selection policy that maximizes the expected total rewards in the future. The Q-table serves as an informative policy matrix, storing the optimal action to be taken given specific condition variables. An example application of RL is the MPPT in wind energy conversion system (WECS) is shown in Figure 13.5. It is important to note here that the wind speed and other related parameters are not needed for its application. Extension of this work can be found through the integration of ANN with the Q-learning. Therefore, it is easy to avoid the challenges of finding the optimal state space [39]. After the learned fine relationship is degraded due to the system's aging

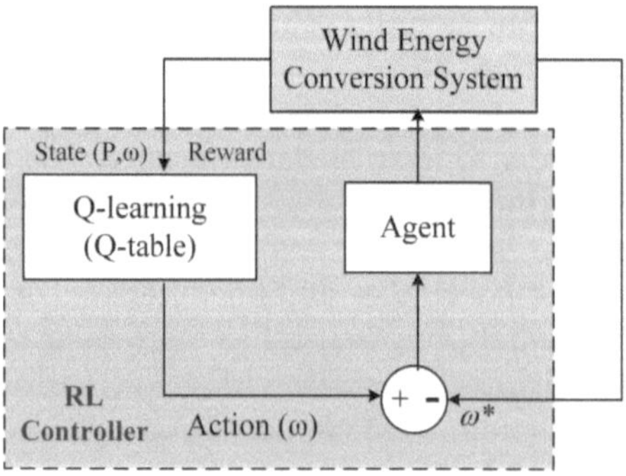

Figure 13.5 Framework for MPPT in wind energy system using reinforcement learning.

characteristics, the online learning method may be reactivated. It is found to significantly enhance the autonomous competence of the wind energy system. It is worth noting that RL obtains experience from interacting with systems rather than relying on existing datasets [40]. This makes RL particularly advantageous in scenarios where the system has limited prior knowledge or formulating its model is challenging.

13.3 MACHINE LEARNING APPLICATION IN FORECASTING

13.3.1 Forecasting solar energy

With the growing number of grid-connected photovoltaic (PV) systems, accurate power production forecasts for grid integration have become increasingly important. The rise in PV systems can be attributed to factors such as reduced investment costs (10–20% decrease in past few years), incentives, regulations on building works technical requirements, and directives. As this trend is estimated to continue, the increased integration of PV systems into the electricity grid can lead to grid instabilities due to sudden weather changes. Many studies utilized load forecasts to enhance system operation. Additionally, research has explored forecasting techniques and accuracy in power grids to improve forecast reliability. It has been noted that weather, as a chaotic system, significantly influences forecasting. General requirements in predicting solar PV generation forecast can be understood from Figure 13.6. Long-term forecasting over extended time scales, such as the next season, becomes impossible. This realization has motivated the development of intelligent techniques based on statistical and stochastic models,

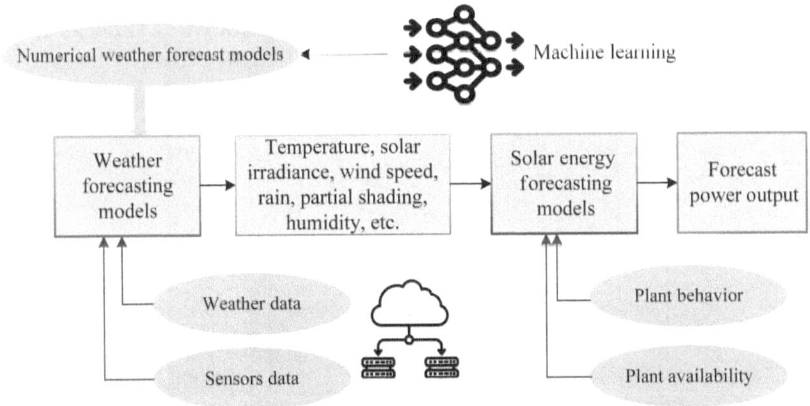

Figure 13.6 Framework explaining the requirement for solar power generation forecast.

enabling long-term planning and providing a comprehensive understanding of the behavior of distributed energy resources and loads.

The liberalization of electricity markets has also introduced spot markets, necessitating accurate output and demand estimation by generators, retailers, large end customers, and communities. These market players heavily rely on forecasting methods for these purposes. However, the proliferation of PV systems has made it challenging for market players to manage their systems effectively, as they face difficulties in accurately forecasting solar irradiance and PV outputs. Consequently, market participants who fail to meet their forecasted output or demand must resort to the balancing market, where they face high prices for imbalances. Therefore, efficient forecasting models are crucial for improving market mechanisms. Early literature on grid management with distributed energy resources focused more on load forecasts rather than forecasting distributed energy resource outputs. ANN with discrete wavelet transform (DWT) for data preprocessing is presented and the flowchart is shown in Figure 13.7 [41]. The model is utilized for estimating solar irradiation using sunshine duration and the average temperature. Similar work for solar PV panel output forecast is developed, which is free from complex calculations [42]. It calculates the required output 24 hours ahead along with utilizing an improved backpropagation technique for a better forecast accuracy. Another work, which provides a comparison between different solar PV energy forecasting techniques based on machine learning, is presented [43]. An assessment and comparison of two commonly used techniques, namely ANN and support vector regression (SVR), for the prediction of energy production from a solar photovoltaic (PV) system in Florida, is proposed [44]. The analysis focuses on forecasting energy production at three different time intervals: 15 minutes, 1 hour, and 24 hours in advance. A hierarchical methodology incorporating the tested machine learning algorithms is developed with the used data of 15-minute mean power measurements acquired throughout the year. To evaluate the accuracy of the models, we employ various error statistics such as mean

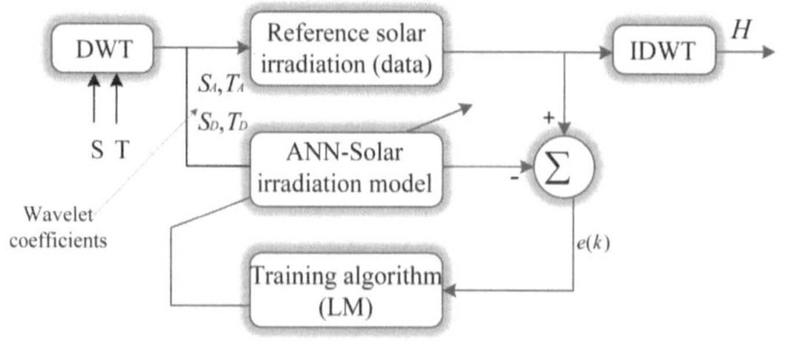

Figure 13.7 Flowchart explaining the implementation of ANN based solar irradiation prediction.

absolute error (MAE), mean bias error (MBE), root mean square error (RMSE), relative mean bias error (rMBE), relative root mean square error (rRMSE), and mean percentage error (MPE).

13.3.2 Forecasting wind energy

Wind energy possesses the capability to generate power consistently throughout the day, making it appropriate for applications that need a continuous energy supply. Furthermore, seasonal variations in wind patterns can be anticipated, enhancing the reliability of wind energy as a sustainable power source [45]. An additional advantage lies in the ability to construct wind turbines on existing farms without encroaching on agricultural land. Nevertheless, the utilization of wind energy comes with numerous challenges [46]. Primarily, the initial investment costs are more compared to traditional power plants. Additionally, due to the immobility of wind turbines, careful analysis of the wind energy prospects in promising geographical locations is essential. Moreover, wind-rich areas are often situated in distant locations, necessitating the installation of transmission lines to connect these areas to the national grid. Last, it is important to consider the potential negative impacts of wind turbines on local wildlife, including noise generation and aesthetic pollution.

Machine learning is an advanced branch of AI, which aims to design algorithms that recognize patterns from the given data, and in case of wind energy applications, it can be applied for potential wind energy forecasts [47]. They can easily depict the behavior of weather and wind speed data, and model the related input features with regard to the projected output. Traditionally, many statistical methods have been used for wind energy forecasts. However, recent trends show an increase in data-driven techniques for wind energy forecasts. Initially, these methods were used for short-term forecasts typically from one hour to some days ahead and gradually, they are progressing toward long-term energy forecasts, which vary from many days to one year. An auto-regressive moving average (ARMA) model was used for wind speed data in time series format and it successfully forecasted short-term data of one hour [48]. Another work for wind speed prediction using a hybrid machine learning technique combining a wavelet transform-based data filtering approach with a soft computing method known as fuzzy ARTMAP (FA) network was proposed [49]. The corresponding framework is represented through Figure 13.8. Further, it is compared extensively with other established methods for wind speed forecasting. The comparative analysis reveals a substantial enhancement in the accuracy of predictions when employing the WT and FA combined models. The developed wind speed forecasting scheme is then applied to actual data obtained from the North Cape wind farm in PEI, Canada. Along with the wind speed, wind power output is also impacted because of commotion and shear. The atmospheric influence on power output was estimated depending on various

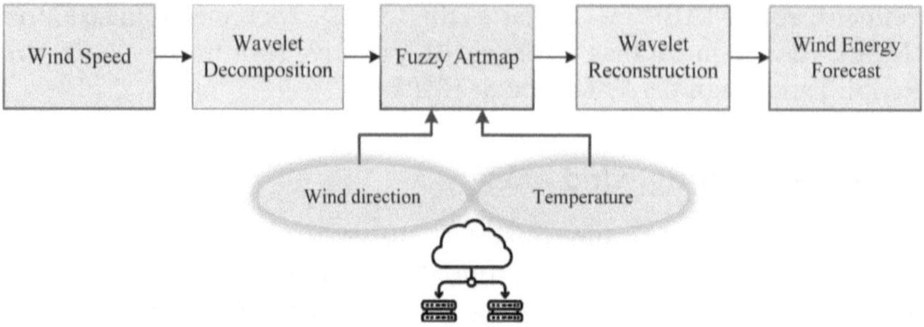

Figure 13.8 Framework providing wind energy forecast based on fuzzy Artmap technique.

factors [50]. One of the most impactful factors is the hub height wind speed, after that the other influencing factor is the hub height commotion strength, and last, wind speed shear at the rotor disk. The data collected is utilized for training regression trees for observing the turbine response given the wind speed, wind shear, and commotion intensity that may be expected at the wind turbine location. This method was found to be free from bias in wind power prediction that may occur due to different commotion and shear at the new location.

13.4 MACHINE LEARNING APPLICATION IN DESIGN

Conventional approaches for analyzing the operational performance of solar panels involve numerical simulations based on equivalent circuit models. These models require parameter identification, which can be achieved through analytical or numerical methods. Analytical methods often rely on assumptions and approximations, leading to model errors. Numerical methods, such as the Newton-Raphson method, non-linear least squares optimization, and pattern search, offer better solutions but are computationally demanding. Markov chains have also been used for parameter identification, but they require extensive data coverage, making them impractical when such data are unavailable. Parameter identification is crucial not only for modeling and simulating PV systems but also for fault diagnosis. Two commonly used models for parameter identification are the single diode model and the double diode model. The RMSE is typically employed as an error metric to optimize solar cell parameters by comparing them to empirical I-V curves. Various optimization algorithms have been developed for parameter identification, including the genetic algorithm, artificial immune system, artificial bee swarm optimization, and artificial bee colony. These algorithms have shown promising results in terms of RMSE compared to other methods in the literature.

Accurate sizing of a solar PV system is vital for ensuring reliable power supply and maximizing cost savings over the system's lifecycle. Traditional non-AI and numerical methods for sizing suffer from data requirements, while intuitive methods lack accuracy. In situations where required data is missing, alternative solutions have been explored. Hybrid approaches combining genetic algorithms and ANNs have been used to optimize sizing coefficients for standalone PV systems. The genetic algorithm optimizes the coefficients by minimizing system costs, and the ANN is trained to determine the optimal coefficients for remote areas. ANNs have also been applied to predict optimal sizing parameters for standalone PV systems. Additionally, the bat algorithm has been adapted for size optimization of grid-connected PV systems, aiming to maximize specific yield. This approach utilizes a database of existing PV modules and their technical specifications, offering faster optimization compared to particle swarm optimization.

13.5 MACHINE LEARNING APPLICATION IN CONTROL

Despite recent progress in various aspects of photovoltaic (PV) utilization, including cost reduction, improved cell efficiency, and enhanced integration with buildings, the limited energy conversion efficiency of PV systems remains a significant obstacle to widespread adoption of PV power generation. Achieving accurate MPPT is crucial in addressing this challenge. Another hurdle in solar power generation lies in its heavy reliance on environmental factors, such as solar irradiance and temperature. Hence, an effective control unit incorporating a proficient MPPT strategy is necessary to extract the maximum energy from PV arrays by appropriately adjusting the duty cycle of the respective DC-DC converter. Taking into account factors such as material efficiency, combination, and organizational configuration, enhancing the MPPT capability emerges as the most cost-effective means to improve the overall efficiency of PV systems. A comprehensive review of various MPPT techniques for solar PV panels is discussed [13].

The control of power electronic converters can be categorized into two modes of operation: (i) grid-connected control and (ii) standalone control as shown in Figure 13.9.

13.5.1 Grid-connected inverter control

In grid-connected control, a conventional controller consisting of a dual cascade loop is used. The outer loop regulates the power and voltage of the inverter, while the inner loop is responsible for current regulation and power quality maintenance [51]. Conventional controllers often employ proportional integral (PI) and proportional-resonant (PR)-based algorithms, but the use of AI algorithms improves controller accuracy and response time

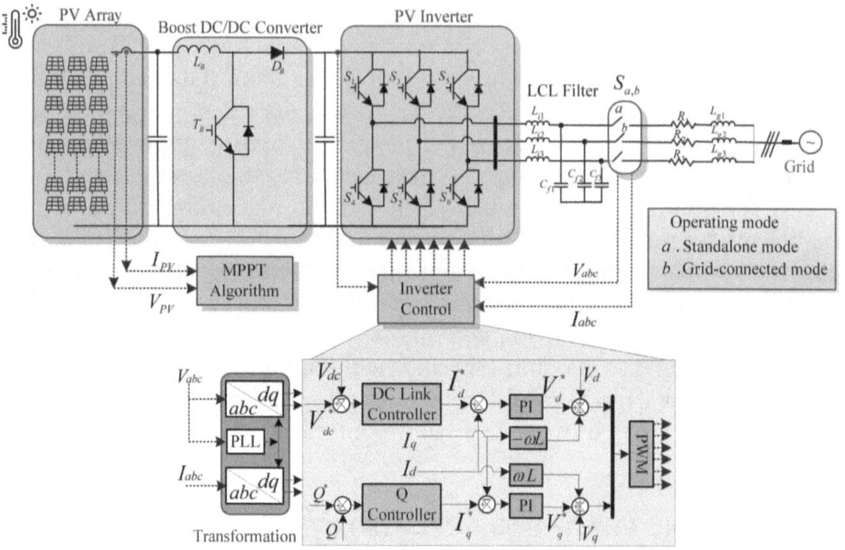

Figure 13.9 PV inverter control in grid-connected and standalone mode.

to transient errors [52]. Fuzzy-based and artificial neural network-based controllers have been simulated, resulting in low total harmonic distortion (THD) output [5]. Anti-islanding protection is crucial to detect abnormalities in the system and disconnect distributed generators (DGs) from the utility grid. Active, passive, and hybrid islanding detection methods are used [53]. Active methods face challenges in multi-inverter systems and can impact power quality. Passive methods monitor system parameters and detect abnormalities when they exceed threshold limits, but false classification is a concern [54]. To address these limitations, AI-based approaches analyze incoming signals, create a database of possible abnormalities, and train classifiers to identify operating conditions in real-time [55]. Low voltage ride through (LVRT) is recommended to maintain grid stability during abnormal grid conditions. PV systems remain connected to the grid and inject reactive current to assist in voltage recovery. LVRT can be achieved using external devices like a flexible alternating current transmission system (FACTS) or by modifying the inverter controller [56].

MPPT is essential to extract maximum power from the PV array. Conventional methods like perturb and observe suffer from limited accuracy [57]. AI-based approaches optimize and regulate MPPT considering variations in the mission profile. Fuzzy logic and genetic algorithms are used to track the maximum operating point. Seamless transition is necessary when faults occur, and the grid is disconnected from DGs. Conventional techniques exhibit response delays and transient voltage issues. AI-based transition techniques enable smooth and transition-free switching between modes. A fuzzy logic-based transition strategy generates a reference trajectory to ensure a smooth transition.

13.5.2 Standalone inverter control

In standalone inverter control, DGs must satisfy local load requirements while regulating voltage and frequency. Conventional controls often fail to optimize operation and reduce THD after transition [58]. AI-based techniques, such as fuzzy logic combined with neural networks and multiple heuristic algorithms, have been proposed to achieve faster recovery and reliable control [59].

13.6 MACHINE LEARNING APPLICATION IN MAINTENANCE

13.6.1 Maintenance in grid-connected PV systems

Grid-connected PV systems operate in complex conditions, making their safety and reliability crucial for efficient operation. Maintenance activities, including condition monitoring, fault diagnosis, and RUL prediction, are employed to improve reliability [60]. Monitoring at both the component and system level is necessary to ensure the intended operation of grid-connected PV systems [61]. A step-by-step procedure for maintenance and health prediction in renewable energy systems utilizing intelligent methods is shown in Figure 13.10.

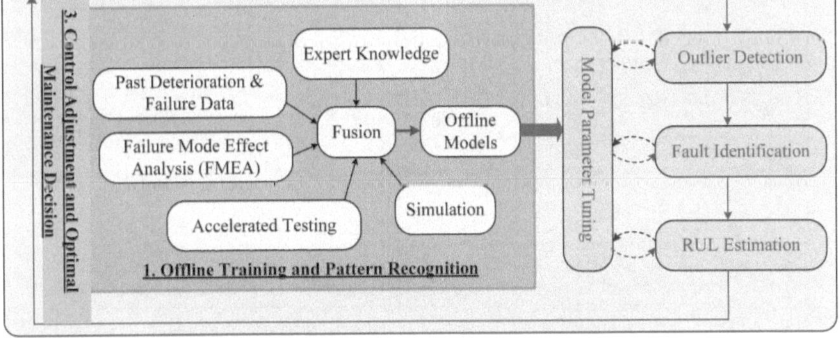

Figure 13.10 Step-by-step procedure for condition monitoring.

13.6.1.1 PV array faults monitoring

Monitoring the status of PV modules is essential to enhance power conversion efficiency. PV panels can experience various faults such as delamination, discoloration, cell cracks, and short circuits due to bypass diodes, snail trails, and glass cracks [62]. Supervised learning-based random forest methods are used for fault diagnosis in PV panels. Simulation data, including array voltage and string current under different solar irradiance and temperature conditions, is pre-processed and used for training. DTs with majority voting are used to quantify prediction accuracy, demonstrating high accuracy and the ability to handle overfitting issues [63]. Other fault diagnosis approaches utilize probabilistic neural networks (PNN), radial basis networks, kernel-based extreme learning machines, and swarm intelligence-based artificial bee colony (ABC) methods [64].

13.6.1.2 Power electronic converter faults monitoring

Machine learning-based intelligent fault classification techniques have proven highly accurate for converter fault diagnosis in grid-connected PV systems. The basic block diagram describing the methodology behind machine learning-based fault diagnosis in power electronic converter with regards to grid-connected PV system is shown in Figure 13.11. ANNs are used for power switch fault identification and classification in multilevel H-bridge inverters. Features such as signal power and energy are extracted using DWT [65]. Radial basis function networks (RBFN) and supervised learning-based PNNs are proposed for fault diagnosis in diode-clamped multilevel inverters [66]. A modified CNN with global average pooling (GAP) is used for inverter switch fault diagnosis [36].

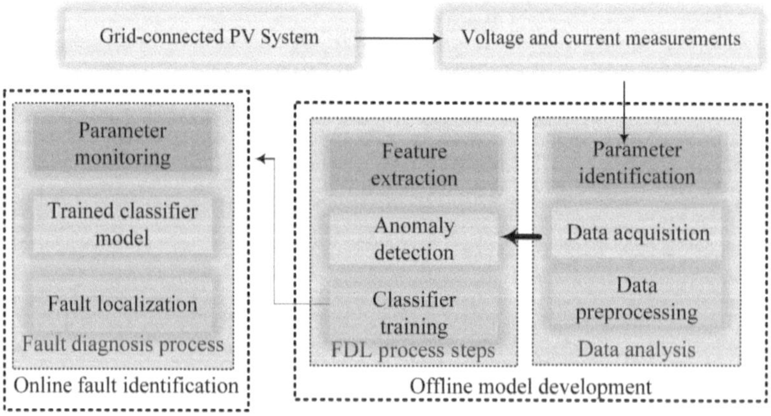

Figure 13.11 Framework for fault diagnosis in grid-connected PV system.

13.6.1.3 *Faults in filters monitoring*

Filters are used in grid-connected PV systems for harmonic attenuation. LCL filters are preferred due to their smaller size, but their capacitors are vulnerable components influenced by operational conditions like temperature and current [67]. Machine learning-based monitoring techniques, including ANN regression, fuzzy-based systems, and neo-fuzzy neuron (NFN) models, are used for estimating the health of capacitors [68]. Recursive least squares (RLS) and SVR methods are compared for capacitor health estimation using offline training [69]. Supervised learning-based ANN approaches are investigated for electrolytic capacitor health monitoring [70].

13.6.1.4 *Battery faults and degradation monitoring*

Batteries are crucial in grid-connected PV systems, and their performance must be ensured [71]. Diagnostic techniques such as state of charge (SOC), state of function (SOF), and state of health (SOH) have been explored for battery health monitoring [72]. Machine learning-based techniques, including Bayesian regression, relevance vector machines (RVM), and adaptive Gaussian mixture models (AGMM), are used for battery fault analysis and RUL estimation [73].

13.6.1.5 *Reliability analysis*

Reliability analysis aims to estimate the lifetime of power electronic devices and provide RUL in case of failure [74]. Analytical methods such as Monte Carlo or Markov analyses are conventionally used, but they may have limitations in terms of rapid convergence [75]. The Monte Carlo-based technique flowchart representation is described in Figure 13.12. Machine learning-based lifetime estimation can overcome these issues by incorporating sudden faults or changes in component stress due to ride-through operations [76].

13.6.2 Maintenance in wind energy applications

The adoption of wind power in the electricity grid is rapidly increasing, accompanied by the continuous growth of individual turbine capacity. To overcome land limitations and tap into greater wind energy potential, wind farms are shifting from onshore to offshore locations [77]. As a result, the impact on the power grid has become highly significant, and the costs associated with maintenance and repairs after occurrences of failures have risen. This necessitates enhanced reliability and resiliency in wind power generation systems to withstand extreme disturbances from the grid and the environment. Reliability is a crucial concern for industrialists and manufacturers of wind power generation systems to ensure a high level of power

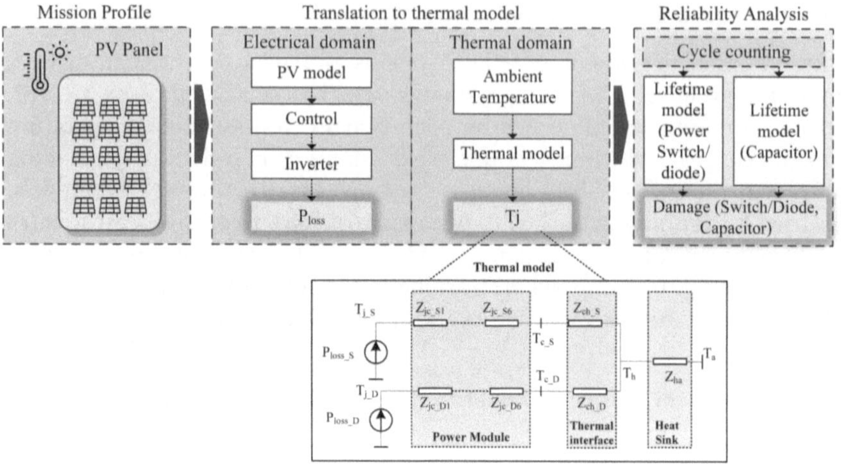

Figure 13.12 RUL Estimation using Monte Carlo method.

availability [78]. Failure drivers in wind power systems include exposure to moisture, dust, vibration, chemicals, temperature, and high voltage [79]. The traditional statistically based methods of reliability in the automotive sector have proven unsatisfactory in attaining elevated safety levels. Consequently, the focus has shifted toward a physics of failure-based approach, which comprises studying the specific phenomena that lead to component failures in these systems. Another condition monitoring approach based on machine learning for wind turbine is proposed [80]. The information in SCADA is used as input, and with the use of a neural network-based method, temperature prediction was done for the gearbox component. Further, with the use of previous data, the difference between the measured and estimated values was analyzed for various wind turbine operating states. The technique was tested with a 2 MW turbine in Sweden for further anomaly detection. An SVM-based fault diagnostic technique for wind energy application is presented as shown in Figure 13.13. Wind turbines may have structural health issues, mostly found in wind turbine blades (WTB). One of the most common issues found in WTBs is delamination, which is caused by disjoining of the layers or through the separation of adhesive bonds [81]. Fault detection in WTB, where guide waves are employed, can be done through machine learning methods. However, it requires the considered ultrasonic signal to be processed and denoised for feeding to the training network as shown in Figure 13.14 [82]. Signal denoising is done through wavelet transform, and the feature coefficients of the studied ultrasonic signal are obtained using the Yule-Walker method. The machine learning classifiers considered are kNN, DT, neural network multilayer perceptron. These classifiers are found to have high accuracy after respective training and testing, and ANN was found to be the best among all. One of the critical components in wind turbines is the

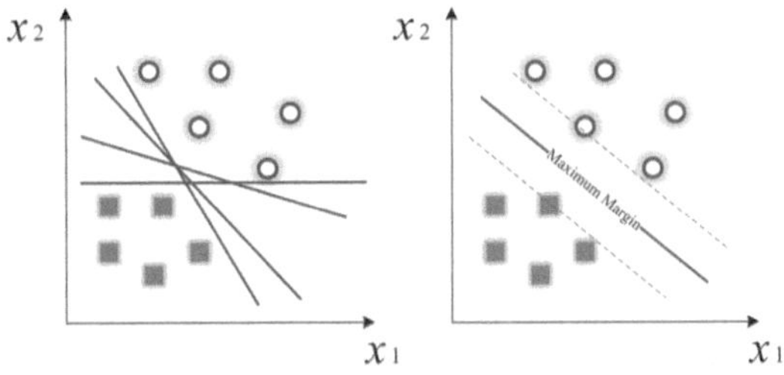

Figure 13.13 SVM technique for wind turbine condition monitoring.

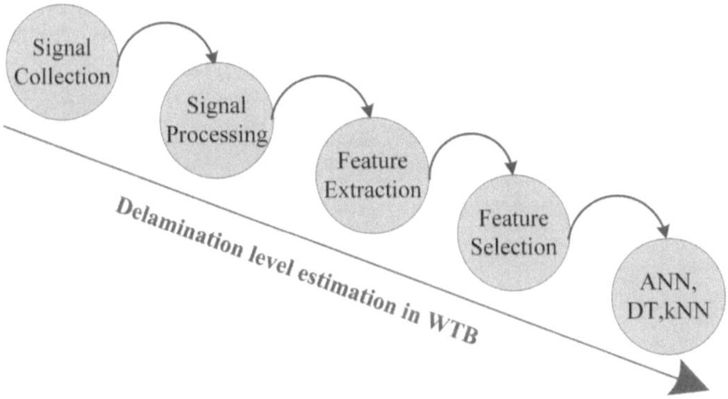

Figure 13.14 Delamination level estimation in wind turbine blades.

high-speed shaft bearing [83]. There are very few studies available related to the gradually progressive defects of shaft bearings. A vibration-based wind turbine shaft bearing fault detection method is presented based on unsupervised learning. Relevant features were extracted using adaptive resonance theory 2(ART2), and the Randall model is employed for bearing testing.

A noteworthy area of research in this field is the utilization of optimized modulation and control techniques to achieve higher junction temperature or temperature deviations under challenging conditions such as LVRT or wind squalls. This approach aims to enhance the understanding and management of thermal stresses, ensuring improved reliability and performance of wind turbine components.

13.7 SUMMARY

To gain a comprehensive understanding of the application of machine learning in diverse lifecycle phases, it is crucial to examine its characteristics and requirements with respect to the smart city applications. This is demonstrated through various areas of renewable energy systems, highlighting the unique demands of machine learning methods in each phase. In general, for applying intelligent control in renewable energy system, real-time errors obtained from the controller, such as voltage and current errors, need to be fed back to the control part for dynamic adjustment in an online mode. Thus, the algorithm's accuracy and speed become crucial considerations. Additionally, the controller's stability must be theoretically confirmed, making interpretability a critical aspect. As the intelligent controller is typically regulated online, there is no need for preparing the dataset that can be used for training the model. In the case of RUL estimation for power-switching devices, the algorithm's speed requirement is adequate due to the gradual degradation of the devices and the satisfactory decision-making time span. RUL prediction requires the degradation model to be prepared in offline mode, and further, can be effectively updated online, resulting in less computational effort. Since the model's accuracy heavily relies on the data, the data size, quality, and label balance in the training dataset become critical considerations. In summary, it is evident that machine learning has tremendous potential in enhancing renewable energy systems, particularly with regard to power electronic converters.

REFERENCES

[1] O. B. Mora-Sanchez, E. Lopez-Neri, E. J. Cedillo-Elias, E. Aceves-Martinez, and V. M. Larios, "Validation of IoT Infrastructure for the Construction of Smart Cities Solutions on Living Lab Platform," *IEEE Trans. Eng. Manag.*, vol. 68, no. 3, pp. 899–908, 2021, doi: 10.1109/TEM.2020.3002250

[2] A. C. Serban and M. D. Lytras, "Artificial Intelligence for Smart Renewable Energy Sector in Europe - Smart Energy Infrastructures for Next Generation Smart Cities," *IEEE Access*, vol. 8, pp. 77364–77377, 2020, doi: 10.1109/ACCESS.2020.2990123

[3] A. Kirimtat, O. Krejcar, A. Kertesz, and M. F. Tasgetiren, "Future Trends and Current State of Smart City Concepts: A Survey," *IEEE Access*, vol. 8, pp. 86448–86467, 2020, doi: 10.1109/ACCESS.2020.2992441

[4] V. Javidroozi, H. Shah, and G. Feldman, "Urban Computing and Smart Cities: Towards Changing City Processes by Applying Enterprise Systems Integration Practices," *IEEE Access*, vol. 7, pp. 108023–108034, 2019, doi: 10.1109/ACCESS.2019.2933045

[5] S. Zhao, F. Blaabjerg, and H. Wang, "An Overview of Artificial Intelligence Applications for Power Electronics," *IEEE Trans. Power Electron.*, vol. 36, no. 4, pp. 4633–4658, 2021, doi: 10.1109/TPEL.2020.3024914

[6] O. H. Abu-Rub, A. Y. Fard, M. F. Umar, M. Hosseinzadehtaher, and M. B. Shadmands, "Towards Intelligent Power Electronics-Dominated Grid Via Machine Learning Techniques," *IEEE Power Electron. Mag.*, vol. 8, no. 1, pp. 28–38, 2021, doi: 10.1109/MPEL.2020.3047506

[7] S. Samal, S. R. Samantaray, and M. S. Manikandan, "A DNN Based Intelligent Protective Relaying Scheme for Microgrids," *in 2019 8th Int. Conf. Power Syst. Transit. Towar. Sustain. Smart Flex. Grids, ICPS 2019*, pp. 1–6, 2019, doi: 10.1109/ICPS48983.2019.9067600

[8] V. S. B. Kurukuru, M. A. Khan, and A. Malik, "Failure Mode Classification for Grid-Connected Photovoltaic Converters," in *Reliability of Power Electronics Converters for Solar Photovoltaic Applications*, Institution of Engineering and Technology, 2021, pp. 205–249.
[9] V. S. B. Kurukuru, A. Haque, M. A. Khan, S. Sahoo, A. Malik, and F. Blaabjerg, "A Review on Artificial Intelligence Applications for Grid-Connected Solar Photovoltaic Systems," *Energies*, vol. 14, no. 15, p. 4690, Aug. 2021, doi: 10.3390/en14154690
[10] M. R. G. Meireles, P. E. M. Almeida, and M. G. Simões, "A Comprehensive Review for Industrial Applicability of Artificial Neural Networks," *IEEE Trans. Ind. Electron.*, vol. 50, no. 3, pp. 585–601, 2003, doi: 10.1109/TIE.2003.812470
[11] B. K. Bose, "Neural Network Applications in Power Electronics and Motor Drives—An Introduction and Perspective," *IECON Proc. (Industrial Electron. Conf.)*, vol. 54, no. 1, pp. 25–27, 2008, doi: 10.1109/IECON.2008.4757921
[12] B. K. Bose, "Artificial Intelligence Techniques in Smart Grid and Renewable Energy Systems - Some Example Applications," *Proc. IEEE*, vol. 105, no. 11, pp. 2262–2273, 2017, doi: 10.1109/JPROC.2017.2756596
[13] M. Seyedmahmoudian et al., "State of the Art Artificial Intelligence-Based MPPT Techniques for Mitigating Partial Shading Effects on PV systems – A Review," *Renew. Sustain. Energy Rev.*, vol. 64, pp. 435–455, 2016, doi: 10.1016/j.rser.2016.06.053
[14] A. Mellit and S. A. Kalogirou, "Artificial Intelligence Techniques for Photovoltaic Applications: A Review," *Prog. Energy Combust. Sci.*, vol. 34, no. 5, pp. 574–632, 2008, doi: 10.1016/j.pecs.2008.01.001
[15] H. Soliman, H. Wang, and F. Blaabjerg, "A Review of the Condition Monitoring of Capacitors in Power Electronic Converters," *IEEE Trans. Ind. Appl.*, vol. 52, no. 6, pp. 4976–4989, 2016, doi: 10.1109/TIA.2016.2591906
[16] L. Duchesne, E. Karangelos, and L. Wehenkel, "Recent Developments in Machine Learning for Energy Systems Reliability Management," *Proc. IEEE*, vol. 108, no. 9, pp. 1656–1676, 2020, doi: 10.1109/JPROC.2020.2988715
[17] K. V. S. Bharath, A. Haque, and M. A. Khan, "Condition Monitoring of Photovoltaic Systems Using Machine Leaming Techniques," in *2018 2nd IEEE International Conference on Power Electronics, Intelligent Control and Energy Systems (ICPEICES)*, Oct. 2018, pp. 870–875, doi: 10.1109/ICPEICES.2018.8897413
[18] V. S. Bharath Kurukuru, A. Haque, R. Kumar, M. A. Khan, and A. K. Tripathy, "Machine Learning Based Fault Classification Approach for Power Electronic Converters," in *9th IEEE Int. Conf. Power Electron. Drives Energy Syst. PEDES 2020*, 2020, doi: 10.1109/PEDES49360.2020.9379365
[19] Y. Chen, Y. Tan, and D. Deka, "Is Machine Learning in Power Systems Vulnerable?," in *2018 IEEE Int. Conf. Commun. Control. Comput. Technol. Smart Grids, SmartGridComm 2018*, pp. 1–6, 2018, doi: 10.1109/SmartGridComm.2018.8587547
[20] F. Aminifar, S. Teimourzadeh, A. Shahsavari, M. Savaghebi, and M. S. Golsorkhi, "Machine Learning for Protection of Distribution Networks and Power Electronics-Interfaced Systems," *Electr. J.*, vol. 34, no. 1, p. 106886, 2021, doi: 10.1016/j.tej.2020.106886

[21] W. Caesarendra and T. Tjahjowidodo, "A Review of Feature Extraction Methods in Vibration-Based Condition Monitoring and Its Application for Degradation Trend Estimation of Low-Speed Slew Bearing," *Machines*, vol. 5, no. 4, 2017, doi: 10.3390/machines5040021

[22] D. Chowdhury, M. Bhattacharya, D. Khan, S. Saha, and A. Dasgupta, "Wavelet Decomposition Based Fault Detection in Cascaded H-Bridge Multilevel Inverter Using Artificial Neural Network," in *RTEICT 2017 - 2nd IEEE Int. Conf. Recent Trends Electron. Inf. Commun. Technol. Proc.*, vol. 2018-Janua, pp. 1931–1935, 2017, doi: 10.1109/RTEICT.2017.8256934

[23] R. B. Dhumale and S. D. Lokhande, "Neural Network Fault Diagnosis of Voltage Source Inverter under Variable Load Conditions at Different Frequencies," *Meas. J. Int. Meas. Confed.*, vol. 91, pp. 565–575, 2016, doi: 10.1016/j.measurement.2016.04.051

[24] M. Manohar, E. Koley, Y. Kumar, and S. Ghosh, "Discrete Wavelet Transform and kNN-Based Fault Detector and Classifier for PV Integrated Microgrid," *Lect. Notes Networks Syst.*, vol. 38, pp. 19–28, 2018, doi: 10.1007/978-981-10-8360-0_2

[25] T. S. Abdelgayed, W. G. Morsi, and T. S. Sidhu, "Fault Detection and Classification Based on Co-Training of Semisupervised Machine Learning," *IEEE Transactions on Industrial Electronics*, vol. 65, no. 2. pp. 1595–1605, 2017, doi: 10.1109/TIE.2017.2726961

[26] W. Yuan, T. Wang, and D. Diallo, "A Secondary Classification Fault Diagnosis Strategy Based on PCA-SVM for Cascaded Photovoltaic Grid-connected Inverter," in *IECON Proc. (Industrial Electron. Conf.)*, vol. 2019-Octob, pp. 5986–5991, 2019, doi: 10.1109/IECON.2019.8927090

[27] W. Gong et al., "A Novel Deep Learning Method for Intelligent Fault Diagnosis of Rotating Machinery Based on Improved CNN-SVM and Multichannel Data Fusion," *Sensors (Switzerland)*, vol. 19, no. 7, 2019, doi: 10.3390/s19071693

[28] G. K. Kumar, E. Parimalasundar, D. Elangovan, P. Sanjeevikumar, F. Lannuzzo, and J. B. Holm-Nielsen, "Fault Investigation in Cascaded H-Bridge Multilevel Inverter through Fast Fourier Transform and Artificial Neural Network Approach," *Energies*, vol. 13, no. 6, 2020, doi: 10.3390/en13061299

[29] B. Cai, Y. Zhao, H. Liu, and M. Xie, "A Data-Driven Fault Diagnosis Methodology in Three-Phase Inverters for PMSM Drive Systems," *IEEE Trans. Power Electron.*, vol. 32, no. 7, pp. 5590–5600, 2017, doi: 10.1109/TPEL.2016.2608842

[30] K. Dhibi et al., "A Hybrid Fault Detection and Diagnosis of Grid-Tied PV Systems: Enhanced Random Forest Classifier Using Data Reduction and Interval-Valued Representation," *IEEE Access*, vol. 9, no. Ml, pp. 64267–64277, 2021, doi: 10.1109/ACCESS.2021.3074784

[31] N. T. Nguyen and H. P. Nguyen, "Fault Diagnosis of Voltage Source Inverter for Induction Motor Drives Using Decision Tree," *Lect. Notes Electr. Eng.*, vol. 398, pp. 819–826, 2017, doi: 10.1007/978-981-10-1721-6_88

[32] Z. He, X. Zhang, C. Liu, and T. Han, "Fault Prognostics for Photovoltaic Inverter Based on Fast Clustering Algorithm and Gaussian Mixture Model," 2020, *Energies*, vol. 13, no. 18, p. 4901, Sep. 2020, doi: 10.3390/en13184901

[33] T. H. Pham, S. Lefteriu, E. Duviella, and S. Lecoeuche, "Auto-Adaptive and Dynamical Clustering for Double Open-Circuit Fault Diagnosis of Power

Inverters," *Conf. Control Fault-Tolerant Syst. SysTol*, pp. 306–311, 2019, doi: 10.1109/SYSTOL.2019.8864777

[34] A. Mellit and S. Kalogirou, "Artificial Intelligence and Internet of Things to Improve Efficacy of Diagnosis and Remote Sensing of Solar Photovoltaic Systems: Challenges, Recommendations and Future Directions," *Renew. Sustain. Energy Rev.*, vol. 143, 2021, doi: https://doi.org/10.1016/j.rser.2021.110889

[35] F. Aziz, A. Ul Haq, S. Ahmad, Y. Mahmoud, M. Jalal, and U. Ali, "A Novel Convolutional Neural Network-Based Approach for Fault Classification in Photovoltaic Arrays," *IEEE Access*, vol. 8, pp. 41889–41904, 2020, doi: 10.1109/ACCESS.2020.2977116

[36] W. Gong, H. Chen, Z. Zhang, M. Zhang, and H. Gao, "A Data-Driven-Based Fault Diagnosis Approach for Electrical Power DC-DC Inverter by Using Modified Convolutional Neural Network with Global Average Pooling and 2-D Feature Image," *IEEE Access*, vol. 8, pp. 73677–73697, 2020, doi: 10.1109/ACCESS.2020.2988323

[37] L. Avila, M. De Paula, I. Carlucho, and C. Sanchez Reinoso, "MPPT for PV Systems Using Deep Reinforcement Learning Algorithms," *IEEE Lat. Am. Trans.*, vol. 17, no. 12, pp. 2020–2027, 2019, doi: 10.1109/TLA.2019.9011547

[38] C. Li, C. Jin, and R. Sharma, "Coordination of PV Smart Inverters Using Deep Reinforcement Learning for Grid Voltage Regulation," in *Proc. - 18th IEEE Int. Conf. Mach. Learn. Appl. ICMLA 2019*, pp. 1930–1937, 2019, doi: 10.1109/ICMLA.2019.00310

[39] J. Sun et al., "An Integrated Critic-Actor Neural Network for Reinforcement Learning with Application of DERs Control in Grid Frequency Regulation," *Int. J. Electr. Power Energy Syst.*, vol. 111, no. March, pp. 286–299, 2019, doi: 10.1016/j.ijepes.2019.04.011

[40] Y. Zhang, "Neural Network Algorithm with Reinforcement Learning for Parameters Extraction of Photovoltaic Models," *IEEE Trans. Neural Networks Learn. Syst.*, vol. PP, pp. 1–11, 2021, doi: 10.1109/TNNLS.2021.3109565

[41] A. Mellit, "Artificial Intelligence Technique for Modelling and Forecasting of Meteorological Data: A Survey," *Atmos. Turbul. Meteorol. Model. Aerodyn.*, vol. 1, no. 1, pp. 293–328, 2011.

[42] M. Ding, L. Wang, and R. Bi, "An ANN-Based Approach for Forecasting the Power Output of Photovoltaic System," *Procedia Environ. Sci.*, vol. 11, no. PART C, pp. 1308–1315, 2011, doi: 10.1016/j.proenv.2011.12.196

[43] D. Su, E. Batzelis, and B. Pal, "Machine Learning Algorithms in Forecasting of Photovoltaic Power Generation," *in SEST 2019 - 2nd Int. Conf. Smart Energy Syst. Technol.*, 2019, doi: 10.1109/SEST.2019.8849106

[44] Z. Li, S. M. Mahbobur Rahman, R. Vega, and B. Dong, "A Hierarchical Approach Using Machine Learning Methods in Solar Photovoltaic Energy Production Forecasting," *Energies*, vol. 9, no. 1, 2016, doi: 10.3390/en9010055

[45] F. Blaabjerg, M. Liserre, and K. Ma, "Power Electronics Converters for Wind Turbine Systems," *IEEE Trans. Ind. Appl.*, vol. 48, no. 2, pp. 708–719, Mar. 2012, doi: 10.1109/TIA.2011.2181290

[46] Y. C. Deng, X. H. Tang, Z. Y. Zhou, Y. Yang, and F. Niu, "Application of Machine Learning Algorithms in Wind Power: A Review," *Energy Sources, Part A Recover. Util. Environ. Eff.*, vol. 00, no. 00, pp. 1–22, 2021, doi: 10.1080/15567036.2020.1869867

[47] H. Demolli, A. S. Dokuz, A. Ecemis, and M. Gokcek, "Wind Power Forecasting Based on Daily Wind Speed Data Using Machine Learning Algorithms," *Energy Convers. Manag.*, vol. 198, no. July, p. 111823, 2019, doi: 10.1016/j.enconman.2019.111823

[48] S. Rajagopalan and S. Santoso, "Wind Power Forecasting and Error Analysis Using the Autoregressive Moving Average Modeling," in *2009 IEEE Power Energy Soc. Gen. Meet. PES '09*, pp. 1–6, 2009, doi: 10.1109/PES.2009.5276019

[49] A. U. Haque, P. Mandal, J. Meng, and M. Negnevitsky, "Wind Speed Forecast Model for Wind Farm Based on a Hybrid Machine Learning Algorithm," *Int. J. Sustain. Energy*, vol. 34, no. 1, pp. 38–51, 2015, doi: 10.1080/14786451.2013.826224

[50] A. Clifton, L. Kilcher, J. K. Lundquist, and P. Fleming, "Using Machine Learning to Predict Wind Turbine Power Output," *Environ. Res. Lett.*, vol. 8, no. 2, 2013, doi: 10.1088/1748-9326/8/2/024009

[51] A. Malik, A. Haque, V. S. Bharath Kurkuru, and R. Kumar, "Improved Stationary Reference Frame for Grid Connected Operation of Single Phase Parallel Inverters," in *2022 IEEE International Conference on Power Electronics, Drives and Energy Systems (PEDES)*, Dec. 2022, pp. 1–6, doi: 10.1109/PEDES56012.2022.10080071

[52] M. A. Khan, A. Haque, and V. S. B. Kurukuru, "Machine Learning Based Islanding Detection for Grid Connected Photovoltaic System," in *2019 Int. Conf. Power Electron. Control Autom. ICPECA 2019 - Proc.*, vol. 2019-Novem, no. 1, 2019, doi: 10.1109/ICPECA47973.2019.8975614

[53] K. Hu, Z. Liu, Y. Yang, F. Iannuzzo, and F. Blaabjerg, "Ensuring a Reliable Operation of Two-Level IGBT-Based Power Converters: A Review of Monitoring and Fault-Tolerant Approaches," *IEEE Access*, vol. 8, pp. 89988–90022, 2020, doi: 10.1109/ACCESS.2020.2994368

[54] M. Ciobotaru, R. Teodorescu, and F. Blaabjerg, "Control of Single-Stage Single-Phase PV Inverter," *EPE J. (European Power Electron. Drives Journal)*, vol. 16, no. 3, pp. 20–26, 2006, doi: 10.1080/09398368.2006.11463624

[55] A. Haque, V. S. B. Kurukuru, M. A. Khan, A. Malik, and F. Fayaz, "Centralized Intelligent Fault Localization Approach for Renewable Energy-Based Islanded Microgrid Systems," *Appl. AI IOT Renew. Energy*, pp. 129–149, Jan. 2022, doi: 10.1016/B978-0-323-91699-8.00007-3

[56] S. Member and X. Wang, "Passivity Enhancement in Renewable Energy Source Based Power Plant with Paralleled Grid-Connected VSIs Frede Blaabjerg," vol. 9994, no. c, pp. 1–9, 2017, doi: 10.1109/TIA.2017.2685363

[57] U. M. Choi, F. Blaabjerg, and K. B. Lee, "Reliability Improvement of a T-type Three-Level Inverter with Fault-Tolerant Control Strategy," *IEEE Trans. Power Electron.*, vol. 30, no. 5, pp. 2660–2673, 2015, doi: 10.1109/TPEL.2014.2325891

[58] A. Z. Fatama, M. A. Khan, V. S. B. Kurukuru, A. Haque, and F. Blaabjerg, "Coordinated Reactive Power Strategy Using Static Synchronous Compensator for Photovoltaic Inverters," *Int. Trans. Electr. Energy Syst.*, vol. 30, no. 6, pp. 1–18, 2020, doi: 10.1002/2050-7038.12393

[59] M. A. Khan, A. Haque, and K. V. S. Bharath, "Droop Based Low voltage Ride through Implementation for Grid Integrated Photovoltaic System." in *2019 International Conference on Power Electronics, Control and Automation (ICPECA)*, Nov. 2019, pp. 1–5, doi: 10.1109/ICPECA47973.2019.8975467

[60] A. Malik, A. Haque, V. S. B. Kurukuru, M. A. Khan, and F. Blaabjerg, "Overview of Fault Detection Approaches for Grid Connected Photovoltaic Inverters," in *e-Prime - Adv. Electr. Eng. Electron. Energy*, vol. 2, no. November 2021, p. 100035, 2022, doi: 10.1016/j.prime.2022.100035

[61] A. Haque, K. V. S. Bharath, M. A. Khan, I. Khan, and Z. A. Jaffery, "Fault Diagnosis of Photovoltaic Modules," *Energy Sci. Eng.*, vol. 7, no. 3, pp. 622–644, 2019, doi: 10.1002/ese3.255

[62] V. S. B. Kurukuru, A. Haque, M. A. Khan, and A. K. Tripathy, "Fault Classification for Photovoltaic Modules Using Thermography and Machine Learning Techniques," in *2019 Int. Conf. Comput. Inf. Sci. ICCIS 2019*, pp. 1–6, 2019, doi: 10.1109/ICCISci.2019.8716442

[63] M. Köntges et al., *Review of Failures of Photovoltaic Modules - IEA-PVPS T13-01:2014*. 2014.

[64] D. C. Jordan, T. J. Silverman, J. H. Wohlgemuth, S. R. Kurtz, and K. T. VanSant, "Photovoltaic Failure and Degradation Modes," *Prog. Photovoltaics Res. Appl.*, vol. 25, no. 4, pp. 318–326, 2017, doi: 10.1002/pip.2866

[65] V. S. B. Kurukuru, A. Haque, A. K. Tripathi, and M. A. Khan, "Condition Monitoring of IGBT Modules Using Online TSEPs and Data-Driven Approach," *Int. Trans. Electr. Energy Syst.*, vol. 31, no. 8, pp. 1–24, 2021, doi: 10.1002/2050-7038.12969

[66] M. Bhattacharya, S. Saha, D. Khan, and T. Nag, "Wavelet Based Component Fault Detection in Diode Clamped Multilevel Inverter Using Probabilistic Neural Network," in *2017 2nd Int. Conf. Converg. Technol. I2CT 2017*, vol. 2017-January, pp. 1163–1168, 2017, doi: 10.1109/I2CT.2017.8226310

[67] D. Zhou, H. Wang, H. Wang, and F. Blaabjerg, "Reliability Analysis of Grid-Interfaced Filter Capacitors," *Chinese J. Electr. Eng.*, vol. 4, no. 3, pp. 21–28, 2019, doi: 10.23919/cjee.2018.8471286

[68] K. W. Lee, M. Kim, J. Yoon, S. Bin Lee, and J. Y. Yoo, "Condition Monitoring of DC-Link Electrolytic Capacitors in Adjustable-Speed Drives," *IEEE Trans. Ind. Appl.*, vol. 44, no. 5, pp. 1606–1613, 2008, doi: 10.1109/TIA.2008.2002220

[69] C. Bhargava, V. K. Banga, and Y. Singh, "An Intelligent Prognostic Model for Electrolytic Capacitors Health Monitoring: A Design of Experiments Approach," *Adv. Mech. Eng.*, vol. 10, no. 10, pp. 1–11, 2018, doi: 10.1177/1687814018781170

[70] V. A. Sankaran, F. L. Rees, and C. S. Avant, "Electrolytic Capacitor Life Testing and Prediction," in *Conf. Rec. - IAS Annu. Meet. (IEEE Ind. Appl. Soc.)*, vol. 2, pp. 1058–1065, 1997, doi: 10.1109/ias.1997.628992

[71] M. F. Samadi and M. Saif, "State-Space Modeling and Observer Design of Li-Ion Batteries Using Takagi-Sugeno Fuzzy System," *IEEE Trans. Control Syst. Technol.*, vol. 25, no. 1, pp. 301–308, 2017, doi: 10.1109/TCST.2016.2549270

[72] J. Yu, "Health Degradation Detection and Monitoring of Lithium-Ion Battery Based on Adaptive Learning Method," *IEEE Trans. Instrum. Meas.*, vol. 63, no. 7, pp. 1709–1721, 2014, doi: 10.1109/TIM.2013.2293234

[73] D. Zhou, Z. Li, J. Zhu, H. Zhang, and L. Hou, "State of Health Monitoring and Remaining Useful Life Prediction of Lithium-Ion Batteries Based on Temporal Convolutional Network," *IEEE Access*, vol. 8, pp. 53307–53320, 2020, doi: 10.1109/ACCESS.2020.2981261

[74] A. Sangwongwanich and F. Blaabjerg, "Monte Carlo Simulation with Incremental Damage for Reliability Assessment of Power Electronics," *IEEE*

Trans. Power Electron., vol. 36, no. 7, pp. 7366–7371, 2021, doi: 10.1109/TPEL.2020.3044438

[75] A. Sangwongwanich, Y. Yang, and S. Member, "Mission Profile-Oriented Control for Reliability and Lifetime of Photovoltaic Inverters," *IEEE Trans. Ind. Appl.*, vol. PP, no. 99, p. 1, 2019, doi: 10.1109/TIA.2019.2947227

[76] Y. Yang, A. Sangwongwanich, and F. Blaabjerg, "Design for Reliability of Power Electronics for Grid-Connected Photovoltaic Systems," *CPSS Trans. Power Electron. Appl.*, vol. 1, no. 1, 2016.

[77] C. Magazzino, M. Mele, and N. Schneider, "A Machine Learning Approach on the Relationship among Solar and Wind Energy Production, Coal Consumption, GDP, and CO_2 Emissions," *Renew. Energy*, vol. 167, pp. 99–115, 2021, doi: 10.1016/j.renene.2020.11.050

[78] A. Arcos Jiménez, L. Zhang, C. Q. Gómez Muñoz, and F. P. García Márquez, "Maintenance Management Based on Machine Learning and Nonlinear Features in Wind Turbines," *Renew. Energy*, vol. 146, pp. 316–328, 2020, doi: 10.1016/j.renene.2019.06.135

[79] A. Stetco et al., "Machine Learning Methods for Wind Turbine Condition Monitoring: A Review," *Renew. Energy*, vol. 133, pp. 620–635, 2019, doi: 10.1016/j.renene.2018.10.047

[80] J. E. Urrea Cabus, Y. Cui, and L. B. Tjernberg, "An Anomaly Detection Approach Based on Autoencoders for Condition Monitoring of Wind Turbines," in *2022 17th Int. Conf. Probabilistic Methods Appl. to Power Syst. PMAPS 2022*, 2022, doi: 10.1109/PMAPS53380.2022.9810575

[81] C. Q. Gómez Muñoz, F. P. García Márquez, and J. M. Sánchez Tomás, "Ice Detection Using Thermal Infrared Radiometry on Wind Turbine Blades," *Meas. J. Int. Meas. Confed.*, vol. 93, pp. 157–163, 2016, doi: 10.1016/j.measurement.2016.06.064

[82] A. A. Jiménez, C. Q. Gómez Muñoz, and F. P. García Márquez, "Machine Learning for Wind Turbine Blades Maintenance Management," *Energies*, vol. 11, no. 1, pp. 1–16, 2018, doi: 10.3390/en11010013

[83] J. Ben Ali, L. Saidi, S. Harrath, E. Bechhoefer, and M. Benbouzid, "Online Automatic Diagnosis of Wind Turbine Bearings Progressive Degradations under Real Experimental Conditions Based on Unsupervised Machine Learning," *Appl. Acoust.*, vol. 132, no. September 2017, pp. 167–181, 2018, doi: 10.1016/j.apacoust.2017.11.021

Index

Pages in **bold** refer to tables.